BASIC ANALYTICAL PETR

BASIC ANALYTICAL PETROLOGY

Paul C. Ragland

Professor of Geology
The Florida State University

New York · Oxford
OXFORD UNIVERSITY PRESS
1989

Oxford University Press

Oxford   New York   Toronto
Delhi   Bombay   Calcutta   Madras   Karachi
Petaling Jaya   Singapore   Hong Kong   Tokyo
Nairobi   Dar es Salaam   Cape Town
Melbourne   Auckland

and associated companies in
Berlin   Ibadan

Library of Congress Cataloging-in-Publication Data
Ragland, Paul C., 1936–
Basic analytical petrology / Paul C. Ragland.
p.   cm.   Bibliography: p. Includes index.
ISBN 0-19-504534-3
ISBN 0-19-504535-1 (pbk.)
1. Rocks—Analysis.   2. Rocks, Igneous—Analysis.   I. Title.
QE438.R34   1989
552'.06—dc 19   89-3051   CIP

9 8 7 6 5 4 3 2 1
Printed in the United States of America
on acid-free paper

To Joan

PREFACE

The adjective ANALYTICAL in the title of this book has two meanings. Use of chemical analyses for igneous rocks and minerals is definitely stressed, but the principal aim of this book is the analysis of igneous rocks in a larger sense. *Analysis* is the separation of any whole into its parts with the intent of determining their nature, thereby learning more about the whole. This book discusses ways by which igneous rocks are separated, literally as well as figuratively, into their component parts and examined chemically. The ultimate purpose, however, is to explore means based on chemistry by which we can learn more about the petrogenesis of igneous rocks. Many students initially look on phase equilibria, basic chemistry, thermodynamics, and calculus as being rather minimally related to the more practical and descriptive aspects of igneous petrology. My intent is to demonstrate quite the opposite!

This book is an outgrowth from parts of three courses, one undergraduate and two graduate. It is not intended as a stand-alone textbook, but is to supplement a more standard book, at the advanced undergraduate or perhaps beginning graduate level. It stresses methodology rather than supplying information, which is available in a number of excellent modern texts. Parts from several books greatly affected my writing (particularly Barker, 1983; Ernst, 1976; McBirney, 1984; and Morse, 1980), but I was most influenced by Cox and others (1979). Several topics covered herein, however, are not included in Cox and others, and this book is generally at a more basic level. The extensive use of example calculations and clarity of writing in a classic book from another field, *Quantitative Zoology* by G. G. Simpson and his colleagues (1960), was also quite influential.

Space constraints dictated that many topics be omitted, e.g., isotopes and physical properties of magmas. Their fundamentals have been covered by other writers (e.g., McBirney, 1984; Cox and others, 1979). Only abbreviated tables of chemical analyses and thermodynamic data are included, but references are given, requiring students to use the library and data tables as learning exercises. No problems are at the ends of chapters, but scores of problems are worked, which hopefully will provide the inspiration for many others.

After the standard disclaimer that all mistakes found herein are of my own making, certainly true in my case, I wish to thank a number of people for their help. This book is dedicated to my wife, Joan, for her constant help and encouragement. It never would have been started, and certainly never completed, without her. I am also indebted to my mother, who in a number of ways provided the wherewithal, and to my sons, for never giving up on me. John Ragland first introduced me to the avocation of computer programming. To all those people, too numerous to list, who initially encouraged my writing and who critiqued parts of the manuscript, I am very grateful. John J. W. Rogers reviewed the entire manuscript and provided many useful comments. Rosemary Raymond and Katherine Milla drafted the figures. Ted Zateslo and David Allison helped with reproduction of the text, and Woody Wise loaned the laser printer. I also wish to thank all the petrology students down through the years who have taught me a great deal. Specifically, I acknowledge the students in my igneous petrology classes during the Fall semesters, 1986 and 1988, for their help. Joe Aylor in my 1988 class was particularly helpful.

Tallahassee, Fla. P. C. R.
March 1989

CONTENTS

1   Basic Tools *3*

1.1   Elements and Oxides *3*
    1.1.1 Nomenclature and Tradition *3*
    1.1.2 Analytical Methods *8*
    1.1.3 Descriptive Statistics *13*
    1.1.4 An Example of the Analytical Method *24*
1.2   Conversions *32*
    1.2.1 Weight/Weight Conversions *32*
    1.2.2 The Equivalent *37*
    1.2.3 Weight/Mole Conversions *38*
    1.2.4 Mineral Formulae *41*
    1.2.5 Norms *44*
1.3   Saturation and Rock Series *52*
    1.3.1 Saturation *53*
    1.3.2 An Introduction to Variation Diagrams *57*
    1.3.3 Silica Saturation and Igneous Rock Series *61*
    1.3.4 Alumina Saturation and Granitoids *66*
1.4   Linear Error Analysis *67*
    1.4.1 Formulation *67*
    1.4.2 Some Practical Examples *70*
1.5   Review of Rock-forming Minerals *72*
    1.5.1 The WXYZ System *72*
    1.5.2 Silicate Structures *75*
1.6   Very Basic BASIC *79*
    1.6.1 BASIC Basics *80*
    1.6.2 Four Examples *83*

2   Classical Thermodynamics *94*

2.1   Fundamental Laws and Free Energy *94*
    2.1.1 Definitions and Conventions *94*

2.1.2 First Law *97*
2.1.3 Enthalpy *99*
2.1.4 Second Law and Entropy *101*
2.1.5 Free Energy *103*
2.2 Preparation for Phase Equilibria *108*
2.2.1 Pressure-Temperature Diagrams *108*
2.2.2 Ideality and Reality *119*
2.2.3 Phase Rule *125*

3 Unary and Binary Systems *132*

3.1 Unary Systems *132*
3.1.1 Water *132*
3.1.2 Silica *136*
3.2 Binary Systems with Immiscible Solids *140*
3.2.1 The Experimental Method *140*
3.2.2 Basic Principles *140*
3.2.3 A P-T-X Diagram *145*
3.2.4 A Simple Binary Eutectic *148*
3.2.5 A Thermal Divide *157*
3.2.6 A Peritectic *159*
3.2.7 Fractional Processes *166*
3.2.8 Liquid Immiscibility *174*
3.3 Binary Systems with Solid Solutions *174*
3.3.1 A Transition Loop *177*
3.3.2 Subsolidus Reactions *183*
3.4 Binary Systems and P-T Diagrams *190*
3.4.1 Preliminaries *191*
3.4.2 Schreinemakers' Rules *194*

4 Ternary and Quaternary Systems *199*

4.1 Ternary Systems *199*
4.1.1 A Simple Ternary Eutectic *199*
4.1.2 Other Ternary Systems with Immiscible Solids *207*
4.1.3 Ternary Systems with Solid Solutions *215*
4.1.4 Quantitative Treatment *226*
4.2 Quaternary Systems *236*
4.2.1 System with a Quaternary Eutectic *236*
4.2.2 A Haplobasalt System: AB-AN-DI-FO *241*
4.2.3 The Haplogranite System OR-AB-AN-Q-H_2O *245*

5 Igneous Rocks *252*

5.1 Melting and Crystallization Experiments *252*
5.1.1 Dry Conditions *253*
5.1.2 Effect of Volatiles *263*
5.2 Treatment of Chemical Analyses *272*
5.2.1 Trace Elements *272*
5.2.2 Variation Diagrams Revisited *286*
5.2.3 Numerical Modeling *313*

5.2.4 Igneous Rock Series Revisited *320*
5.3 **Igneous Rocks and Plate Tectonics** *329*
5.3.1 Approaches *330*
5.3.2 Comparisons *337*
5.3.3 Conclusion *339*

APPENDICES *340*

REFERENCES *355*

SUBJECT INDEX *363*

BASIC ANALYTICAL PETROLOGY

1

BASIC TOOLS

This chapter could be entitled "MISCELLANEA." It is an eclectic mix of six main topics, only the first two of which are obviously related. In fact, they are all related because they are necessary tools to obtain maximum benefit from this book. One of the principal aims of the book is the manner in which theoretical calculations, experimental results, and chemical analyses of igneous rocks are all connected. These analyses do not appear miraculously in the hands of the petrologist. They represent a great deal of work on someone's part and, once they are obtained, are used for a variety of purposes and reported in a number of ways. Many different techniques and conventions have evolved over the years regarding their use.

We begin with some of these conventions and continue with chemical analytical methods, descriptive statistics, and chemical conversions. Then we consider the concept of saturation, enabling us to relate igneous rock suites with experimental phase equilibria. The next topic on linear error analysis is included because petrologists are required to make calculations in which a knowledge of the error (precision or reproducibility) for their data allows them to constrain their hypotheses. Silicate mineralogy is briefly reviewed as most igneous rocks are mainly comprised of silicates. Finally, a short lesson in BASIC computer programming is included because most numerical calculations in this book can be much more easily solved using simple BASIC programs.

1.1 ELEMENTS AND OXIDES

1.1.1 Nomenclature and Tradition

Igneous rocks are aggregates of one or more minerals, which in turn are inorganic chemical compounds with definite, ordered crystal structures and either fixed or predictably varying compositions. Only 14 chemical elements make up over 99 percent of these naturally occurring compounds, and thus igneous rocks themselves. These elements, in order of increasing atomic number, are: H, C, O, Na, Mg, Al, Si, P, S, K, Ca, Ti, Mn, and Fe.

1.1.1.1 Relative Abundances

Of the 103 elements listed on any periodic table, 90 occur naturally on earth. In general, elements with relatively small atomic (Z) numbers are much more abundant, by orders of magnitude, than those with large Z numbers. For instance, O (Z = 8) is about 10^9 times more abundant in our solar system than U (Z = 92; Appendix C). Iron, the heaviest of the 14 most common elements listed, only has a Z number of 26.

TABLE 1.1   A Typical Chemical Analysis of a Basalt

a	b	c	d	e	f
SiO_2	48.7	Ba	62	La	8.1
TiO_2	1.29	Co	73	Ce	20
Al_2O_3	16.6	Cr	170	Sm	2.2
Fe_2O_3	2.05	Cu	110	Eu	0.70
FeO	8.29	Nb	12	Tb	0.58
MnO	0.16	Ni	230	Yb	2.3
MgO	6.63	Rb	35	Lu	0.33
CaO	10.7	Sr	350		
Na_2O	2.83	Th	3.3		
K_2O	0.47	U	1.2		
H_2O+	0.81	V	79		
H_2O-	0.67	Y	37		
P_2O_5	0.20	Zn	49		
CO_2	0.09	Zr	66		
Total	99.49				

a: major and minor elements expressed as oxides
b: concentrations of oxides in Column a in weight percent
c: trace elements
d: concentrations of trace elements in Column c in parts per million (ppm)
e: rare-earth trace elements
f: concentrations of rare-earth trace elements in Column e in parts per million (ppm)

Notes: This list of trace elements is only representative and many other trace elements can be included. The rare-earth trace elements are included together as a group; of the 14 rare-earth elements, these are the seven most commonly analyzed.

Reasons for these differences in abundance involve the origin of our solar system, formation of the elements in stars, and even the origin of the universe itself. These topics are beyond the scope of this book, but general discussions can be found in most introductory geochemistry texts (for example, see Mason and Moore, 1982; Richardson and McSween, 1989). The important fact here is that light elements are generally more abundant than heavy ones, for whatever reasons.

Notable exceptions exist, however, so the preceeding observation is useful only in a broad sense. As an example, Li, Be, and B ($Z = 3\text{-}5$) should be more abundant than C, N, and O ($Z = 6\text{-}8$). In fact, the latter three elements are much more abundant, by about seven orders of magnitude (Appendix C), the reasons for which are beyond our concern.

1.1.1.2 Major, Minor, and Trace Elements

Patterns in overall abundance of elements in the universe are reflected by those in igneous rocks, in which major and minor elements on average have lower Z numbers than do trace elements. A typical example of a chemical analysis of an igneous rock, in this case a basalt, is given in Table 1.1. Note that the major and minor elements are expressed as weight percent oxides. Weight percent is equivalent to any weight measure per 100 measures, such as grams per hundred grams. Trace elements, on the other hand, are expressed in parts per million of the element (commonly referred to as *ppm*, equivalent to grams per million grams; one weight percent of an element is 10,000 ppm). Reasons for expressing major and minor elements as oxides and trace elements simply as elements are discussed in Section 1.1.1.4. The major and minor oxides in Table 1.1 do not total exactly 100 percent. This is because of errors in the individual analyses (Section 1.4.2.1) and the fact that some oxides not analyzed, such as SO_2, and the trace elements are not included in the summation.

Major and minor elements are reported to three significant figures as oxides in Table 1.1 and Appendix E, whereas only two are reported for trace elements, because the precision and accuracy of analysis (Section 1.1.4.5) is generally better for major and minor elements. Some researchers prefer to report all major and minor elements to two decimal places. This means that any oxide whose concentration is over 10 percent, such as SiO_2 and Al_2O_3 (or commonly CaO, MgO, and FeO in mafic rocks), is reported to four significant figures, which is not justified based on the precision of most analyses.

We have referred several times to the terms *major, minor,* and *trace* elements. For purposes of convenience, elements in igneous rocks are divided into these three groups based upon their abundances. Limits on these groups are strictly arbitrary, but most petrologists at least tacitly acknowledge the following:

Major element	>1.0 weight percent of the rock or mineral
Minor element	0.1 - 1.0 weight percent
Trace element	<0.1 weight percent (<1000 ppm)

Some workers prefer to include minor elements with majors and think in terms of only two groups.

These limits are only rough approximations. Typically, MnO, P_2O_5, and TiO_2 are considered minor elements, despite the fact that they may not fall within 0.1 - 1.0 weight percent. According to these limits, TiO_2 in Table 1.1 is a major element. Likewise, K_2O is generally thought to be a major element, although its concentration is commonly less than 1.0 percent in mafic rocks (as in Table 1.1). It is important to realize just how arbitrary these conventions are. Consider a typical rock analysis of 0.11 percent MnO as an example. This analysis can be easily converted to 0.085 percent Mn (Section 1.2.1.2), which is equivalent to 850 ppm Mn. Manganese, then, can be validly expressed either as a minor oxide or a trace element. Only tradition requires that it be expressed as the former. In fact, for those who prefer to think in terms of only two groups, it is a major oxide.

Note that O^{2-} is the only anion in Table 1.1. Silicates are by far the most common igneous rock-forming minerals, followed by oxides, and then by sulfides, carbonates, and phosphates (not necessarily in that order). Only the sulfides contain a simple anion other than O^{2-} in any quantity. *Fluorite* (CaF_2) is occasionally found in granitic igneous rocks, but it is generally quite rare.

1.1.1.3 Oxidation States and Volatiles

Of the cations bonded to O^{2-} as oxides in Table 1.1, only three (Ti, Mn, and Fe) can have more than one valence state under natural conditions. It is assumed that Ti is generally in the +4 state and Mn is in the +2, thus their oxides are reported as TiO_2 and MnO. These assumptions are not strictly correct but are sufficient for our purposes. In contrast, Fe clearly exists in two valence states, +2 (ferrous) and +3 (ferric), so it is normally reported as FeO and Fe_2O_3.

Because Fe exists in both the oxidized (ferric) and reduced (ferrous) states, some conventions with regard to reporting Fe analyses have been adopted. Many modern instrumental techniques cannot distinguish between Fe^{2+} and Fe^{3+}, so they actually analyze the total Fe content of the rock. Ferrous Fe is determined separately, normally either by titration or colorimetry (Section 1.1.2.2), and Fe^{3+} is obtained by difference. If both FeO and Fe_2O_3 are reported, there is no confusion. If only total Fe is analyzed, however, several different terms are used to report it. FeO^*, FeO_T, and ΣFeO all mean total Fe as FeO; comparable terms are used for total Fe as Fe_2O_3. Although they lead to confusion, all six terms are perfectly acceptable and all are widely used. Only total Fe is analyzed in minerals using the electron microprobe.

The volatile constituents reported in Table 1.1 are H_2O+, H_2O-, and CO_2. Sometimes SO_2 is analyzed as well. H_2O+ is generally referred to as "structural H_2O" and exists primarily as the hydroxyl ion (OH^-) in hydrous minerals such as the micas and amphiboles. H_2O+ cannot be determined in a mineral analysis using the electron microprobe. H_2O-, or "adsorbed H_2O," is the water adsorbed on grain surfaces that can be driven off in a drying oven at temperatures of 90-110°C.

Occasionally the analyst does not determine the volatile constituents separately, but rather reports *LOI*. The term refers to "loss on ignition" and is a measure of the total volatile content of the rock. LOI is determined by placing the sample in a furnace at several hundred °C (commonly 800°C) and measuring the weight loss. Some analysts measure it before H_2O- is driven off, whereas others measure it afterward. Its chief drawback, aside from the fact that it does not provide information about specific volatiles, is that the sample actually takes on some oxygen in the furnace due to oxidation of some Fe^{2+} to Fe^{3+}. Consequently, the reported value for LOI is lower than the actual volatile content. If LOI is reported, then total Fe should be reported as $Fe_2O_3^*$ (or Fe_2O_{3T} or ΣFe_2O_3).

1.1.1.4 Traditions

Chemists commonly express concentration units in molarity, molality, or normality. Low-temperature geochemists, particularily those who analyze water samples, do so as well. Although petrologists occasionally use some variation on molar units (Section 1.2.3), they generally quote weight percent oxides and ppm elements. The reasons go back to the early days of rock analysis.

Analytical chemists began quantitatively analyzing rocks and minerals in the early days of chemistry. For example, the pioneering German chemist, von Bunsen (1853) analyzed some basalts from Iceland. The only analytical techniques available at that time were gravimetric and volumetric (Section 1.1.2.1). Not only do many of the chemical reactions involved in classical volumetric and gravimetric analyses proceed in our oxygen-bearing atmosphere, but also the cations are generally bonded to O^{2-} in mineral lattices. The result is that products of these reactions were either oxides or could be expressed as combinations of oxides, so the tradition of quoting major elements as oxides was established. These methods were incapable of analyzing trace elements, the technology for which was not developed until the early to mid 1900s.

xxxxxxxxxxxxxxxxxxxxxxxxxxxxxxxxxxxxxxxxxxxxxxxxxxxxxxxxxxxxxxxxxx

Table 1.1 is arranged in the customary order in which the oxides are tabulated for publication, which is more or less with decreasing valence of the cation: Si^{4+}, Ti^{4+}, Al^{3+}, Fe^{3+}, Fe^{2+}, Mn^{2+}, Mg^{2+}, Ca^{2+}, Na^+, K^+, H^+, P^{5+}, and C^{4+}. Although one can argue that CO_2 is not out of place because all volatiles are listed at the end, that argument does not apply for P_2O_5. Moreover, the other two minor oxides, TiO_2 and MnO, are in their "proper" positions. Apparently P_2O_5, a minor element, could not be accorded the distinction of being at the top of the list.

xxxxxxxxxxxxxxxxxxxxxxxxxxxxxxxxxxxxxxxxxxxxxxxxxxxxxxxxxxxxxxxxxx

Because many of these oxides had and still have many industrial and everyday uses, they are referred to by common as well as chemical names:

SiO_2	silica
TiO_2	titania
Al_2O_3	alumina
Fe_2O_3	ferric oxide
FeO	ferrous oxide
MnO	manganese oxide
MgO	magnesia
CaO	lime
Na_2O	soda
K_2O	potash
H_2O+	structural water
H_2O-	adsorbed water
CO_2	carbon dioxide

Terms like *silica, lime, potash, soda,* and *milk of magnesia* are familiar to most people, far more so than the names of most common rocks or minerals.

In the early days trace elements were analyzed only by electronic instrumental methods, primarily *emission spectroscopy* and *colorimetry* (Section 1.1.2.1). Emission and absorption, respectively, of electromagnetic energy was measured. Because the concentration of an element is related to the intensity change of the energy recorded, the element was analyzed quantitatively. Elements were analyzed rather than oxides, and ppm were convenient for reporting purposes, therefore the conventions were established.

1.1.2 Analytical Methods

Chemical analyses of igneous rocks and minerals provide the main database for analytical petrology. Before one can use these data to place constraints on petrogenetic processes, it is necessary to have some idea about how these analyses are obtained. This section provides a brief overview. For a more detailed summary of all modern techniques, in addition to a brief review of classical methods, refer to Potts (1987).

1.1.2.1 Early Techniques

The only analytical method available in the 1800s and early 1900s was classic volumetric and gravimetric analysis (also known colloquially as *wet chemical analysis*). It was as much of an art as a technology and was a very slow, tedious procedure; the best analysts became skillful only after years of experience. Furthermore, it was only capable of determinations for major and minor elements and could take a day or more to analyze a single sample. Maxwell (1968) provides a compendium of volumetric and gravimetric analytical methods. All the chemical steps involved in a classic analysis are beyond our purposes here, but a brief summary is informative. The procedure is still in use today, although only rarely.

The sample was first decomposed by some fusion technique, such as with Na_2CO_3 as a flux. Silica was then determined gravimetrically by evaporation and dehydration from HCl. A series of precipitations, first with NH_4OH and then with NH_4 oxalate, was then made. The first precipitates concentrated the R_2O_3 *group*, which principally included Al, Fe, Ti, and Mn. After Fe, Ti, and Mn were analyzed separately from the R_2O_3 group, Al was determined by difference. The NH_4 oxalate precipitates concentrated Ca and Mg. Calcium was determined gravimetrically as $CaCO_3$ and Mg was titrated. In later years K and Na were analyzed by an instrumental method, flame photometry (discussed later). Similarly, colorimetry (another instrumental method) was later used for Ti, Mn, and P. Ferrous iron was determined by titration.

Trace-element analysis had to await the development of instrumental techniques. Two early instrumental methods that were used for trace-element analysis are emission spectroscopy and colorimetry. The use of emission spectroscopy for trace analysis of rocks and minerals was pioneered by Goldschmidt in Norway and Fersman in the U.S.S.R. during the 1920s and 1930s. Later, Ahrens in South Africa continued this pioneering work. A classic text on spectrochemical analysis is by Ahrens and Taylor (1961). Colorimetry was developed as a viable analytical technique for trace elements by the analytical chemist Sandell (1959). These instrumental methods, particularly colorimetry, are still used today, although they are not as popular as earlier.

1.1.2.2 Rapid Analysis

Before discussing various instrumental methods, the *rapid analysis* method for major and minor oxides developed by the U.S. Geological Survey must be mentioned. This technique is still being used by the USGS. It is actually a combination of instrumental and wet chemical methods that has proven to be quite durable and has been widely adapted in other laboratories. The equipment necessary to perform rapid analyses is relatively inexpensive. The procedure is not so rapid as some modern, automated methods, but it is a vast improvement over earlier techniques. If performed properly, it also provides high quality analytical data.

Two versions of the rapid analysis procedure have been published: Shapiro and Brannock (1962) and Shapiro (1975). The two variations are similar except for analysis of alkalis and alkaline earths. The older method used titration for CaO and MgO, whereas Na_2O and K_2O were determined by flame photometry (see later discussion of these instrumental methods). The newer method uses atomic absorption for all four elements. SiO_2, TiO_2, Al_2O_3, FeO^*, MnO, and P_2O_5 are all analyzed colorimetrically by both methods. Likewise, both methods use titration for FeO, gravimetric analysis for H_2O+ and H_2O-, and volumetric analysis for CO_2.

1.1.2.3 General Types of Analytical Methods

The instrumental methods most widely used today for bulk rock samples are frequently referred to by the following acronyms: XRF, AA, NAA, and ICP (Table 1.2). The *electron probe* is used for in situ chemical analyses of minerals and the *mass spectrometer* is used for isotope analyses. The theoretical basis for these and the older analytical methods will be briefly discussed below (refer to Potts, 1987, for more details).

Most chemical analytical methods can be classified into two groups: electrochemical and spectrochemical. The pH meter is a familiar example of an electrochemical method, but electrochemical methods are seldom used in chemical analyses of rocks and minerals. Spectrochemical methods used in geochemistry and analytical petrology can be divided into three classes, depending on what part of the atom is undergoing either emission or absorption of electromagnetic energy or matter (Table 1.2). For this classification the parts of the atom are the nucleus, "inner" electrons, and "outer" electrons.

The electromagnetic spectrum of energy is critical in this regard. Relatively low (long wavelength) energies, in the ultraviolet (UV), visible (VIS), or infrared (IR) parts of the spectrum, originate with changes in energy levels of the outer electrons. X-rays are involved in energy level transitions within the inner electrons, whereas gamma rays are related to instability (spontaneous disintegration, i.e., radioactivity) of the nucleus. In comparison with energies associated with outer electrons, gamma-rays and X-rays have higher energies, with relatively short wavelengths. Expulsion of matter, such as alpha particles (He nucleii -- two protons and two neutrons) or beta particles (essentially electrons), from the nucleus is also associated with radioactivity.

1.1.2.4 Methods Involving Outer Electrons

Starting with the bottom of Table 1.2, analytical methods that involve the outer electrons are classified into two groups that involve either the absorption or emission of electromagnetic energy. All the classic instrumental methods for chemical analysis fall into one of these two groups. These instruments all have a *monochrometer*, which isolates certain wavelengths of light involved in the analysis. A glass prism is a simple example of a monochrometer, although it is seldom used in modern analytical instruments. Modern monochrometers are either mechanically ruled or holographically recorded diffraction gratings.

TABLE 1.2
Analytical Methods used in Rock and Mineral Analysis

I. Electrochemical

II. Spectrochemical

    A. The nucleus
        1. radioactivity
            a. natural -- alpha, beta, gamma spectrometers
            b. artificial -- neutron activation (NAA)
        2. mass differences -- mass spectrometry

    B. "Inner" electrons
        1. X-ray fluorescence (XRF)
        2. X-ray diffraction (XRD)
        3. electron probe

    C. "Outer" electrons
        1. emission of electromagnetic energy
            a. flame photometry
            b. emission spectroscopy
            c. inductively coupled plasma-atomic emission
               spectrometry (ICP-AES or simply ICP)
        2. absorption of electromagnetic energy
            a. atomic absorption spectrophotometry (AA)
            b. colorimetry

    Atomic absorption spectrometry and colorimetry both involve absorption of energy in the UV-VIS or near-IR part of the electromagnetic spectrum. In colorimetry the sample is a colored aqueous solution, the color being achieved by the addition of a chemical that forms a colored complex ion with the element to be analyzed. The deeper the color, the more absorption of light, and the higher is the concentration of the element in the sample. Each element and thus each colored complex is wavelength specific; that is, it absorbs light in only one or more specific parts of the spectrum. The sample is placed in a glass or quartz *cuvette* (a test tube can be used as a simple cuvette) through which a beam of light is directed.
    In atomic absorption (AA), the solution is aspirated into a flame, commonly a burning mixture of acetylene and either air or N_2O. Another AA technique is to excite a sample in a *graphite furnace,* which produces better

detection limits, but is slower and less precise than using a flame. A beam of monochromatic light is directed through the flame or graphite furnace. The purpose of the flame or furnace is to reduce compounds and complex ions in solution to the atomic state; only atoms can absorb the monochromatic radiation of the light beam. The more atoms of a particular element in the flame that absorb that specific wavelength, the greater will be the absorption of light. Atomic absorption is still widely used because of its relatively modest initial cost and its capability for analyzing a wide variety of major and trace elements.

Two classic instrumental methods, flame photometry and emission spectroscopy, both use the emission of light, again in the UV-VIS or near-IR part of the spectrum. Flame photometry, an optional addition for many modern AA spectrometers, is still widely used for determination of the alkali metals. In flame photometry, as in AA, an aqueous solution containing the sample is aspirated into a flame. From this point, however, the techniques are quite different. Heat from the flame excites the outer valence electrons into higher energy states. As they "fall back" to their original ("ground") state, they emit energy. The familiar yellow Na light at 589 nm upon striking a match is an example. The light intensity is proportional to concentration of the element emitting the energy. The chief drawback of this method is that the flame is only energetic enough to excite atoms that are easily ionized, primarily the alkali metals.

Emission spectroscopy is quite similar to flame photometry, except energy input is much greater, thus more electrons are excited and more elements can be analyzed. The sample, in powdered form, is packed into the hollowed end of a graphite electrode. A direct current (DC) electric arc, or occasionally alternating current (AC) spark, at very high voltage from a nearby electrode provides energy input that excites the electrons. This energy eventually passes through a monochrometer, where individual wavelengths are isolated. Intensities of spectral lines are measured by photomultiplier tubes or photographically on glass plates. The DC arc is sensitive and versatile (i.e., it has low detection limits for many elements), but controlling the arc temperature is quite difficult, so reproducibility (precision; Section 1.1.4.5) can suffer. Temperature, and thus precision, are much more easily controlled in the AC spark, but detection limits are considerably higher.

The inductively coupled plasma (ICP) technique is relatively new and has become widespread only since the 1970s. Although it is akin to both emission spectroscopy and flame photometry, it has several distinct advantages. The most important is that simultaneous determinations of many elements with reasonable accuracy and precision can be achieved. It is an extremely sensitive technique, much more so than most other analytical methods. It, along with some types of NAA, produces the most data of reasonable quality in the least amount of time. It can analyze more samples, each for at least twice as many elements, in a day than classic wet chemistry can analyze in a month.

In ICP the samples are first dissolved and the resultant solutions are then simultaneously mixed with argon gas and aspirated into a miniature radio frequency generator, where an *inductively coupled plasma* is created. Atoms in this plasma are in their excited states. This is a much more effective means of excitation than those employed by either flame photometry or emission spectroscopy. Each emission line produced, characteristic of a particular element, is first isolated with a holographic diffraction grating (monochrometer) and then detected with a photomultiplier tube. The intensity from each tube, proportional to the concentration of that element in the solution, is then fed into an on-board computer for data reduction.

1.1.2.5 Methods Involving Inner Electrons

Three methods involving inner electrons are widely used in analytical petrology (Table 1.2) and they all involve the utilization of X-rays and Bragg's Law:

$$n\lambda = 2d\sin\theta \tag{1.1}$$

where λ is the wavelength of incident X-radiation; d is distance between lattice planes of a crystal; θ is the angle of incidence, which equals the diffraction angle; and n is the *order* (not explained herein). The value for n is always a whole number, usually 1, 2, or 3.

   X-ray fluorescence (XRF) differs from X-ray diffraction (XRD) in that the former is used for quantitative chemical analysis and the latter is useful for identification and semiquantitative analysis of minerals in a rock. In terms of Bragg's Law, for X-ray diffraction (XRD), λ is constant (the specific wavelength of the X-ray tube), whereas d is a function of $\sin \theta$. A powdered sample is bombarded with primary X-rays from the tube. Different minerals in the sample diffract the X-rays according to Bragg's Law. One can then determine a number of *d-spacings* from a *diffractogram,* which collectively serve as a "fingerprint" for a particular mineral.

   In contrast, for X-ray fluorescence d is constant and λ is a function of $\sin \theta$. For both the XRF unit and electron probe, we discuss the more widely used *wavelength dispersive,* rather than newer *energy dispersive,* types. In XRF the d-spacing is that of an *analyzing crystal,* which acts as a monochrometer and allows certain wavelengths of interest to be isolated. Primary X-rays from the tube bombard a powdered or fused rock sample, producing secondary (or fluorescent) X-rays from individual elements in the sample, which are diffracted according to Bragg's Law. The analyzing crystal then isolates X-rays of the analyte element, which are then counted by some type of radiation detector, such as a *scintillation counter.* Because the X-rays of interest have a particular wavelength (λ), they occur at a certain θ angle. Concentration of the analyte is proportional to number of counts per second (cps) recorded by the detector. X-ray fluorescence is one of the most widely used modern instrumental techniques and is used for both major- and trace-element analysis.

   An electron probe is somewhat similar to an XRF unit, except that it has one distinct advantage. In XRF electrons are "boiled" off the cathode of the tube, are accelerated across a very large electric potential gradient (up to 100,000 volts), and bombard a metal anode or "target." Primary X-rays are produced, which in turn impinge upon a sample, producing secondary X-rays. This is not an efficient means of energy production because considerable energy is dissipated as heat.

   The electron probe is much more efficient than XRF because a sample is the target for a beam of electrons, so primary rather than secondary X-rays are produced. They are then treated in much the same way as secondary X-rays from an XRF unit; they are diffracted by an analyzing crystal according to Bragg's Law and detected by a radiation detector. Because of increased efficiency, an electron probe can analyze very small samples, down to diameters of 1 micron or less in a thin section. It is not as useful, however, as XRF for trace-element analysis. Consequently, the electron probe is used for in situ mineral analyses, while an XRF unit is used for whole-rock analyses of much larger samples (commonly several grams).

1.1.2.6 Methods Involving the Nucleus

Methods that involve the nucleus are concerned with either measurement of radioactivity or mass differences. Radioactivity can be natural or artificially induced. Neutron activation analysis (NAA) requires artificially induced radioactivity. If a sample is placed in a nuclear reactor and is bombarded with neutrons, artificial radioisotopes are produced. A radiation detector, somewhat similar to the scintillation counter used for X-rays, can then be used to count gamma radiation emitting from one or more radioisotopes of interest. Because the intensity of this radiation is proportional to the amount of the original element in the sample, quantitative analysis can be performed. The advantage of this method is that many elements can be analyzed simultaneously and detection limits are quite low.

Natural radioactivity is normally only measured for Th, U, and their many radioactive daughters, or K. Potassium has one radioactive isotope, ^{40}K, which decays to ^{40}Ar. Three types of spectrometers are used, depending on the type of radiation involved: alpha, beta, or gamma. Today Th and U are widely analyzed by NAA, whereas K can be easily analyzed by many methods -- XRF, AA, ICP, and flame photometry. Consequently, measurement of natural radioactivity as a means of determining these elements is not as common now as in earlier years. When a method is used for igneous rocks, it is normally gamma-ray spectrometry. It operates much as NAA, except natural rather than artificial radioactivity of K, U, and Th (or their radioactive daughters) is used.

Finally, the most precise and accurate of all methods is mass spectrometry. It is the primary method used in geochronology, where extremely accurate and precise data for individual isotopes are required. Except for investigations of natural radioactivity, it is the only method used for isotopic studies of igneous rocks. Because it can differentiate different mass numbers, isotopes can be distinguished from one another.

In mass spectrometry ions from the sample are introduced into a very high vacuum chamber, where they are accelerated and their paths are curved by a large magnet. Just as automobiles rounding a curve, the heavier ions (those with larger mass numbers) tend to drift to the outside of the curve relative to the lighter ions. Thus individual isotopes are isolated and are detected by some device such as a *Faraday cup*. The disadvantages of mass spectrometry are its high initial cost and the procedures, compared with those used for other methods, are quite time consuming. For most isotopic studies, however, one has no choice.

1.1.3 Descriptive Statistics

Statistical manipulation of data can be considered as two types: *descriptive* or *exploratory statistics*, as opposed to *inferential* or *confirmatory statistics*. Descriptive statistics are sufficient for perceiving a relationship between variables, but to demonstrate the validity of that relationship requires inferential statistics. Either type can be considered on the basis of whether we are dealing with one, two, or many variables. These are referred to respectively as uni-variate, bivariate, and multivariate methods. This book does not discuss multi-variate techniques, or deal with inferential statistics, which is concerned with confidence levels and probabilities. Any basic statistics book covers this area (e.g., Dowdy and Wearden, 1983). Le Maitre (1982) provides a particularly good general survey of geostatistics in igneous petrology, including multivariate procedures. For elementary statistical theory explained in an especially literate fashion, the book by Simpson and others (1960) is recommended.

Variables in petrology are either *discrete* or *continuous*. A variable representing the number of samples we have collected or the number of zircon grains in a thin section is a discrete variable. We may have collected 22 or 23 samples, but not 22.5743 . . . samples. Chemical or mineralogical compositions are continuous variables, as the example of the true quartz content of a granitic rock being 22.5743 . . . percent. Continuous variables, however, appear to be discrete. For instance, we report the quartz content mentioned above to be 22.57, 22.6, or even 23 percent, depending upon the number of decimal places quoted. The true amount of quartz, 22.5743 . . . percent, is a value that we will never know. Data such as 22.5, 22.6, 22.7 appear to be discrete variables when in fact they are continuous.

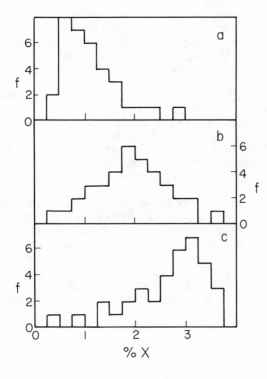

Figure 1.1 Three histograms showing *a.* positively skewed (approximately log normal) distribution, *b.* apparently normal distribution, and *c.* negatively skewed distribution. Number of analyses, or frequency, is *f.*

1.1.3.1 Distributions

Continuous variables are commonly normally distributed. A *normal* (or *Gaussian* or *bell-shaped)* distribution means that the variables are symmetrically distributed around the *arithmetic mean* and their *histogram* follows a particular mathematical function. Figure 1.1 shows three histograms, the middle of which approximates a normal distribution. The other two distributions are *positively skewed* and *negatively skewed*.   Histograms of chemical or mineralogical analyses can exhibit any   gradation between these three distributions.   Most *parametric* (i.e., classic) inferential methods require that variables be normally distributed, but this is not so important for simple descriptive statistics.

A histogram of a trace element is very likely positively skewed and approximates a *log-normal* pattern. This means that if one were to take logs of all data, the histogram of logs would be normally distributed. Figure 1.2 is based on logs of data used to construct the positively skewed histogram of Figure 1.1. As negative concentrations are impossible, trace elements frequently exhibit log-normal distributions because their arithmetic means are quite near zero and any spread in the data must be toward larger values.   This spread toward larger values is a positive skew, which can approximate a log-normal distribution.

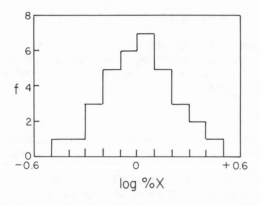

Figure 1.2 Histogram of logs for data used to plot Figure 1.1*a*.

In contrast, SiO_2 is probably the most common element whose histogram exhibits a negative skew.   The silica content of any igneous rock not affected by secondary processes normally cannot exceed about 75 percent SiO_2. If we are dealing with granitic rocks, the arithmetic mean may be near the maximum of 75 percent.   Hence, any spread in data is toward lower values for SiO_2, causing a negative skew in the histogram.   To summarize, if the mean of a variable is near its lower limit, a log-normal distribution is probable; if near the upper limit, the distribution is more likely negatively skewed.   The remaining oxides commonly have distributions that approximate normality, although many exceptions exist.

To be strictly correct, one should not calculate any statistical parameter defined below unless a normal distribution can be at least approximated for the variable involved. If, for example, the histogram appears to be log-normal, one can simply take logs of all data for that variable and perform all statistical calculations on those logs. This process is called *normalizing* the data set. If parametric, inferential statistics are being used, this normalization is essential. If only descriptive statistics are involved, it is not so critical.

1.1.3.2 Univariate Parameters

Basic univariate, descriptive statistical parameters are of two types, those that are measures of *central tendency* and those that measure *dispersion*. Any one of the former, of which there are several, can be loosely called an *average*. These include the arithmetic mean, median, mode, geometric mean, and harmonic mean, but we are only concerned with the arithmetic mean. Any standard introductory statistics text defines the remaining terms. The arithmetic mean, or simply mean ($\bar{x}$), is defined by a familiar equation:

$$\bar{x} = \Sigma X_i / n \qquad\qquad (1.2)$$

where ΣX_i is the summation of all values and n is the number of values for the variable. Thus $\bar{x}$ for the seven apparently discrete variables 1, 3, 2, 5, 4, 1, and 5 is:

$$\bar{x} = (1 + 3 + 2 + 5 + 4 + 1 + 5) / 7 = 21 / 7 = 3$$

Four measures of dispersion of data about the mean are of interest. If we define a deviation *(d)* as being equal to the difference between an individual value and the mean value, then $d = X_i - \bar{x}$. The first measure of dispersion is the *variance* (s^2) of a sample, which is defined as:

$$s^2 = \Sigma d^2 / (n - 1) \qquad\qquad (1.3)$$

Because each deviation is squared, s^2 will always be a positive quantity. In a statistical context, a *sample* refers to a group of individual measurements from a much larger *population* of measurements and is assumed to be representative of that population. In petrology, a sample is normally used to mean an individual specimen. The denominator of Equation 1.3 is the *degrees of freedom (df)*, which are defined as the number of values minus the number of statistical parameters required to explain those values. Because $\bar{x}$ is the only parameter required, for Equation 1.3 $df = n - 1$.

The second measure of dispersion, the sample *standard deviation,* is simply the square root of variance:

$$s = [\Sigma d^2 / (n - 1)]^{1/2} \qquad\qquad (1.4)$$

A standard deviation has the same units as its arithmetic mean. If a mean is in ppm, the standard deviation will also be in ppm. We square each deviation before summing the squares, rather than squaring the sum, e.g.:

X_i	d	d^2
1	-2	4
3	0	0
2	-1	1
5	2	4
4	1	1
1	-2	4
5	2	4

$$\bar{x} = 3 \quad \Sigma d^2 = 18 \quad n = 7$$

$$s^2 = 18 / 6 = 3 \quad \therefore \quad s = 1.732$$

The *percent deviation* or *coefficient of variation (V)* is the third measure of dispersion:

$$V = 100s / \bar{x} \qquad\qquad (1.5)$$

For the above example:

$$V = 100 \times 1.732 / 3 = 58 \ \%$$

No matter what the original units, V is always in units of percent.

We seldom use this basic formula for s^2, but rather one that seemingly is more complicated. This second equation has an advantage over the first in that it is easily solved by calculators and computers. It is algebraically equivalent to Equation 1.5. The equation is:

$$s^2 = [\Sigma X^2 - ((\Sigma X)^2 / n)] / (n - 1) \qquad\qquad (1.6)$$

Hand calculators that have "hard-wired" statistical functions normally use this formula for sample variance and its square root for sample standard deviation. The ΣX^2 term requires that we square each variable before we sum the squares, whereas the $(\Sigma X)^2$ requires the opposite -- the variables are first summed and then the sum is squared. Using Equation 1.6 to solve the previous example:

$$\Sigma X^2 = 1 + 9 + 4 + 25 + 16 + 1 + 25 = 81$$
$$(\Sigma X)^2 = 21^2 = 441$$
$$n = 7$$

$$s^2 = [81 - (441 / 7)] / 6$$
$$= 18 / 6 = 3$$

A BASIC computer program to calculate the arithmetic mean, standard deviation, and coefficient of variation is given in Section 1.6.2.2.

The arithmetic mean and standard deviation are related to the normal distribution. If we represent a histogram of apparently normally distributed data by a smooth curve (Fig. 1.3), the mean will be at the center and divide the curve into two symmetrical halves. This is why the mean is referred to as a measure of central tendency. About 68 percent (68.26, to be exact) of the area under a normal curve is within $\pm$ 1 standard deviation ($\pm$ 1s) of the mean, whereas about 95 percent will be within $\pm$ 2 standard deviations ($\pm$ 2s). The greater a value is for s, the "wider" will be the normal curve, hence standard deviation is a measure of dispersion about the mean. If, using the example calculations above, our sample is normally distributed and has $\bar{x} \pm s = 3.0 \pm 1.7$ ppm Rb, then about 68 percent of the area under the normal curve for this sample should fall between 1.3 and 4.7 ppm. About 95 percent theoretically falls between 0 (actually, -0.4) and 6.4 ppm.

These percentages under the normal curve for $\pm$ 1s and $\pm$ 2s probably do not exactly hold for a histogram of our 3.0 $\pm$ 1.7 ppm Rb sample. It would be sheer coincidence if they did. This is because our sample was drawn from a much larger population. In the case of a large igneous body, such as a batholith, effectively an infinite number of rock specimens could be collected and Rb analyses performed. Our statistical sample probably comprises, at most, a few tens or hundreds of petrological samples. The smooth normal curve represents a theoretical histogram for the entire population, a histogram for which we never know its exact shape, but can approximate it with the sample histogram. Just how well we approximate it is the realm of inferential statistics and is not dealt with here.

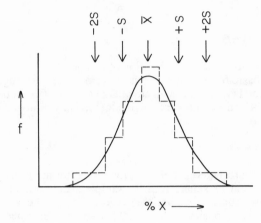

Figure 1.3 A smooth normal distribution curve superimposed on a histogram.

Rather than collecting individual specimens from a batholith, suppose we had grouped our specimens into a number of statistical samples. We could determine $\bar{x}$ for each sample and then calculate the standard deviation of sample means. Standard deviation of the arithmetic means must be smaller than standard deviation of all individual specimens. This is because we have more confidence in our knowledge of a mean value than an individual value and, therefore, dispersion of the means should be smaller than dispersion of individual values. Standard deviation of the means is referred to as the *standard error of estimate* of the mean, or simply the *standard error*. This is the fourth measure of dispersion. We can estimate the standard deviation of means without actually having made all the measurements by calculating the standard error, which is defined as:

$$e_s = s / n^{1/2} \tag{1.7}$$

where s is the standard deviation and n is number of measurements for the sample. This equation clearly indicates that standard error e_s will always be smaller than s for all samples of more than one specimen. If we wish to consider the dispersion of individual measurements, we must use standard deviation, but if we are concerned with dispersion of mean values, standard error must be used. For the preceeding example the standard error can be calculated as:

$$e_s = 1.732 / 2.646 = 0.655$$

This, then, is an estimate of the standard deviation of means taken from multiple samples of the entire population, although we never literally calculated all those means.

If we can calculate the standard error of estimate for an arithmetic mean, we can clearly do so for other statistical parameters. For example, standard error for the coefficient of variation (e_V) can be estimated by $V/(2n)^{1/2}$. A similar expression can be used to estimate standard error for the standard deviation

(e_{ss}), i.e., $s/(2n)^{1/2}$. Notice that standard error for the standard deviation can be referred to as "standard deviation of the standard deviations." Again, using the above example:

$$e_V = 58 / (14)^{1/2} = 58 / 3.742 = 15.5 \%$$
$$e_{ss} = 1.732 / 3.742 = 0.463$$

1.1.3.3 Correlation

Bivariate statistics are particularly useful in petrology because of our use of variation diagrams (Sections 1.3.2 and 5.2.2). The simplest of all variation diagrams is an *X-Y scattergram* (Fig. 1.4). Chemical or mineralogical data are often plotted on such diagrams to characterize the two variables. Petrologists and geochemists are forever searching for "trends" on these diagrams. A trend is considered to exist if the two variables apparently vary together in some systematic way. The intensity of such a relationship between two variables can be measured by the *correlation coefficient,* whereas the mathematical relation-ship itself is described by *regression coefficients.*

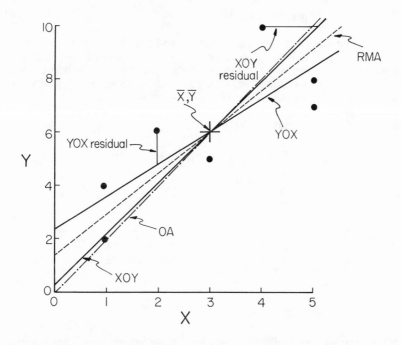

Figure 1.4 An *X-Y* scattergram. x̄ and ȳ are arithmetic means for variables *X* and *Y*. *Dots* represent the actual data points. Lines labeled XOY and YOX were obtained by regressing *X* on *Y* and *Y* on *X*, respec-tively. The line OA was drawn through the point representing the two means and the origin of the graph. The line RMA was fit by the reduced major axis method.

If two variables vary in such a way that one apparently increases or decreases as the other increases, we say that these two variables are correlated. The term *correlate* is informative if we think of it as *co-relate,* a measure of how well two variables relate to one another. If both variables increase, this is referred to as a positive correlation; if one increases while the other decreases, we call this a negative correlation. This correlation can be linear; that is, allowing for some scatter, the plotted points appear to approximate a straight line. The correlation can also approximate a curve, but this discussion is only concerned with linear correlations. The intensity of a linear correlation can be measured by a *linear correlation coefficient,* also known as the *Pearson* or *product moment* correlation coefficient. In correlation, as opposed to some types of regression, no assumptions have to be made about which variable is independent and which is dependent. Section 1.1.3.4 discusses regression as well as independent and dependent variables.

Before defining the linear correlation coefficient, however, we must introduce *covariance.* Covariance is to bivariate statistics as variance is to univariate statistics. We can define a deviation for an X-variable (plotted on the horizontal axis of the scattergram) as $d_X = X_i - \bar{x}$. A deviation for a Y-variable (vertical axis) is $d_Y = Y_i - \bar{y}$. Covariance *(C)* is then defined as:

$$C = \Sigma(d_X d_Y) / (n - 1) \tag{1.8}$$

Notice the similarity between Equations 1.3 and 1.8. In this case n refers to the number of pairs of X,Y data. Because one pair is required to plot one point, n also refers to the number of points on the scattergram. With reference to the definition of degrees of freedom, as n equals the number of pairs of variables, only one pair of means is required ($\bar{x}$ and $\bar{y}$), so $df = n - 1$.

A simple example illustrates the concept of covariance. The following data are used to plot points on Figure 1.4:

X	d_X	Y	d_Y	$d_X d_Y$
1	−2	2	−4	8
3	0	5	−1	0
2	−1	6	0	0
5	2	8	2	4
4	1	10	4	4
1	−2	4	−2	4
5	2	7	1	2

$\bar{x} = 3$ $\bar{y} = 6$ $\Sigma d_X d_Y = 22$

$\therefore C = 22 / 6 = 3.667$

In this example Y generally increases with X, so we can state that Y and X are "positively correlated." As a result, positive deviations in X are generally paired with positive deviations in Y; negative deviations are likewise paired. The result is a positive covariance. Had negative deviations been paired with positive deviations, the result would be a negative covariance and thus negative correlation. A negative correlation, therefore, does imply a relationship between two variables, where one variable generally increases as the other decreases. Using Equations 1.3 and 1.4, $s_Y^2 = 7.0$ and $s_Y = 2.646$. The calculations above indicate that $s_X^2 = 3.0$ and $s_X = 1.732$.

As in the case of variance, an equation for C algebraically equivalent to Equation 1.8 and convenient for solution on computers and calculators is:

$$C = [\Sigma XY - ((\Sigma X \Sigma Y) / n)] / (n - 1) \qquad\qquad (1.9)$$

For our example of covariance:

$$\Sigma XY = 2 + 15 + 12 + 40 + 40 + 4 + 35 = 148$$
$$\Sigma X \Sigma Y = 21 \times 42 = 882$$
$$n = 7$$

$$C = [148 - (882 / 7)] / 6 = 22 / 6 = 3.667$$

We can now define a linear correlation coefficient *(r)*. It is covariance divided by the product of standard deviations for X and Y:

$$r = C / (s_X s_Y) \qquad\qquad (1.10)$$

For the preceeding data:

$$r = 3.667 / (1.732 \times 2.646) = +.80$$

Because s_X and s_Y will always be positive, the algebraic signs on C and r are the same.

It follows from the previous discussion that if all data points fall exactly on a straight line and Y increases with increasing X, (i.e., the line has a positive slope), this is a perfect positive correlation and $r = +1.0$ (Fig. 1.5c). In this situation the product of the standard deviations equals the covariance. If, as stated earlier, Y decreases with increasing X, then the covariance will be negative. In this case, if all points plot exactly on a line with a negative slope, $r = -1.0$ (Fig. 1.5d). If no correlation, and thus no relationship, exists between X and Y (colloquially referred to as a "shotgun" plot), $r = 0$ (Fig. 1.5a). Small but non-zero values of r indicate a questionable relationship between X and Y. The correlation coefficient, then, can vary continuously between +1.0 and -1.0. Some plots with intermediate values for r are also shown (Figs. 1.5e-f). Any set of points whose best-fit line is either horizontal or vertical will have $r = 0$ (Fig. 1.5b), even if all points fall on the line, because C will be zero.

Occasionally a *coefficient of determination (D)* will be quoted rather than r. The coefficient of determination is nothing more than r^2. Because it is the square of r, it will always be a positive quantity and, therefore, cannot reflect a negative correlation. Its usefulness comes from the fact that, when expressed as a percent, it approximates the percentage of variance in one variable explained by the other. Hence if $r = 0.90$, $D = 0.81$, and roughly 81 percent of the variance in the dependent variable Y is explained by the independent variable X (see the following discussion).

1.1.3.4 Regression

The correlation coefficient or coefficient of determination quantifies the intensity of a mathematical relationship between two variables, but tells us nothing about the relationship itself. Stating this in practical terms, we may need some way to calculate an intercept and slope of a best-fit straight line through the data points on our scattergram. We do this through calculation of regression coefficients, which are simply this intercept and slope. As in treatment of correlation, we deal only with straight lines, equations for which are called *first-order polynomials*. The equation is $Y = mX + b$ where m is the slope and b is the intercept on the Y axis. We could fit logarithmic curves or curves defined by higher order polynomials, such as a parabola, which requires a "second order" or *quadratic* equation, $Y = a + bX + cX^2$. Principles are the same for either a linear or quadratic fit, so we will be content with straight lines. A program for fitting either a straight line or a parabola, which includes univariate statistics, a linear correlation coefficient, and coefficients of determination, is included in Appendix D.

Suppose we have a scattergram (Fig 1.4) on which a best-fit line may pass through the origin of the graph. Let us further suppose that we have some good a priori reason why the best-fit line should go through the origin. The simplest way to fit a straight line through these points is to calculate the mean for X ($\bar{x}$) and for Y ($\bar{y}$) and plot this point on the graph (marked by a large plus sign on Fig. 1.4). The best-fit line using this simple method passes through this point and the origin of the graph (line OA on Fig. 1.4). The slope, m, of this line will be $\bar{x}$ / $\bar{y}$ and the intercept, b, is zero.

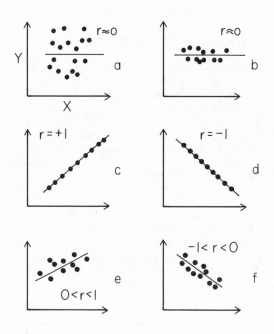

Figure 1.5 Some examples of correlation coefficients: *a.* and *b.* no correlation; *c.* and *d.* perfect correlation; *e.* and *f.* intermediate level of correlation.

For most scattergrams the best-fit line does not necessarily pass through the origin, so how do we fit a line through data points in the more general case? In fact, a close examination of the distribution of points on Figure 1.4 suggests that perhaps a best-fit line should intersect the Y-axis rather than go through the origin. Many methods for fitting a line through points on a scattergram have been proposed, but only two are considered.

The most widely used method is called *least-squares regression*. The term *least squares* is derived from the fact that when the summation of all squared *residuals* is at a minimum, the line so defined must be the best-fit line. In this case, a residual is simply the difference between an actual value and its value determined from the best-fit line. If residuals are measured parallel to the Y-axis (Fig. 1.4), this is referred to as "regressing Y on X"; Y becomes the dependent (or "response") variable and X is the independent (or "predictor") variable. If residuals are measured parallel to the X-axis (Fig. 1.4), the opposite is true -- regression of X on Y means that X is dependent and Y is independent. The more common procedure is to regress Y on X; this is the technique "hard-wired" into many hand calculators. A common definition for *regression* in a dictionary is a "going back." Hence we can look upon regression analysis as going back from the predictor variable to the response variable, that is, esti-mating the response variable based on a value for the predictor variable.

Herein lies a problem. If the data plotted on the scattergram are some-what scattered ($r < +1.0$ or > -1.0), regressing Y on X will yield a different best-fit line from regressing X on Y (Fig. 1.4). Furthermore, frequently we cannot decide which is the dependent and which is the independent variable. In theory, the independent variable has no error (precision) associated with its measure-ment, whereas the dependent variable does have error. If, for example, the X-variable were distance down a core hole or along a sampling profile, and Y were some mineralogical or chemical percentage, X could be considered independent and Y dependent (although some small error will be associated with making the distance measurement). Both X and Y are commonly mineralogical or chemical compositions, so neither can be rightly called independent or dependent. The differentiation (or fractionation) index (see Section 5.2.2.2) is generally plotted on the X-axis, but it cannot be considered as a truly independent variable.

The second method provides a solution to this problem; it is called the *reduced major axis* (or RMA) technique. One assumes no independent or dependent variables. Residuals parallel to X and Y are both minimized by RMA. Using RMA, the best-fit line bisects an angle formed by the two best-fit lines determined by regressing Y on X and X on Y by least-squares (Fig. 1.4). As r approaches +1.0 or -1.0, the acute angle formed by the Y-on-X (YOX) and X-on-Y (XOY) regression lines becomes increasingly small until these two lines and the RMA best-fit line merge into one line for a perfect correlation. If a best-fit line is defined by $Y = mX + b$, then, using RMA, m and b are defined as:

$$m_{RMA} = s_Y / s_X \quad \text{and} \quad b_{RMA} = \bar{y} - m_{RMA}\bar{x} \qquad (1.11)$$

where m_{RMA} is the slope and b_{RMA} is the intercept of the best-fit line. All remaining terms are defined above. A word of warning is required here -- as both s_Y and s_X are positive quantities, the slope m_{RMA} can never reflect a negative correlation. If r is negative, m_{RMA} must be made negative. We have already calculated all the data necessary to determine m_{RMA} and b_{RMA}. Using data from the example above:

$$m_{RMA} = 2.642 / 1.732 = 1.525$$
$$b_{RMA} = 6 - (1.525 \times 3) = 1.425$$

Because $r = +0.80$, $m_{RMA} = +1.525$. Thus the formula for the best-fit line using the reduced major axis technique (line labeled RMA on Fig. 1.4) is :

$$Y = 1.525X + 1.425$$

Equations 1.11 can be compared with those necessary to calculate the least-squares regression coefficients (regressing Y on X):

$$m_{YOX} = C / s_X^2 \quad \text{and} \quad b_{YOX} = \bar{y} - m_{YOX}\bar{x} \qquad (1.12)$$

The equation for b in Equations 1.12 is identical to that for b in Equation 1.11, except the latter requires m_{YOX} and the former specifies m_{RMA}. Note that C is involved in the calculation of m_{YOX} but not m_{RMA}. As $C > 0$ for a positive correlation and $C < 0$ if the correlation is negative, the sign on m_{YOX} will be correct. We have also already calculated all the parameters necessary to determine these YOX coefficients. Continuing to use our example and regressing Y on X:

$$m_{YOX} = 3.667 / 3 = 1.222$$
$$b_{YOX} = 6 - (1.222 \times 3) = 2.334$$

$$\therefore Y = 1.222X + 2.334$$

The line resulting from this equation is labeled on Figure 1.4 as YOX. Similarly, an X on Y regression results in:

$$m_{XOY} = C / s_Y^2 \quad \text{and} \quad b_{XOY} = \bar{x} - m_{XOY}\bar{y}$$
$$m_{XOY} = 3.667 / 7 = 0.524$$
$$b_{XOY} = 3 - (0.524 \times 6) = -0.144$$

$$\therefore X = 0.524Y - 0.144$$

The best-fit line based on this regression is labeled as XOY on Figure 1.4.

We can now characterize a set of points on a scattergram by r (or D), m, and b. These parameters are all interrelated, especially if one also considers $\bar{x}$, $\bar{y}$, s_X, s_Y, and C. To determine all these statistical parameters, we need calculate only the following summations: ΣX, ΣY, ΣX^2, ΣY^2, ΣXY, and n. These calculations are all very easily done by a fairly short computer program if arranged in the proper order (Appendix D). After calculating the six summations, the best order is: (1) $\bar{x}$, $\bar{y}$, s_X, and s_Y; (2) C; and (3) m, b, r, and D. If some reason exists for the line to go through the origin of the graph, simply calculate $\bar{x}$ and $\bar{y}$, then draw the line through this point and the origin. If no reason exists, and if neither an X nor Y variable is independent and the other dependent, use the RMA method. If one variable clearly is independent and the other dependent, then a choice can be made to regress Y on X or X on Y. Traditionally, whenever possible, Y is regressed on X.

1.1.4 An Example of the Analytical Method

Section 1.1.2 summarizes various analytical methods used in igneous rock and mineral analysis. This section takes as an example one analytical instrument, the atomic absorption (AA) spectrophotometer, and goes through various steps necessary to produce a chemical analysis of an igneous rock. The discussion is based on an AA unit with a burner system, rather than one with a graphite furnace. Multielement techniques, such as ICP, NAA, and energy dispersive XRF, are quite common now, but the broad principles are similar and a great deal can be learned from a simple example.

1.1.4.1 Sample Preparation

Atomic absorption spectrophotometry requires that a sample be in solution, so the first step is to crush the rock to a powder. One must be careful that the sample is large enough to be representative and that it is not contaminated during crushing. Fist- or even smaller-sized specimens of fine-grained volcanic rocks are adequate, but the coarser grain sizes of plutonic rocks require larger specimens to be representative. Consequently, adequate samples of pegmatites, which are extremely coarse-grained rocks, generally have to be extremely large, on the order of many kilograms.

After chipping with a rock hammer or trimming off weathered material with a diamond saw, crushing is normally done in two stages: first crush to gravel-sized material and then crush to a powder. After the first crush the sample is split to obtain an unbiased smaller sample for the second crush, which is necessary because the relatively fine material is compositionally different from the relatively coarse. A common technique to achieve an unbiased sample is to *cone and quarter* the material. This is done by pouring all the sample onto a large sheet of clean paper in a pile and then quartering the "cone" with a stiff sheet of plastic or cardboard. This procedure is repeated until a sufficiently small sample has been obtained for the second crush. This final crush should be done in some type of apparatus in which the grinding container is composed of some extremely hard material that will not contaminate the sample, such as W carbide, Si carbide, or Al_2O_3 ceramic. Tests should be run to assure that samples are not being contaminated. This can be done by grinding pure quartz and analyzing it for the appropriate elements. The final powder should at least pass through a -100-micron mesh diameter screen.

The next step is to "open" the sample, that is, dissolve a known weight of it in a known volume of aqueous, acidic solution, one of which is normally HF. An alternative is to fuse the sample with some flux, such as lithium metaborate, before dissolution. Generally one can dissolve the crushed powder directly. Any of a number of acids can be used with HF, the most common of which are HCl - HNO_3, H_2SO_4, or H_3BO_3. If SiO_2 is not to be analyzed, the samples can be dissolved in open containers with moderate heat. Silica escapes as the gas SiF_4. Teflon beakers and a water bath or bank of heat lamps work well. Recently analysts have turned to microwave ovens for opening samples, in which case the sample containers must be tightly sealed. An HF - H_3BO_3 mixture in sealed polycarbonate flasks or teflon "bombs" opens the sample.

1.1.4.2 Standards

After dissolution the samples are ready to be analyzed, but first some *standards* must be prepared. Standards are of two types: (1) natural rock standards analyzed many times, preferably in a number of different laboratories, such that their compositions are well known, and (2) artificially prepared solutions, commonly called *salt standards,* that contain known amounts of the elements to be analyzed. If an adequate supply of rock standards is available, salt standards are normally not required. We confine our discussion to the use of rock standards. Fortunately, a large number of igneous rock standards are available from various laboratories around the world. Recent compilations of "preferred values" of elements for these standards, as well as sources of the standards themselves, can be obtained from the journal *Geostandards Newsletter.* Pioneering work in this area has been done by several U. S. Geological Survey geologists, notably Fairbairn and others (1951) in their classic study of the

USGS rock standards G-1 and W-1, as well as Flanagan (see his 1976 paper for earlier references). A recent compilation of preferred values for elements in rock standards from around the world is by Abbey (1983).

1.1.4.3 Complications

Atomic absorption depends on standards, as do most instrumental methods used for the analysis of igneous rocks. Ultimately, in AA we determine the composition of an unknown sample (i.e., a sample of unknown concentration, or *unknown*) by comparing the degree to which it absorbs monochromatic light with that same measure for several standards. This is not always as simple as it appears because there are usually some complications, two of which are worthy of mention.

The first complication is due to chemical or spectral interferences from other chemical species in the solution. We make no attempt to distinguish between these two types of interferences, but rather to consider them collectively as *matrix effects*. If the total matrix effect is to decrease the absorbance of light by an atom in the flame relative to its absorbance in the absence of interfering species, this is referred to as *depression* (absorbance is a measure of light absorption by atoms in the flame). Conversely, *enhancement* occurs when the matrix effect increases absorbance over the value that would exist had no interfering species been present. Depression lowers the efficiency of light absorption, whereas enhancement raises it.

Interferences in AA are not as serious a problem as they are in many other techniques, but they still occur. Depression is generally more common than enhancement. The simplest way to solve this problem is to be certain that matrices (i.e., collective compositions of possible interfering ions) are matched as closely as possible in all rock standards and unknowns. For example, if mafic rocks are unknowns, use mafic rocks as standards. Another precaution is to be sure that rock standards and unknowns are opened by exactly the same procedure.

The second complication concerns instrumental drift. During analysis sensitivity of the instrument changes with time. If one analyzes the same sample numerous times while analyzing the unknowns and standards, almost invariably the replicate absorbances for this sample changes. This change is not always systematic. The absorption for this "drift sample" may drift up (higher sensitivity) or, more often, down (lower sensitivity). This drift may be due to either electronic drift or, more likely, slight changes in the aspiration rate of the solutions.

Several cures for instrumental drift are possible: (1) instrumental settings can be changed as the drift sample is monitored so that it always yields about the same absorbance; (2) a different calibration curve (discussed later) can be used for different absorbances of the drift sample; or (3) rather than plotting absolute absorbances on the X axis of the calibration curve, a ratio of absorbance of a standard or unknown to the absorbance of the nearest (i.e, aspirated closest in time) drift sample is plotted.

1.1.4.4 The Calibration Curve

The next step is to prepare a *calibration curve* (also called "standard curve" or "working curve;" Fig. 1.6). Some AA instruments allow for direct calibration by entering concentrations of standards directly into the memory of the "on-board" computer, but frequently not enough standards can be entered. To prepare this calibration curve, one simply plots absorbance or a ratio of absorbances on the X-axis and concentration on the Y-axis for a number of standards. If an ample number of rock standards is available, then one calculates the formula for the "best-fit" curve through these data points by regression. Only for quite low concentrations or short path lengths will the best-fit line be linear, although the path length can be shortened by revolving the burner. In general, a second-order polynomial (parabola) provides the best fit. Knowing absorbance for each unknown, one can simply solve this regression equation for its concentration.

A word of warning is required here. Never extrapolate a calibration curve beyond its highest standard in AA, because the curve very likely is not linear; that is, at least one rock standard should have a higher concentration than any of the unknowns. This is less of a problem at the lower end of the calibration curve, because the curve either passes through the origin of the graph or some Y-intercept value at zero concentration.

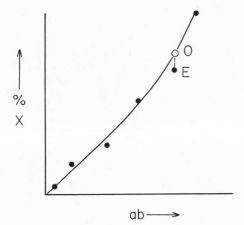

Figure 1.6 A calibration curve for hypothetical element *X* based on six standards (absorbance, *ab)*. The point *O* is the obtained value for that particular standard, while point *E* is the expected (accepted) value. The difference between *O* and *E* *(O - E)* is the residual.

Simply plotting and calculating regression coefficients for a calibration curve are not enough. We must have some way to evaluate reliability of the curve, based on the assumption that concentrations of standards are accurately known. This evaluation is generally done by calculation of a correlation coefficient r for a linear calibration curve, or a coefficient of determination D for a second-order curve (Appendix D). A correlation coefficient (r) of at least 0.98 and preferably >0.99 is normally recommended. Alternatively, χ^2 can be calculated for either a linear or non-linear curve. The quantity χ^2 is defined as:

$$\chi^2 = \Sigma[(O - E)^2 / E] \qquad\qquad (1.13)$$

where O is the observed concentration of a standard determined from the calibration curve and E is its expected (or accepted) concentration (Fig. 1.6). Because the magnitude of χ^2 depends in part on the number of standards analyzed, no absolute value of χ^2 exists above which the calibration curve is not acceptable. The lower χ^2 is, the more reliable is the calibration curve.

A common situation is for the calibration curve to be unacceptable because r (or D) and χ^2 are too high. This is perhaps more common for other instrumental methods than for AA, but it can happen for any method. Assuming (1) published concentrations for rock standards are accurate (certainly not true for some trace elements in some standards); (2) the sample preparation procedure is adequate; and (3) good laboratory procedures are used, the most likely problem is some type of elemental interference. A simple procedure for determining which element is the culprit is to plot residuals versus several elements that might cause the interference. A hypothetical example is shown in Figure 1.7. Clearly too much scatter exists around the calibration curve for element A (Fig. 1.7a), but a plot of the Y-residuals versus element B reveals that apparently it is responsible (Fig. 1.7b). Because of the negative correlation between the concentration of B and the residuals, B must depress the absorbance of A.

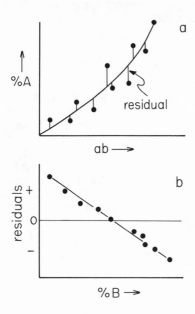

Figure 1.7 *a.* Calibration curve for concentration of element *A* plotted against absorbance *(ab). b.* Plot of residuals measured from calibration curve against concentration of element *B.*

A simple solution to this problem is to develop a correction factor based on the degree of depression of A by B. This correction factor is related to the slope of the line on Figure 1.7b. About a 10 percent increase in the amount of B leads to an approximate 2 percent decrease in the *apparent* concentration of A. Each of the unknowns must then be analyzed for B and the correction factor applied to each analysis of A. Fortunately, if unknowns and standards are closely matched, differences in matrix effects are minimized and this technique is usually not required. As an example, if reliable mafic rock standards are used for mafic unknowns, typically there is little scatter about the calibration curve.

1.1.4.5 Accuracy and Precision

Once our standards have been analyzed, how do we assess *accuracy* and *precision* of the analytical method? Although both contribute to the quality of the analyses, they are quite different. The difference between them can best be understood by examining Figure 1.8. We can make the analogy between firing at a target and performing a chemical analysis. Target *a* represents good accuracy and good precision; the shots are all clustered in the "bull's eye." Target *b* is an example of good precision but poor accuracy; the shots are tightly clustered but considerably away from the bull's eye. The opposite situation is shown on target *c;* the average shot is near the center of the bull'e eye but there is considerable scatter. This is indicative of good accuracy but poor precision. Target *d* obviously represents the worst situation, poor accuracy and precision.

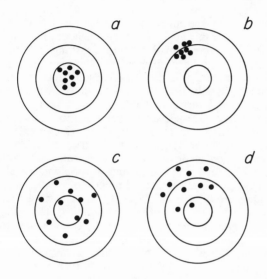

Figure 1.8 An explanation of accuracy and precision using targets. *a.* good accuracy, precision; *b.* good precision, poor accuracy; *c.* good accuracy, poor precision, *d.* poor accuracy, precision.

Thus precision is a measure of how well we can reproduce our results, whereas accuracy measures how closely we approach the correct value. If chemical analyses are not of high quality, target *b*, representing good precision but poor accuracy, is probably the most common analog. In some studies only relative differences are important, because the data will be used to compare chemical compositions of samples for each single study. In such cases good precision is quite important but good accuracy is less so. If data from one investigation are to be compared with those from other studies, however, good accuracy is also necessary. The safe course is always to strive for both good accuracy and precision. Hence our goal is to achieve analyses comparable to target *a,* but we do not always achieve them. How, then, do we quantify precision and accuracy?

Precision is normally easier to assess than accuracy. Perhaps this is why target *b* is a common situation for analyses of poor quality. Precision actually consists of three parts: (1) sampling precision, (2) precision involved in sample preparation, and (3) instrumental (including operator) error. We use the terms *reproducibility, precision,* and *error* synonymously. An equation can be written for total precision in terms of *V* (coefficient of variation; Equation 1.5):

$$V_T{}^2 = V_S{}^2 + V_P{}^2 + V_I{}^2 \qquad\qquad (1.14)$$

where V_T is total error, V_S is sampling error, V_P is sample preparation error, and V_I is instrumental plus operator error.

In the case of AA, instrumental (plus operator) precision can be evaluated by repeatedly analyzing the same prepared solution and calculating V. For many modern instrumental methods, V is less than 1-2 percent. Instrumental plus sample preparation error can be assessed in one of two ways. The first is by dissolving multiple samples of the same rock powder (ten dissolutions is a common number), analyzing each, and calculating V. Typically, V is less than 3-5 percent. A more conservative approach is to saw and separately crush multiple samples of the same hand specimen, then dissolve and analyze each sample, and finally calculate V. For this approach, V also depends on homogeneity of the hand specimen, as well as error involved in the crushing and grinding.

To assess total error, multiple hand specimens of an apparently homogeneous lithologic unit in outcrop must be collected. Each hand specimen is then prepared and analyzed separately before V is calculated, which can be shockingly large in some cases. If instrumental error, preparation error, and sampling error are 1, 3, and 7 percent, respectively, then Equation 1.14 indicates that total error is 7.7 percent. This is a typical situation in which total error is dominated by sampling error. When analytical error is quoted in the literature, most commonly it is a combination of preparation plus instrumental error, but does not include sampling error. Reporting instrumental error alone is really not fair, because it gives an overly optimistic estimate of precision.

Rather than evaluating error by multiple analyses of a single sample, some analysts prefer using duplicate analyses of several samples. The logic here is that no one sample can represent the entire range in composition for all samples and an average error is more meaningful. It is an observational fact that as concentrations approach the lower detection limit (the lower concentration limit of the analyte below which the AA unit records no absorbance), the error drastically increases. As a result, an evaluation of error based on duplicate analyses of several samples that represent the entire compositional range should be more realistic. Before V can be calculated the data must be "standardized." An example of standardization for five duplicate MgO analyses is:

% MgO	$\bar{x}$	f	$f \cdot \%$ MgO
0.72	0.755	3.466	2.50
0.79			2.74
1.39	1.350	1.939	2.70
1.31			2.54
2.59	2.645	.9894	2.56
2.70			2.67
3.41	3.485	.7509	2.56
3.56			2.67
4.92	4.850	.5396	2.65
4.78			2.58

$$\bar{x}_G = 2.617 \qquad f = \bar{x}_G / \bar{x}$$

The grand mean, $\bar{x}_G$, is simply the mean of the five means. The last column contains the standardized MgO analyses on which the error calculations are made. The coefficient of variation, V, for these 10 values is 3.02 percent, so this is the reported precision for the MgO analyses of this study.

Accuracy can also be rigorously evaluated. A common procedure is to report expected (accepted values based on published averages in the literature) and obtained concentrations for elements in one or more rock standards *not used in the regression calculations for the standard curve*. Reporting values for a standard used to determine the calibration curve is not legitimate. If several elements have been analyzed, as is normally the case, the χ^2 function (Equation 1.13) can be used to quantify collective accuracy of the analyses. The χ^2 function is used differently here than for the calibration curve discussed previously. In this case, we are dealing with one standard and several elements. For the calibration curve, χ^2 is determined on the basis of one element and several standards. Another common practice when reporting accuracy is to calculate the percent deviation of expected and obtained values for each analyte in a standard. An example of the use of χ^2 and percent deviation in quantifying accuracy for major elements in the USGS rock standard BCR-1 is given in Table 1.3.

TABLE 1.3 Estimation of Accuracy using Chi-Squared

	O	E	c	p
SiO_2	55.4	54.5	0.0149	1.65
Al_2O_3	13.2	13.6	0.0118	−2.94
TiO_2	2.31	2.20	0.0055	5.00
FeO^*	12.3	12.1	0.0033	1.65
MgO	3.29	3.46	0.0084	−4.91
CaO	6.74	6.92	0.0047	−2.60
Na_2O	3.39	3.27	0.0044	3.67
K_2O	1.63	1.70	0.0029	−4.12
Σ			0.0559	

O: values obtained for USGS rock standard BCR-1
E: expected values for USGS rock standard BCR-1
$c = (O - E)^2 / E$
$p = [(O - E) \times 100] / E$ = percent deviation

$$\chi^2 = \Sigma[(O - E)^2 / E] = \Sigma c = 0.0559$$

In conclusion, our analytical goal is to minimize V (as a measure of error or precision) and χ^2 (an estimate of accuracy). For a major-element analysis of around ten oxides, normally V for each oxide should be < 3-4 percent and χ^2 should be < 0.10-0.15 for a single comparison standard. For SiO_2 even better precision is required; V should be < 2 percent. We wish to optimize accuracy and precision while developing a method that allows us to analyze as many samples as possible in the shortest time. Frequently these two goals are in conflict. Good accuracy and precision may not be possible by a rapid, simple technique. As in many other areas, ultimately we look for acceptable compromises.

1.2 CONVERSIONS

When we change from one set of chemical concentration units to another we make the tacit assumption that rock-forming minerals are *stoichiometric compounds*. A stoichiometric mineral is one in which, within our limits of measure, the true composition is quite close to the formula composition. Take quartz (SiO_2) as an example. If quartz is stoichiometric, the atomic ratio of Si (plus any additional cations that substitute in the Si lattice site, such as trace quantities of Al or Ge) to O must be very close to 0.5.

Nonstoichiometry in crystals is caused by various types of lattice defects, such as vacant lattice sites or ions in interstitial positions rather than true lattice sites. Quartz and most other rock-forming silicates and oxides are generally fairly stoichiometric. Sulfides are notoriously non-stoichiometric. Stoichiometry, which assumes that simple considerations of valence and gram formula weight govern the calculations, is assumed throughout this book. The term *gram-formula weight (gfw)* is used to represent both molecular weight and atomic weight. One gfw represents the number of grams in one mole of an element or compound. Water, for example, has a gfw of 18.0 g/mol.

1.2.1 Weight/Weight Conversions

These conversions are frequently necessary during rock analysis. They are also useful in converting major oxides to atoms so that major/trace element ratios can be calculated. Other uses include converting weight percent oxides or atoms to weight percent minerals and vice versa.

1.2.1.1 Simple Conversions

To take the simplest type of conversion first, we convert weight percent Mg to MgO and back again:

Given: 4.55 % Mg, gfw(Mg) = 24.3, gfw(MgO) = 40.3

MgO = 4.55 x 40.3 / 24.3 = 7.55 %

Obviously, the second conversion is the inverse of the first:

Mg = 7.55 x 24.3 / 40.3 = 4.55 %

The weight percent Mg in MgO is simply:

Mg = 24.3 x 100 / 40.3 = 60.3 %

So the remainder, 39.7 percent, must be O.

We are not restricted to conversions to and from elements and oxides. Supposing we decide to convert the 4.55 percent Mg to enstatite ($MgSiO_3$, abbreviated EN; gfw = 100.4):

$$EN = 4.55 \times 100.4 / 24.3 = 18.80 \%$$

If our original data are expressed in oxides, we might prefer:

$$EN = 7.55 \times 100.4 / 40.3 = 18.80 \%$$

Petrologists frequently express mineral formulae in "oxide form" rather than conventional form; EN would be $MgO \cdot SiO_2$ as well as $MgSiO_3$. Using the types of calculations above it would contain, by weight, 24.2% Mg, 28.0% Si, and 47.8% O (or 40.1% MgO and 59.9% SiO_2).

In the previous example, all three chemical formulae, Mg, MgO, and $MgSiO_3$, contain one mole Mg. Any two formulae can be converted from one to another in the manner described above if the same number of "critical moles" (in the above case, moles of Mg) exists in both species. If the number of critical moles differs in the two formulae, the conversion calculation is slightly, but only slightly, more complicated. A handy rule of thumb is, in the case where the number does differ, multiply or divide one formula or the other by a number such that the number of critical moles is the same in both formulae. An example is the conversion of K_2O to K:

Given: 5.12 % K_2O, gfw(K) = 39.1, gfw(K_2O) = 94.2

gfw($KO_{.5}$) = 94.2 / 2 = 47.1
K = 5.12 x 39.1 / 47.1 = 4.25 %

Alternatively:

gfw(2K) = 39.1 x 2 = 78.2
K = 5.12 x 78.2 / 94.2 = 4.25 %

This calculation is apparent if we consider the following reaction:

$$2K^+ + O^{2-} = K_2O$$

Two moles of K^+ are required to make one mole of K_2O, which is exactly the premise for the calculation above. If any doubt exists about converting from one unit of concentration to another, simply write the reaction between the units. In the elementary case above little opportunity exists for confusion, but every conversion is not so straightforward.

More complex formulae can be handled with equal ease, as in the case of phlogopite ($KMg_3AlSi_3O_{10}(OH)_2$, abbreviated PH; gfw = 417.3):

Given: 7.55 % MgO or 4.55 % Mg

gfw(3Mg) = 3 x 24.3 = 72.9
PH = 4.55 x 417.3 / 72.9 = 26.0 %

gfw(3MgO) = 3 x 40.3 = 120.9
PH = 7.55 x 417.3 / 120.9 = 26.0 %

1.2.1.2 Conversion Factors

Ratios of gfws (Appendix C) and EWs (for explanation of EWs, see Section 1.2.2) as used in the calculations above are commonly referred to as *conversion factors*. A number of books tabulate these factors, but perhaps the most complete list can be found in the Appendix of the classic geochemistry book by Rankama and Sahama (1950). A more recent compilation is tabulated in Le Maitre (1982). Factors for the major oxides and $CaCO_3$ are calculated in Table 1.4. For example, to convert from 75.5 weight percent SiO_2 to weight percent Si:

$$Si = 75.5 \times 0.4675 = 35.3 \%$$

To convert from 8.37 weight percent Al to weight percent Al_2O_3:

$$Al_2O_3 = 8.37 / .5292 = 15.8 \%$$

TABLE 1.4 Weight/Weight Conversion Factors from Oxides to Atoms

From	To	a	b	c
SiO_2	Si	28.09	60.09	0.4675
TiO_2	Ti	47.90	79.90	0.5995
Al_2O_3	Al	26.98	50.98	0.5292
Fe_2O_3	Fe	55.85	79.85	0.6994
Fe_2O_3	FeO	71.85	79.85	0.8998
FeO	Fe	55.85	71.85	0.7773
FeO	Fe_2O_3	79.85	71.85	1.1113
MnO	Mn	54.94	70.94	0.7745
MgO	Mg	24.31	40.31	0.6031
CaO	Ca	40.08	56.08	0.7147
Na_2O	Na	22.99	30.99	0.7419
K_2O	K	39.10	47.10	0.8301
H_2O+	H	1.01	9.01	0.1121
H_2O+	OH^-	17.01	18.01	0.9445
P_2O_5	P	30.97	70.97	0.4364
CO_2	C	12.01	44.01	0.2729
CO_2	$CaCO_3$	100.09	44.01	2.2743

a: gfw for atom or EW (EW = equivalent weight; section 1.2.2) for oxide in second column
b: EW (equivalent weight) for oxide in first column
c = a / b = conversion factor

To convert from a value expressed in units of the first column to its equivalent in the second column, multiply by the appropriate factor. To convert any unit in the second column to its equivalent in the first, divide by the appropriate factor.

TABLE 1.5
Conversion of Data in Table 1.1 to Weight Percent Atomic Units

	a	b	c	
SiO_2	48.7	0.4675	22.8	Si
TiO_2	1.29	0.5995	0.77	Ti
Al_2O_3	16.6	0.5292	8.78	Al
Fe_2O_3	2.05	0.6994	1.43	Fe^{3+}
FeO	8.29	0.7773	6.44	Fe^{2+}
MnO	0.16	0.7745	0.12	Mn
MgO	6.63	0.6031	4.00	Mg
CaO	10.7	0.7147	7.65	Ca
Na_2O	2.83	0.7419	2.10	Na
K_2O	0.47	0.8301	0.39	K
H_2O+	0.81	0.9445	0.76	OH^-
P_2O_5	0.20	0.4364	0.09	P
CO_2	0.09	0.2729	0.02	C
Σ	99.49		55.35	
			44.1	O

a: weight percent oxide
b: weight conversion factor (from Table 1.4)
c = a x b = weight percent atom

Note: Oxygen obtained by subtracting total atoms from total oxides.

Petrologists often calculate ratios between elements, such as K/Rb, K/Ba, Ca/Sr, Ti/Y, and Ti/Nb. Potassium, Ca, and Ti are normally expressed as weight percent oxides, whereas Rb, La, Y, and Nb are reported in ppm element. How do we calculate these ratios? A percentage can be considered as so many parts per hundred (pph), so to convert to ppm oxide simply multiply by 10^4. Then convert ppm oxide to ppm element using a conversion factor from Table 1.4. These two steps can be accomplished in one by thinking of the conversion factors in Column C of Table 1.4 as having been multiplied by 10^4; e.g., the factor for $SiO_2 \longrightarrow$ ppm Si would be 4675, and for $TiO_2 \longrightarrow$ ppm Ti, 5995. An example is:

Given: 10.23 % CaO, 328 ppm Sr

Ca/Sr = 10.23 x 7147 / 328 = 223

The major-oxide analysis given in Table 1.1 is converted from oxides to atoms in Table 1.5. Oxygen content is determined by subtracting the sum of all the cations from the original total. Oxygen can be analyzed quite accurately by other means and the degree to which the two methods agree measures the validity of the assumption of stoichiometric minerals. Unfortunately, this separate oxygen determination is seldom performed.

A careful inspection of Table 1.5 reveals that H_2O+ is not handled in the same manner as the oxides. H_2O- is normally not involved in these conversions as it is primarily adsorbed water. Reasons for dealing with H_2O+ differently becomes obvious in Section 1.2.3, but recall that most H_2O+ exists as the OH^- ion in mineral lattices. Consequently, for sake of the calculations in Table 1.5, H_2O+ can best be considered as the hydroxide HOH rather than an oxide. Thus OH^- is listed as a separate anion in Table 1.5. Conversion of H_2O+ is handled differently for different types of conversions and must be given special attention.

1.2.1.3 FeO, Fe_2O_3, and Total Fe

Another topic dealing with weight/weight conversions concerns FeO, Fe_2O_3, and total Fe (as FeO^*, $Fe_2O_3^*$, or whatever; Section 1.1.1.3). If FeO and Fe_2O_3 are reported separately in the analysis and FeO^* is desired, use the conversion factors in Table 1.4. An example is:

Given: 8.29 % FeO, 2.05 % Fe_2O_3

Fe_2O_3 as FeO = 2.05 x .8998 = 1.84 %
FeO^* = 8.29 + 1.84 = 10.13 %

If $Fe_2O_3^*$ is required, simply multiply 8.29 by 1.1113 (the appropriate factor from Table 1.4) and add it to 2.05, obtaining 11.26 percent.

For norms (Section 1.2.5) and some other computations, FeO and Fe_2O_3 must be reported separately, yet most modern instrumental methods only analyze for total Fe. Assuming total Fe is reported as FeO^*, an assumption about the FeO/FeO^* oxidation ratio must be made so values for the individual oxides can be estimated. For fresh, unaltered basaltic rocks this ratio is commonly taken to be 0.85 or 0.90. An alternative is simply to assign a value of 1.5% to Fe_2O_3 for basaltic rocks and the remainder, after conversion, is assigned to FeO. The FeO/FeO^* ratio decreases due to oxidation associated with most types of secondary alteration, such as weathering. Given a basalt with 12.0% FeO^* and assuming a ratio of 0.85:

FeO = 12.0 x 0.85 = 10.2 %
Fe_2O_3 = (12.0 - 10.2) x 1.1113 = 2.00 %

Unfortunately, the ratio varies widely for even unaltered felsic rocks. Irvine and Baragar (1971) have offered the only simple solution to this problem, assuming analyzing for ferrous iron is impractical. They suggest that the following equation approximates the Fe_2O_3 content for a range in volcanic rock compositions:

% Fe_2O_3 = % TiO_2 + 1.5

with the excess Fe_2O_3 being converted to FeO (see Section 5.2.4.3). TiO_2 and Fe_2O_3 commonly exhibit similar patterns of variation in unaltered igneous rocks because the distribution of oxides such as magnetite and ilmentite strongly affects both of them.

1.2.2 The Equivalent

The *equivalent* is a frequently used concentration unit in many different types of chemical calculations. A one mole per liter (1 M) aqueous solution is referred to as a "one molar" solution. Similarly, a one equivalent per liter (1 N) solution is called a "one normal" solution. Just as a 1M solution is equivalent to 1 gfw/L, a 1 N solution equals 1 gew/L. Obviously, gew represents "gram equivalent weight." Refer to any standard introductory chemistry text for a discussion of gew and normality.

"Equivalents" of a different sort were popularized a number of years ago in Europe by Niggli (1954) and Barth (1962). They have not been widely adopted outside continental Europe, but they are extremely useful in a number of different petrologic calculations. They are referred to repeatedly throughout this book.

For sake of simplicity, terminology in this book is slightly different from that used by Niggli and Barth. Moreover, use of the term gew seems inappropriate, because the equivalent as defined below is seldom if ever used in chemistry. We use the term *EW* for gram equivalent weight as defined in this manner: EW of a compound is simply its gfw divided by the number of cations in its formula. Some examples are:

$$EW(SiO_2) = gfw(SiO_2) / 1 = 60.1$$
$$EW(Al_2O_3) = gfw(Al_2O_3) / 2 = 51$$
$$EW(KAlSi_3O_8) = gfw(KAlSi_3O_8) / 5 = 55.7$$

Thus 1 mole SiO_2 equals 1 equivalent SiO_2 but 1 mole $KAlSi_3O_8$ (orthoclase or OR) equals 5 equivalents OR. A compound has one equivalent for each cation in its formula.

Equivalents are very useful and avoid confusion in making the weight/weight conversions discussed in Section 1.2.1. To return to the example of K_2O:

Given: 5.12 % K_2O, gfw(K) = 39.1, $EW(KO_{.5})$ = 47.1

% K = 5.12 x 39.1 / 47.1 = 4.25

When using equivalents, chemical equations are not balanced in the usual manner. A good rule of thumb is that the total number of equivalents must be the same on both sides of the equation. First, an equation is balanced in the usual fashion:

$$Mg_2SiO_4 + SiO_2 = 2MgSiO_3 \tag{1.19}$$
$$FO + Q = 2EN$$
$$\text{forsterite} + \text{quartz} = 2 \text{ enstatite}$$

Using equivalents, the balanced equation is:

$$3FO + Q = 4En \tag{1.20}$$

There are three equivalents (cations) in FO, one in Q, and four (2 moles x 2) in EN, so the equation balances.

1.2.3 Weight/Mole Conversions

Two types of weight/mole conversions are the most common: weight percent oxide - mol percent oxide and weight percent oxide - mol percent atom (frequently called *atomic percent*). A variation on this latter type is *cation* percent. The values for gfws of oxides are used exclusively in the former and the EWs of oxides are most useful in the latter two conversions.

1.2.3.1 Why Use Molar Units?

Why do we use mol percents, rather than weight percents? In later chapters of this book different diagrams are introduced, some of which are more easily constructed and interpreted if molar quantities are used. In Sections 1.2.4 and 1.2.5 mineral formulae and norms are considered, both of which require the use of molar quantities.

TABLE 1.6 Conversion of Data in Table 1.1 to Mol Percent Oxides

	a	b	c	d
SiO_2	48.7	60.1	0.8103	51.5
TiO_2	1.29	79.9	0.0161	1.02
Al_2O_3	16.6	102.0	0.1627	10.4
Fe_2O_3	2.05	159.6	0.0128	0.81
FeO	8.29	71.8	0.1155	7.34
MnO	0.16	70.9	0.0023	0.15
MgO	6.63	40.3	0.1645	10.4
CaO	10.7	56.1	0.1907	12.1
Na_2O	2.83	62.0	0.0456	2.90
K_2O	0.47	94.2	0.0050	0.32
H_2O+	0.81	18.0	0.0450	2.86
P_2O5	0.20	142.0	0.0014	0.09
CO_2	0.09	44.0	0.0020	0.13
Σ	99.49		1.5739	100.02

a: weight percent oxide
b: gfw of oxide
c = a / b = mol proportion oxide
d = (c x 100) / Σc = mol percent oxide

Another reason for converting to mol percent has to do with balancing chemical reactions. Chemical equations written in terms of moles or equivalents are generally easily balanced. Coefficients for simple equations involving pure compounds, for instance, are commonly integers (see Equations 1.19 and 1.20).

This is not the case for equations balanced in weight units. As an example, balanced on the basis of weight units, Equations 1.19 and 1.20 become:

$$140.7FO + 60.1Q = 200.8EN \qquad (1.21)$$

where the coefficients are the gfws multiplied by the molar coefficients (or the EWs multiplied by the equivalent coefficients). In more complex equations, particularly when solid solutions are involved, working with molar or equivalent units is far easier.

TABLE 1.7 Conversion of Data in Table 1.1 to Atomic and Cation Percent

	a	b	c	d	e	f	
SiO_2	48.7	60.1	0.8103	1.6206	17.94	46.13	Si
TiO_2	1.29	79.9	0.0161	0.0322	0.36	0.92	Ti
Al_2O_3	16.6	51.0	0.3255	0.4882	7.21	18.53	Al
Fe_2O_3	2.05	79.8	0.0257	0.0386	0.57	1.46	Fe^{3+}
FeO	8.29	71.8	0.1155	0.1155	2.56	6.57	Fe^{2+}
MnO	0.16	70.9	0.0023	0.0023	0.05	0.13	Mn
MgO	6.63	40.3	0.1645	0.1645	3.64	9.36	Mg
CaO	10.7	56.1	0.1907	0.1907	4.22	10.86	Ca
Na_2O	2.83	31.0	0.0913	0.0456	2.02	5.20	Na
K_2O	0.47	47.1	0.0100	0.0050	0.22	0.57	K
H_2O	0.81	9.0	0.0900	0.0450	1.99	----	OH^-
P_2O_5	0.20	71.0	0.0028	0.0070	0.06	0.16	P
CO_2	0.09	44.0	0.0020	0.0040	0.04	0.11	C
					59.11j		O
Σ	99.49		1.8467	2.7592	99.99	100.00	

a: weight percent oxide
b: EW (equivalent weight) oxide
c = a / b = mol proportion ion
d: mol proportion O
e = (c x 100) / g = atomic percent
f = (c x 100) / i = cation percent
g = Σc + h = 1.8467 + 2.6692 = 4.5159
h = Σd - (mol prop OH^-) = 2.7592 - 0.0900 = 2.6692
i = Σc - (mol prop OH^-) = 1.8467 - 0.0900 = 1.7567
j = atomic percent O = 59.11 = (h x 100) / g

In Section 1.2.1.1 it was shown that EN ($MgO \cdot SiO_2$) contains 40.1 percent MgO and 59.9 percent SiO_2 by weight. Because there is 1 mole of each oxide in EN, the mineral must contain 50 mol percent MgO and 50 mol percent SiO_2. Mol percents are much more convenient and minerals such as EN are more easily dealt with when they are written in oxide form.

1.2.3.2 Mol Percent Oxides

The major oxide analysis given in Table 1.1 has been converted to mol percent oxides in Table 1.6. The procedure is quite straightforward -- first divide each oxide by its gfw and obtain its *mol proportion*. Then sum these mol proportions and recalculate to 100 percent. Notice that H_2O+ is treated as an oxide. Oxides are divided by their respective gfws and the proper units are maintained. If the units on weight percent are considered as being g/100 g and those on moles as g/mol, then units on mol proportions are:

$$(g/100\ g) / (g/mol) = moles/100\ g$$

By recalculating these mol proportions to 100 percent, the units become moles/100 moles, or mol percent. A check of this type to determine if the units cancel properly is commonly referred to as *dimensional analysis*. These checks are extremely helpful in avoiding mistakes.

If the analysis is already expressed in mol percent oxides and weight percent oxides are required, the process described above is simply reversed. Each mol percent oxide is multiplied by its appropriate gfw; these weight proportions are then summed and recalculated to 100 percent.

1.2.3.3 Atomic and Cation Percents

Atomic percents are calculated in a similar manner to mol percent oxides, except that EWs are used rather than gfws. Major oxide data from Table 1.1 have been converted to atomic percents in Table 1.7. Each oxide is divided by its EW to determine its atomic (more correctly, ionic) mol proportion (Column c). The mol proportion O (Column d) associated with each oxide is calculated along with that for the cation (e.g., for Al_2O_3, mol prop O = 1.5 x mol prop Al). Hydroxyl ion (OH^-) is obtained from H_2O^+. The mol proportions of O are then summed and this total, along with mol proportions for OH and the cations, are recalculated to 100 percent (Column e). The calculation is considerably simpler if H_2O+ is not included. Dimensional analysis indicates that the final units are moles/100 moles. Notice that in this typical basalt about 59 of 100 atoms (ions) are oxygen, whereas only about 18 are silicon.

Cation percents, later used to calculate molecular norms (Section 1.2.5), are closely related to atomic percents. One simply ignores and does not calculate mol proportion O or OH. As a consequence, the mol proportions of only cations are recalculated to 100 percent. These values are referred to as cation percents. A BASIC computer program to calculate cation percents is given in Section 1.6.2.4.

1.2.3.4 Molar Ratios

Recalculating an entire weight percent analysis if only a few elements are of interest is not necessary. For instance, the *Mg number* (or M-ratio) is a useful indicator of the degree of crystal fractionation of a basalt magma. The Mg # is defined as the mol ratio $Mg^{2+} / (Mg^{2+} + Fe^{2+})$. It can be calculated by:

Given: 6.63 % MgO, 8.29 % FeO
EW(MgO) = 40.3, EW(FeO) = 71.8

$$Mg\# = \frac{100(6.63/40.3)}{6.63/40.3 + 8.29/71.8} = 58.8$$

Either EWs or gfws are obviously satisfactory in the above calculation because they are numerically identical. In general, however, working with EWs for similar conversions is safer.

It is not necessary first to convert to weight percent atom and then mol proportion atom in any conversion from weight percent oxide to atomic percent. This is because the EWs for Fe and Mg cancel in the mol proportion terms, e.g.:

$$mol\ prop\ Mg = (6.63 \times 24.3) / (40.3 \times 24.3)$$

A partial analysis can be converted to a mol percent oxide ratio just as easily. An important ratio for delineating igneous rock suites is mol percent $Al_2O_3 / (CaO + Na_2O + K_2O)$ *(A/CNK)*. An example calculation is:

Given: % Al_2O_3 = 14.27, % CaO = 2.05
% Na_2O = 3.62, % K_2O = 4.41
gfw(Al_2O_3) = 102, gfw(CaO) = 56.1
gfw(Na_2O) = 62, gfw(K_2O) = 94.2

$$A/CNK = \frac{14.27/102}{2.05/56.1 + 3.62/62 + 4.41/94.2} = 0.99$$

As in the example in Table 1.6, gfws rather than EWs are used.

1.2.4 Mineral Formulae

Chemical formulae of minerals, especially those of solid solutions, are much more complicated than formulae of the pure compounds (such as $MgSiO_3$) discussed previously. Because compositions of real minerals are needed for some petrologic calculations, a method has been devised to convert weight percent oxide analyses to mineral formulae.

1.2.4.1 The Anion Number

The method for calculating mineral formulae is a variation on the theme of calculating atomic percents outlined in Section 1.2.3.3. The principal difference is that rather than recalculating to 100 percent, a ratio of cations to a certain number of O^{2-} + OH^- ions is calculated (if analyzed, F^- and Cl^- may also be included with these anions). If the mineral is anhydrous, only O^{2-} is used. The number of anions in the denominator of the ratio depends on the mineral group. This number, referred to herein as the *anion number,* for each major mineral group is:

amphiboles		24	garnets	24
epidotes		13	micas	24
feldspars		32	olivines	4
feldspathoids:	nepheline	32	pyroxenes	6
	leucite	6	spinels	32

Reasons for these numbers have to do with silicate structures and become apparent if Section 1.5.2.1 is consulted.

1.2.4.2 Two Examples

Two examples, analyses of olivine and biotite, are given to illustrate the manner in which mineral formulae are calculated. The olivine analysis is the simpler and contains no H_2O or Fe_2O_3, and Al_2O_3 was not detected (Table 1.8). Complications normally not involved with an olivine analysis include the manner in which H_2O, Al_2O_3, and anions other than O^{2-} are handled. These are addressed in the biotite analysis. Additional examples, including some that deal with these and other complications, can be found in Deer and others (1966 Appendix 1).

TABLE 1.8 Calculation of Mineral Formula for a Typical Olivine

	a	b	c	d	e	
SiO_2	39.6	60.1	0.6589	1.3178	0.99	Si
TiO_2	0.02	79.9	0.0002	0.0004	0.00	Ti
FeO^*	13.9	71.8	0.1936	0.1936	0.29	Fe
MnO	0.25	70.9	0.0035	0.0035	0.01	Mn
MgO	45.5	40.3	1.1290	1.1290	1.70	Mg
CaO	0.31	56.1	0.0055	0.0055	0.01	Ca
Na_2O	0.03	31.0	0.0010	0.0005	0.00	Na
K_2O	0.01	47.1	0.0002	0.0001	0.00	K
Σ	99.62			2.6504		

a: weight percent oxide
b: EW (equivalent weight) oxide
c = a / b = mol proportion cation
d: mol proportion O
e = c x N = number cations per 4 oxygen ions
N = 4 / Σd = 4 / 2.6504 = 1.5092

Note: Mineral formula: $(Mg_{1.70}Fe_{.29}Mn_{.01}Ca_{.01})Si_{.99}O_4$

For the olivine analysis (Table 1.8), weight percent oxides in Column a are divided by EWs in Column b to obtain mol proportion cations in Column c. The mol proportion O associated with each cation is then tabulated in Column d; these are summed and the summation (2.6504) is then divided into the anion number (4 in the case of olivine) to obtain the factor 1.5092 (N). This factor is then multiplied by each mol proportion cation in Column c to obtain the number of cations per four oxygens tabulated in Column e. The mineral formula given at the bottom of Table 1.8 can be written directly from the data in Column e. This form is convenient to include in chemical equations. If the chemical analyses are done accurately and this particular olivine approximates a stoich-iometric compound, the mineral formula should have the form:

$$(Mg,Fe,Mn,Ca)_2SiO_4$$

(Section 1.5.2.3). Within limits of analytical error, the formula given in Table 1.8 does indeed approach an ideal stoichiometric formula.

TABLE 1.9 Calculation of Mineral Formula for a Typical Biotite

	a	b	c	d	e	
SiO_2	35.0	60.1	0.5824	1.1647	5.33	Si
TiO_2	2.51	79.9	0.0314	0.0628	0.29	Ti
Al_2O_3	20.4	51.0	0.4000	0.6000	3.66	Al
Fe_2O_3	1.13	79.8	0.0142	0.0212	0.13	Fe^{3+}
FeO	20.4	71.8	0.2841	0.2841	2.60	Fe^{2+}
MnO	0.03	70.9	0.0004	0.0004	0.00	Mn
MgO	7.05	40.3	0.1749	0.1749	1.60	Mg
CaO	0.19	56.1	0.0034	0.0034	0.03	Ca
Na_2O	0.92	31.0	0.0297	0.0148	0.27	Na
K_2O	8.66	47.1	0.1839	0.0919	1.68	K
H_2O+	3.64	9.0	0.4044	0.2022	3.70	OH^-
Σ	99.93			2.6204		

a: weight percent oxide
b: EW (equivalent weight) oxide
c = a / b = mol proportion ion
d: mol proportion O + OH
e = c x Q = number ions per 24 (O + OH) ions
Q = 24 / Σd = 24 / 2.6204 = 9.1589

Note: Mineral formula:

	W	$(Ca_{.03}Na_{.27}K_{1.68})$
	X and Y	$(Mg_{1.6}Fe^{2+}_{2.6}Fe^{3+}_{.13}Ti_{.29}Al_{.99})$
	Z	$(Al_{2.67}Si_{5.33})_8$
	anions	$(OH_{3.7}O_{20.3})_{24}$

A more complex example, a hydrous aluminosilicate, is one that involves dealing with H_2O as well as partitioning Al between the tetrahedral and octa-hedral lattice sites. A sample calculation for a typical biotite mica is given in Table 1.9. The calculation is performed in a manner similar to that for olivine (Table 1.8), except for some complexities mentioned previously and discussed more fully later. The chemical formula is expressed at the bottom of Table 1.9 in a form comparable with the general formula for biotite given in Section

1.5.2.3. Cations are grouped according to whether they occupy 12-fold (W), six-fold (X and Y), or four-fold (Z) coordination lattice sites. Refer to Section 1.5.1.2 for a discussion of coordination.

The following comments specifically pertain only to a mica analysis, but they are generally applicable to any mineral formula for a hydrous alumino-silicate. If analyzed, F^- or Cl^-, or both, would be included in the calculations and grouped with the O^{2-} + OH^- anions in the chemical formula. Hydroxyl ion is determined from H_2O+ and added to oxygen calculated from the oxides to obtain the total anions (Column d). Eight cations are present in the Z site in biotite, so first all Si is placed in this site; the site is then filled with Al, and the remaining Al is placed in the (X + Y) site. Although X- and Y-position ions typically are not separated in a mineral formula, the major X-position ions are Mg, Fe^{2+}, and Mn, whereas the major Y-position ions are Fe^{3+}, Ti, and octahedral Al (Section 1.5.1.2).

If an electron probe were used, no H_2O+, Fe^{2+}, or Fe^{3+} could be analyzed, so some assumptions must be made. The first is to assume an O:OH ratio of 20:4. The second is either to assume an Fe^{2+}:Fe^{3+} ratio or assume all Fe is Fe^{2+}. The latter is normally adequate because (1) X- and Y-position ions are not separated; (2) most of the Fe in a typical biotite is Fe^{2+} (Table 1.9); and (3) EWs for Fe_2O_3 and FeO are similar, 79.8 and 71.8, respectively. Little error is introduced in assuming all Fe is Fe^{2+}.

1.2.5 Norms

The final type of conversion we consider is calculation of the *norm*. The term norm, like the term *hard rock,* has a very different meaning to petrologists than to other people. A norm is a list of hypothetical minerals, in either weight percent or mol percent, that is calculated from the chemical analysis of a rock. It has been said that the norm is a list of minerals that might have crystallized from the magma, whereas the *mode* is the group of minerals that actually did crystallize. Norms are very widely used as concentration units by petrologists. Calculating a norm combines principles of stoichiometry with those of business accounting. Major and minor elements are apportioned into a predetermined group of normative minerals (the norm) according to a fixed set of rules.

1.2.5.1 Why Calculate a Norm?

Given this definition of a norm, reasonable questions are: "Why calculate a norm? Why doesn't a mode suffice?" Norms were first determined in the early 1900s and used to estimate the mineralogy of rocks that were so fine-grained their minerals could not be identified with a petrographic microscope. With the development of X-ray diffraction, this use of the norm was less necessary.

In modern times the norm generally is used for different reasons. Through the years a plethora of chemical classifications of igneous rock suites have been proposed (see Section 5.2.4), many of which are based on norms. If we wish to compare our rocks with those discussed in the literature, one way to do so is with norms.

A second, and equally important, use is to "bridge the gap" between lab-oratory studies on synthetic systems and investigations of naturally occurring rocks. Stating this differently, normative minerals are commonly used to proxy for components in experimentally determined phase diagrams of synthetic systems, which allows chemical compositions of natural rocks to be plotted on these diagrams. Because we have some knowledge of the physical and chemical conditions under which the synthetic systems formed, we can speculate about those conditions for the rocks themselves.

1.2.5.2 CIPW and Molecular Norms

As mentioned earlier, norms are calculated in either mol percent or weight percent. Rules for calculating weight percent norms were devised first, in 1903, by four American geologists named Cross, Iddings, Pirsson, and Washington. As a result, these norms are known as *CIPW norms* and are very widely used. Like equivalents, which form the basis for their computation, molecular norms were popularized in Europe by Niggli (1954) and Barth (1962). Thus molecular norms are occasionally referred to as *Niggli norms* or *Niggli-Barth norms*. In this book we demonstrate how to calculate a molecular norm and then convert it to the equal of a CIPW norm. A set of rules for directly computing a CIPW norm is given in Cox and others (1979).

TABLE 1.10 Standard Normative Minerals

	Name	Formula	EW
ab	albite	$NaAlSi_3O_8$	52.5
ac	acmite	$NaFeSi_2O_6$	57.8
an	anorthite	$CaAl_2Si_2O_8$	55.7
ap	apatite	$Ca_5P_3O_{12}(OH)$	62.8
c	corundum	Al_2O_3	51.0
cc	calcite	$CaCO_3$	50.0
di	diopside	*wo* + (*en*+*fs*)	----
en	enstatite	$MgSiO_3$	50.2
fa	fayalite	Fe_2SiO_4	67.9
fo	forsterite	Mg_2SiO_4	46.9
fs	ferrosilite	$FeSiO_3$	66.0
hm	hematite	Fe_2O_3	79.8
hy	hypersthene	*en* + *fs*	----
il	ilmenite	$FeTiO_3$	75.8
kp	kaliophilite	$KAlSiO_4$	52.7
lc	leucite	$KAlSi_2O_6$	54.6
mt	magnetite	Fe_3O_4	77.1
ne	nepheline	$NaAlSiO_4$	47.4
ol	olivine	*fo* + *fa*	----
or	orthoclase	$KAlSi_3O_8$	55.7
pl	plagioclase	*ab* + *an*	----
q	quartz	SiO_2	60.1
ru	rutile	TiO_2	79.9
sp	sphene	$CaTiSiO_5$	65.4
wo	wollastonite	$CaSiO_3$	58.1

Why calculate a molecular norm first? There are many reasons, but five immediately come to mind:

1. Molecular norms are easier to calculate and use, because units are in mol percent rather than weight percent.
2. Molecular norms are quite flexible in that nonstandard normative minerals can be easily determined.
3. Molecular norms are more easily converted to the equivalent of CIPW norms than vice versa.
4. Molecular norms have become more popular in recent years for classification of igneous rock suites (e.g., Irvine and Baragar, 1971). A recent igneous petrology textbook by McBirney (1984) emphasizes molecular norms.
5. Mol percentages expressed as norms are more similar to volume percentages of the actual minerals present (i.e., the mode) than are weight percentages expressed as norms.

1.2.5.3 Normative Minerals

Before actual calculation of a norm a few words are necessary about normative minerals. The common normative minerals, along with their standard chemical formulae and EWs, are given in Table 1.10. There are no complex minerals such as augite, hornblende, or biotite, and no hydrous minerals, with the exception of *ap*. Normative minerals typically are considered as end-members of solid solution series, except *ol, hy, di,* and *pl,* all of which have variable EWs. For these reasons most normative minerals are useful to proxy for components on phase diagrams. The normative mineral *pl*, not in most norms, is included herein.

xxxxxxxxxxxxxxxxxxxxxxxxxxxxxxxxxxxxxxxxxxxxxxxxxxxxxxxxxxxxxxxxxxx

In Table 1.10 the mineral enstatite (MgSiO$_3$) is abbreviated en, whereas in Section 1.2.1.1 EN was used. Why this difference? Introductory petrology students are often confused when dealing with phase diagrams where reference is being made to a particular component or mineral, both of which have the same name (in this case, enstatite). To add to the confusion, the normative mineral enstatite exists as well. The convention adopted throughout this book is to abbreviate accordingly:

component	EN
actual mineral (phase)	En
normative mineral	en

Confusion possibly still could exist for quartz (q) and corundum (c), but normally their meaning should be clear when taken in context.

xxxxxxxxxxxxxxxxxxxxxxxxxxxxxxxxxxxxxxxxxxxxxxxxxxxxxxxxxxxxxxxxxxx

A few minerals not given on Table 1.10 are occasionally included in a norm. Additional minerals exist as well, but these are representative:

pyrite *(py)*	FeS_2
fluorite *(fr)*	CaF_2
zircon *(z)*	$ZrSiO_4$
halite *(hl)*	$NaCl$
chromite *(cm)*	$FeCr_2O_4$
sodium metasilicate *(ns)*	Na_2SiO_3
potassium metasilicate *(ks)*	K_2SiO_3

The first five of these can be calculated if some element normally in trace amounts is in sufficient quantities to warrant inclusion of its mineral in the norm (step 4, Appendix A). The latter two minerals are rarely needed, except in rocks extremely Al-undersaturated (Section 1.3.4.2; Appendix A, steps 8B and 8D).

1.2.5.4 Haplogranite - A Simple Example

As mentioned earlier, calculation of a norm is as much an exercise in accounting as in geology or chemistry. A comparison can be made to a family's personal finances, where income from various sources is apportioned out to pay for the family's different expenses. Likewise, chemical constituents from the various oxides are apportioned out to "make" different normative minerals. In fact, a table similar to a spreadsheet used for financial accounting is extremely useful in calculating a norm.

TABLE 1.11 Calculation of a Molecular Norm for a Haplogranite

	A	B	C	*or*	*ab*	*an*	*c*	*q*
Si	73	1.2146	67.98	17.82	16.26	4.00	--	29.90
Al	17	0.3333	18.66	5.94	5.42	4.00	3.30	--
Ca	2	0.0357	2.00	--	--	2.00	--	--
Na	3	0.0968	5.42	--	5.42	--	--	--
K	5	0.1062	5.94	5.94	--	--	--	--
	100		100.00	29.70	27.10	10.00	3.30	29.90

A: weight percent oxide
B: mol proportion (see Table 1.4, column b, for EWs)
C: cation percent

A simple example of calculating a molecular norm for a *haplogranite,* an idealized granite consisting of only quartz and feldspars (i.e., no ferromagnesian minerals or oxides), is very informative (Table 1.11). First, the weight percent oxide analysis must be converted to cation percent according to the procedure outlined in Table 1.7. Then the cations Si, Al, Ca, Na, and K are apportioned among the normative minerals *or, ab, an, c,* and *q.* Because the formula for *or* is $KAlSi_3O_8$, these three cations must be apportioned according to K:Al:Si = 1:1:3 in *or.* Similarly, *ab,* with a formula $NaAlSi_3O_8$, must be calculated according to

Na:Al:Si = 1:1:3. Because *an* has a formula $CaAl_2Si_2O_8$, it must be made according to Ca:Al:Si = 1:2:2. Excess Al is expressed as *c* and excess Si as *q*. By excess we are referring to any remaining Al or Si after all three feldspars are calculated. Had there been a deficiency in either Al or Si, a more complex calculation would be required, which is described in Appendix A and the more involved example below.

A few notes are necessary regarding Table 1.11:

1. Mol percents for the different normative minerals are given as the last five values in the bottom row.
2. The two decimal places carried throughout the table are not necessary for the final mol percents (one decimal place will suffice). Two places are carried to demonstrate that the initial weight percents, cation percents, and total of normative minerals add to 100.00. This provides a check to determine if the calculatiuons were performed correctly.
3. Mol percent *an* can be used to demonstrate how the data in Table 1.11 can be related to the information in the previous paragraph. Conversion of 2.00 weight percent CaO to cation percent resulted in no change, so 2.00 cation percent Ca is available to make *an*. This 2.00 percent is combined with 4.00 percent Al and 4.00 percent Si according to Ca:Al:Si = 1:2:2 to make 10.00 mol percent *an*.
4. After making all feldspars, 3.30 percent of the original 18.66 percent Al is left and reported as *c*. Likewise, 29.90 percent of the original 67.98 percent Si remains and is reported as *q*. This haplogranite is considered to be "oversaturated" in Si and Al (Section 1.3.1).

1.2.5.5 Basalt: A More Complex Example

We have now learned the principles of a molecular norm calculation through a simple example, so let us continue by considering a full major-element analysis, a considerably more complex situation. The data in Table 1.1 have been converted to cation percent in Table 1.7 (Column f), which is repeated here so that the steps below can be more easily followed:

Si	46.1	Mg	9.36
Ti	0.92	Ca	10.9
Al	18.5	Na	5.20
Fe^{3+}	1.46	K	0.57
Fe^{2+}	6.57	P	0.16
Mn	0.13	C	0.11

In the sample computation, refer to Appendix A (numbers below are keyed to step numbers in Appendix A). Most calculations are shown as equations. For this analysis, as well as any other, all steps in Appendix A are not required. Numbers in italics are final percentages; they add to 99.98 percent, equal to the sum of the above cation percentages.

1. $Fe^{2+} = 6.57 + 0.13 = 6.70$ %

2. 0.16 % P + 0.27 % Ca = *0.43 % ap*

3. 0.11 % C + 0.11 % Ca = *0.22 % cc*

5. $0.92 \% \ Ti + 0.92 \% \ Fe^{2+} = 1.84 \% \ il$

6. $0.57 \% \ K + 0.57 \% \ Al + 1.71 \% \ Si = 2.85 \% \ or$

7. $5.20 \% \ Na + 5.20 \% \ Al + 15.60 \% \ Si = 26.00 \% \ ab$

9. $10.52 \% \ Ca + 21.04 \% \ Al + 21.04 \% \ Si = 52.60 \% \ an$ (provisional)

10. $18.5 \% \ Al - 0.57 \ or - 5.20 \ ab - 21.04 \ prov. \ an = -8.31 \% \ Al$

11. $20.78 \% \ an = 4.16 \% \ Ca + 8.31 \% \ Al + 8.31 \% \ Si$

      $an = 52.60 - 20.78 = 31.82 \%$

12. $4.16 \% \ Ca + 4.16 \% \ Si = 8.32 \% \ wo$

13. $1.46 \% \ Fe^{3+} + 0.73 \% \ Fe^{2+} = 2.19 \% \ mt$

15. $5.05 \% \ Fe^{2+} + 5.05 \% \ Si = 10.10 \% \ fs$

16. $9.36 \% \ Mg + 9.36 \% \ Si = 18.72 \% \ en$

17. $Mg \ / \ (Mg + Fe^{2+}) = 18.72 \ / \ (18.72 + 10.10) = 0.65$

18. $di = 8.32\% \ wo + 8.32\% \ (en + fs) = 16.64 \%$

      $hy = 28.82 - 8.32 = 20.50 \%$ (provisional)

19. $46.1 \% \ Si - 1.71 \ or - 15.60 \ ab - 21.04 \ prov. \ an$
      $+ 8.28$ from step 11 $- 4.14 \ wo - 5.05 \ fs - 9.36 \ en$
      $= -2.51 \% \ Si$

      $10.04 \% \ hy = 7.53 \% \ ol + 2.51 \% \ Si$

      $hy = 20.50 - 10.04 = 10.46 \%$

      $ol = 7.53 \%$

27. $pl = 26.00 + 31.82 = 57.82 \%$

1.2.5.6 From Molecular to "CIPW Norm"

Computation of a molecular norm is tedious but straightforward. Conversion to
the equivalent of a CIPW norm is equally simple (Table 1.12). Multiply the
percentage of each molecular mineral by its appropriate EW, sum these weight
proportions, and recalculate to 100 percent. One word of caution is required. Be
certain that the solid solutions *hy, di,* and *ol* are expressed in terms of their end-
member compositions (*en, fs, wo,* etc.) and these end-member normative minerals are multiplied by their EW's. These end-member compositions can be
determined by using the Mg/Mg+Fe ratio (step 17 above and in Appendix A)
and remembering that one equivalent *wo* equals one equivalent (*en + fs*). In
most cases molecular percentages of normative minerals are not greatly different
from weight percentages; greatest differences are for *mt* and *il.*

1.2.5.7 Tailor-made Norms

To this point we have only considered conventional norms that contain the standard normative minerals given in Table 1.10. A "tailor-made" norm can be created, which contains one or more nonconventional minerals. This can be done in two ways: convert some standard minerals in a conventional norm to minerals of interest or create an entirely new set of minerals.

TABLE 1.12 Conversion from a Molecular to a "CIPW Norm"

		a	b	c	d	
	ap	0.43	62.8	27.00	0.48	
	cc	0.22	50.0	11.00	0.20	
	il	1.84	75.8	139.47	2.50	
	mt	2.19	77.1	168.85	3.03	
	or	2.85	55.7	158.75	2.85	
pl	ab	26.0	52.5	1365.00	24.5	56.3
	an	31.8	55.7	1771.26	31.8	
di	wo	8.32	58.1	483.39	8.67	17.0
	en	5.41	50.2	271.58	4.87	
	fs	2.91	66.0	192.06	3.44	
hy	en	6.80	50.2	341.36	6.12	10.4
	fs	3.66	66.0	237.60	4.26	
ol	fo	4.89	46.9	229.34	4.11	7.32
	fa	2.64	67.9	179.26	3.21	
Σ		99.96		5575.92	100.04	

a: mol percent normative mineral
b: EW (equivalent weight) normative mineral
c = a x b = weight proportion normative mineral
d = (c x 100) / Σc = weight percent normative mineral

An example of the first method is the formation of a normative hydrous mineral in a granitic rock, such as muscovite ($KAl_2AlSi_3O_{10}(OH)_2$; *mu*). This calculation could be done in a variety of ways, but probably the most straightforward is to include H_2O^+ in the calculation of cation percent (Table 1.7) and write the following formula in equivalents:

$$2c + 5or + 2OH^- = 9mu \tag{1.22}$$

Take an example of a rock that contains 2 percent OH⁻. The modified norm would contain 18 percent *mu;* 4 percent *c* and 10 percent *or* are used up in making *mu*. The amount of *mu* that can be made obviously depends on the amount of H_2O+ or *c* available, because in a granitic rock sufficient *or* normally will be present. Granitic rocks that contain modal muscovite generally have *c* in their conventional norms and are Al-oversaturated (Section 1.3.4.2). Niggli (1954) tabulates many equations that can be used to modify a standard norm.

TABLE 1.13 Minerals for Eclogite Norm

	Name	Formula
ap	apatite	$Ca_5P_3O_{12}(OH)$
cc	calcite	$CaCO_3$
ru	rutile	TiO_2
mt	magnetite	Fe_3O_4
ph	phlogopite	$KMg_3AlSi_3O_{10}(OH)_2$
jd	jadeite	$NaAlSi_2O_6$
gt	garnet	$(Mg,Fe,Ca,Mn)_3Al_2Si_3O_{12}$
px	pyroxene	$(Ca,Mg,Fe,Mn)Si_2O_6$
om	omphacite	*jd* + *px*
q	quartz	SiO_2

An eclogite norm is an example of creating a new set of normative minerals. *Eclogite* is roughly the chemical equivalent of a basalt, but it crystallized under extremely high pressures (commonly by metamorphism of a basalt), so its mineralogy is different from that of a basalt. The most essential minerals in an eclogite are *omphacite* (a Na-rich clinopyroxene) and a Mg-rich garnet. The Na component of omphacite is called *jadeite* ($NaAlSi_2O_6$), so jadeite *(jd)* is made rather than albite, and garnet *(gt)* plus pyroxene *(px)* rather than anorthite and olivine. Pyroxene is made rather than the more specific *di* and/or *hy* because no accounting is made of the ratio Ca:(Mg+Fe^{2+}+Mn). If H_2O+ is present, K_2O is converted to phlogopite *(ph),* a Mg-rich biotite. Under these high pressures rutile is more likely stable than ilmentite, so all TiO_2 is placed in rutile. The minerals in this eclogite norm are given in Table 1.13.

According to Table 1.13, five cations of Ca and three of P are required to make apatite, and so on for the remaining minerals. The cation percent data in Column f on Table 1.7 are recalculated to an eclogite norm using the formulae from Table 1.13. Results are shown in Table 1.14, which is in the form of a sample "spreadsheet." For the example of apatite, all available P is combined with Ca in a ratio of Ca:P = 5:3 to make 0.43 percent *ap*. All other normative minerals are made by similar calculations.

Several points should be made regarding Table 1.14:

1. Olivine is present in the conventional norm for this rock, indicating that it is SiO_2-undersaturated (Section 2.1.1.1). The eclogite norm indicates that the rock is SiO_2-oversaturated, because *q* is present. Thus the degree of

SiO_2-saturation depends on which normative minerals are made.

2. After *ap, cc, mt,* and *ph* are made, Fe^{2+}, Mn, Mg, and Ca are combined to make first *gt* (using the remaining Al), and then *px.*

3. Phlogopite contains OH^-, but H_2O+ was not included in the determination of cation percents. The ratio of OH^- to K in *ph* is 2:1 (Table 1.13), so adequate OH^- is available to make *ph* (Table 1.7, Column E).

4. If the basalt originally introduced in Table 1.1 were metamorphosed to an eclogite, Table 1.14 implies that it should contain about 50 percent garnet, 30 percent omphacite, 10 percent quartz, and 10 percent assorted other minerals. In fact, it would probably contain less garnet and quartz but more omphacite. This is because some Al will be in the omphacite structure, raising its content at the expense of garnet and quartz. Moreover, the Al would be in *Ca Tschermak's molecule* ($CaAlAlSiO_6$) in omphacite, which would reduce the Ca component in the garnet to a more realistic amount. We must make some assumptions about mineral compositions to partition Al and Ca between omphacite and garnet.

TABLE 1.14 Worksheet for Calculation of Eclogite Norm

	ap	*cc*	*ru*	*mt*	*ph*	*jd*	*gt*	*px*	*q*
Si					1.71	10.4	19.1	5.07	9.82
Ti			.92						
Al					.57	5.20	12.7		
Fe3				1.46					
Fe2				.73					
Mn							19.1	5.07	
Mg					1.71				
Ca	.27	.11							
Na						5.20			
K					.57				
P	.16								
C		.11							
Norm	.43	.22	.92	2.19	4.56	20.8	50.9	10.1	9.82

Notes: Fe^{2+}, Mn, Mg, and Ca are grouped together when *gt* and *px* are calculated. Cation percent from Column f, Table 1.7.

1.3 SATURATION AND ROCK SERIES

Igneous rocks can be grouped according to which igneous rock *series* (also referred to as *association* or *suite*) they belong. All rocks of a series are thought to be genetically related in some way, such as all having formed from derivative magmas of compositionally similar parental magmas through fractional crystallization. As an example, a granodiorite is typically in the *calcalkaline* series, which together with the *tholeiitic* series form the main two subgroups of the *subalkaline* series. Igneous rock series are normally defined on the basis of chemical (including normative) criteria, although petrographic criteria can play a role. Calcalkaline rocks, for instance, are generally quartz bearing.

1.3.1 Saturation

The term *saturation* as it is used in chemistry has a particular meaning. An oversaturated aqueous solution of NaCl can, if prompted by some means such as agitation, undergo precipitation of salt. Conversely, an undersaturated solution can retain an even larger quantity of NaCl and dissolution will occur if additional salt is available. Thus, in theory, an exactly saturated solution cannot undergo additional precipitation or dissolution.

1.3.1.1 Silica Saturation

Saturation, as used in igneous petrology, has a similar but still unique connotation. If a magma is oversaturated with the component SiO_2, theoretically it will precipitate quartz. On the other hand, a magma undersaturated in SiO_2 (at low pressure with a low water content) should crystallize a "silica-deficient" silicate, such as a feldspathoid or an olivine. The meaning of the term *silica-deficient* becomes apparent when enstatite ($MgSiO_3$ or $MgO{\cdot}SiO_2$) is compared with its silica-deficient counterpart, forsterite (Mg_2SiO_4 or $2MgO{\cdot}SiO_2$). Enstatite has 50 mol percent SiO_2, whereas forsterite has only 33 percent. Thus we can look on feldspathoids and olivines as silica-deficient analogs of feldspars and orthopyroxenes, respectively:

$$NaAlSiO_4 + 2SiO_2 = NaAlSi_3O_8 \qquad (1.23)$$
nepheline + 2 quartz = albite

$$(Mg,Fe)_2SiO_4 + SiO_2 = 2(Mg,Fe)SiO_3 \qquad (1.24)$$
olivine + quartz = 2 orthopyroxene

A similar reaction can be written for leucite, quartz, and orthoclase. Carmichael and others (1974) have shown that the concept of saturation can be treated in terms of activity or chemical potential of SiO_2, or any other species such as Al_2O_3, in the melt.

A useful way to consider the concept of saturation in igneous rocks is by means of norms. The calculation of norms is based on the assumption that most igneous rock-forming minerals behave approximately as stoichiometric compounds. If so, the magma presumably must have a stoichiometric "excess" of SiO_2 over the amount of SiO_2 required to make other silicates (i.e., be oversaturated with respect to SiO_2) in order for quartz to crystallize. Hence, any rock that contains normative quartz is said to be "SiO_2 oversaturated." It follows that rocks with olivine or feldspathoids in the norm are "SiO_2 undersaturated." Ideally a rock that contains none of these minerals in the norm is exactly SiO_2 saturated. In fact, however, if the percentages of olivine, feldspathoids, or quartz are less than about 2 percent, a rock can be considered approximately SiO_2 saturated. Two percent is about the limit of error for calculation of a normative mineral, considering the precision of the original chemical analysis.

Some petrologists prefer to broaden the range of silica-saturated rocks. Their definitions of the three groups are:

SiO_2-oversaturated $q + hy$ in the norm
SiO_2-saturated $ol + hy$ in the norm
SiO_2-undersaturated $ol + ne$ in the norm

If taken to extreme, this definition of silica saturation would mean that most peridotites, which are ultramafic rocks, are SiO_2-saturated rocks. This is contrary to the way in which they are normally considered. In this book *ol-ne* normative rocks are considered to be "critically" SiO_2 undersaturated and *ol-hy* normative rocks simply undersaturated.

1.3.1.2 Other Oxides and Incompatible Normative Minerals

Silica is not the only component in a rock to which the concept of saturation can be applied. Alumina saturation is almost equally important, especially with respect to felsic rocks. If a rock contains corundum in the norm, it is considered to be Al_2O_3 oversaturated; if diopside is present as a normative mineral, the rock is Al_2O_3 undersaturated. If normative acmite is present, it is critically undersaturated in Al_2O_3.

The presence of other diagnostic normative minerals is indicative of saturation with respect to other components, such as TiO_2, Fe_2O_3, or CaO. A rock with free *wo* (i.e., *wo* not tied up in *di*) can be considered CaO oversaturated. Likewise, if a rock contains *he*, it is Fe_2O_3 oversaturated, or *ru*, TiO_2 oversaturated.

TABLE 1.15 Examples of Incompatible Normative Minerals

	Incompatible minerals		Incompatible minerals
ab	kp, lc	lc	ab, hy, q
ac	an, c	mt	sp, ns, ru
an	ac, ns	ne	hy, kp, q
c	ac, di, wo, ns, sp	ns	an, c, mt, hm
di	c, ru	ol	q, wo
fs	hm, ru, sp	or	kp
hm	il, ns, fs	q	kp, lc, ne, ol
hy	kp, lc, ne, wo	ru	mt, fs, wo, di
kp	ab, hy, or, q	sp	mt, fs, c
		wo	c, hy, ru, ol

Any mineral indicative of oversaturation for a particular component never appears in a norm with a mineral indicating undersaturation of that same component. Table 1.15 lists examples of incompatible normative minerals, which never appear together in the same norm. For instance, *kp* and *lc* can never appear in a norm with *ab*. Likewise, because they are diagnostic of radically different degrees of saturation, *c* and *ac* never appear in the same norm, nor do *ne* and *q*. The mineral *wo* in Table 1.15 refers to the separate normative mineral that is made because of CaO oversaturation, not the "component" of *di*.

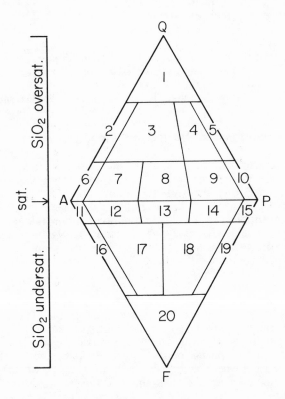

Figure 1.9 The QAPF double triangle for the modal classification of plutonic igneous rocks (simplified after Streckeisen, 1976). Q, quartz; A, alkali feldspar; P, plagioclase; F, feldspathoids. Fields are: 1, quartz-rich granitoids; 2, alkali granite; 3, granite; 4, granodiorite; 5, quartz diorite; 6 and 11, alkali syenite; 7 and 12, syenite; 8 and 13, monzonite; 9 and 14, monzodiorite and monzogabbro; 10 and 15, diorite and gabbro; 16, foyaite; 17, plagifoyaite; 18, essexite; 19, theralite; 20, foidite. An analogous double triangle exists for volcanic rocks. Rocks whose compositions plot in fields 6-15 can be considered as approximately SiO_2 saturated.

1.3.1.3 A Common Petrographic Classification

Most petrographic classifications of individual rock types at least implicitly acknowledge the concept of silica saturation. The most widely used petrographic classification for igneous rock types was originally developed by an IUGS Subcommission (Streckeisen, 1976). It was the culmination of a half-century's attempts to refine ways in which igneous rocks are classified and is now the international standard. Excluding ultramafics, it is based on four variables arranged in a double triangle (Fig. 1.9). The exact manner in which this diagram is used is discussed in the next section. The variables (volume percentages in the mode) are:

Q percent quartz
A percent alkali feldspar
P percent plagioclase
F percent feldspathoids

The concept of SiO_2 saturation is implicit in Figure 1.9. According to these double triangles, quartz and feldspathoids cannot coexist in the same rock. Thus rocks in the upper triangle are SiO_2 oversaturated; in the lower triangle, undersaturated; and near the boundary between the two triangles, saturated.

1.3.1.4 Correspondence of Norms and Modes

In the earlier discussion it is tacitly assumed that at least a crude correspondence exists between the norm and the mode. If a rock contains normative quartz, does it also literally contain quartz (referred to as modal quartz)? The answer depends on the degree to which the overall modal mineralogy (the actual minerals present) corresponds to the standard normative mineralogy. In unaltered mafic rocks that crystallized at moderate to shallow levels the correspondence is usually quite close; in most other types it frequently is not. Any of several reasons can account for these differences:

1.   Standard normative minerals do not include extremely common real minerals such as biotite, hornblende, or muscovite. For instance, muscovite-bearing granitoids are generally, but not always, c normative. The mineral corundum only rarely appears in igneous rocks. Of course, it is possible to calculate a tailor-made norm that more closely corresponds to the mode.
2.   The norm is based on the present-day Fe oxidation ratio rather than the ratio that existed when the rock crystallized. Because weathering, or even sample preparation for chemical analysis, typically oxidizes some of the Fe^{2+} to Fe^{3+}, more mt is calculated. This reduces the amount of SiO_2 required to make ferromagnesian silicates and increases q. For SiO_2-undersaturated norms, oxidation increases normative orthopyroxene (hy) relative to ol. An example is given in Table 1.16.
3 .   The alkali and alkaline earth metals, including the major elements Na, K, and Ca, are quite mobile during secondary processes, such as weathering or hydrothermal alteration. Calcium and its trace counterpart Sr, for example, are normally the first cations to be leached out during chemical weathering in a warm, humid climate. The concentration of these ions in the present-day rock may be quite different from the original content. Small changes in Na, K and Ca have profound effects on q and c; the higher the alkali content, the more SiO_2 and Al_2O_3 are required to make ab, or, and an; thus less q is made. Likewise, a higher alkali content will increase the ol/hy ratio, to the point that ne rather than hy may be present in the norm (Table 1.16).
4.   Standard norms apply to rocks that crystallized under moderate to low pressures and in nonhydrous conditions. If the rocks under investigation formed in a different environment, a standard norm may be totally different from the mode and a more appropriate norm can be calculated.

1.3.2 An Introduction to Variation Diagrams

Before further discussion of saturation and igneous rock series, a brief look at *variation diagrams* is necessary. These diagrams are used throughout this book as one of the most common means by which chemical analyses of igneous rocks are displayed. The simplest type of variation diagram is one in which one variable (element, oxide, normative mineral, or any numeric combination) is plotted against another on an *X-Y scattergram* using an arithmetic scale. An example is the plot of SiO_2 on the X-axis versus $Na_2O + K_2O$ on the Y-axis shown in Figure 1.10.

TABLE 1.16 Effect of Oxidation and Alkali Content on Norms

	A	B	C
Si	46.1	46.1	46.1
Ti	0.92	0.92	0.92
Al	18.5	18.5	18.5
Fe^{3+}	1.46	3.21	1.46
Fe^{2+}	6.57	4.82	6.57
Mn	0.13	0.13	0.13
Mg	9.36	9.36	9.36
Ca	10.9	10.9	10.9
Na	5.20	5.20	6.24
K	0.57	0.57	0.68
P	0.16	0.16	0.16
C	0.11	0.11	0.11
ap	0.40	0.40	0.40
cc	0.22	0.22	0.22
il	1.84	1.84	1.84
mt	2.19	4.81	2.19
or	2.85	2.85	3.40
ab	26.0	26.0	30.1
an	31.9	31.9	29.0
di	16.5	16.5	19.0
hy	10.6	15.3	--
ol	7.44	--	14.0
q	--	0.11	--
ne	--	--	0.65

A: cation percent and molecular norm for basalt of Table 1.1 and 1.7. This norm is calculated in section 1.2.5.5.
B: identical to A, except Fe is more oxidized
C: identical to A, except higher alkali content

1.3.2.1 A Simple Triangular Diagram

Another common type of variation diagram is a triangular diagram, which requires that the three variables be recalculated to 100 percent before plotting (Fig. 1.11). As an example, the following data were used to plot point X on the CaO-Na_2O-K_2O triangular diagram of Figure 1.11:

%CaO = 1.73	1.73 x 100 / 8.66	=	20.0
%Na$_2$O = 2.41	2.41 x 100 / 8.66	=	27.8
%K$_2$O = 4.52	4.52 x 100 / 8.66	=	52.2
Total = 8.66			100.0

Only two of the three recalculated values are necessary to plot a point; the third can be used as a check. Gridded triangular graph paper is commercially available. Triangular variation diagrams have their limitations and pitfalls (see Section 5.2.2.1), but they are quite useful.

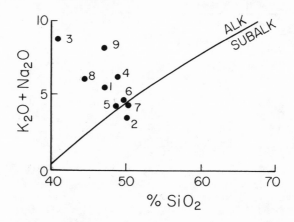

Figure 1.10 An alkali-SiO_2 scattergram. All data are in weight percent. According to this diagram, rocks whose compositions plot at points *7* and *2* are in the subalkaline *(SUBALK)* rock series; the remainder are in the alkaline *(ALK)* series.

1.3.2.2 The QAPF Double Triangle

A variation on triangular diagrams must be understood to plot data on the double triangles of Figure 1.9. Some of the grid lines on Figure 1.9 radiate from the points Q and F, rather than remain parallel to the sides of the triangles as they are on Figure 1.11. Each radiating grid line is a line with a constant P/A ratio. The horizontal grid lines yield values for percent quartz (Q; top triangle) or percent feldspathoids (F; "foids"; bottom triangle).

Hence it is not necessary to recalculate three values to 100 percent. One needs only calculate the percent plagioclase (P) of the total feldspars ($P + A$); that, combined with either percent quartz or foids in the rock, will enable an igneous rock to be named. As an example, a plutonic rock with 30 percent quartz and 75 percent of the total feldspar as plagioclase is classified as a granodiorite.

1.3.2.3 A More Complex Triangular Diagram

Figure 1.12a is a typical triangular plot of normative data. A similar diagram has been used by Irvine and Baragar (1971) to classify volcanic rocks (see Section 5.2.4.3). Data cannot be plotted as easily on this figure as on a simple equilateral triangle (Fig. 1.11). This is because the equilateral triangle of *ne-ol-q* in Figure 1.12a is divided into the three scalene triangles *ne-ab-ol, ab-ol-hy*, and *ab-q-hy*. Because *ne* and *q*, or *ol* and *q*, do not coexist in the same norm, data must be either recalculated to 100 percent and plotted within each nongridded, scalene triangle or recast so that they can be plotted on the equilateral triangle *ne-ol-q*.

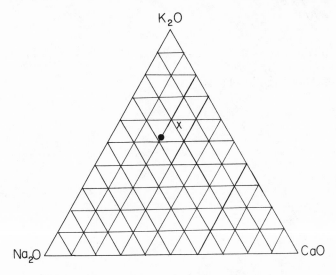

Figure 1.11   Position of point *X*, as calculated in the text, on an alkali triangular variation diagram.

This latter method is not difficult if one is dealing with molecular norms as plotted on Figure 1.12a. Note that *hy* is plotted on Figure 1.12a at the position *ne*0-*ol*75-*q*25 and that *ab* plots at *ne*60-*ol*0-*q*40. Equations 1.23 and 1.24 are written and balanced in terms of moles (gfws). In order to understand how data are converted to be plotted on Figure 1.12a, however, it is necessary to recast Equations 1.23 and 1.24 using EW values. These equations become:

$$3ne + 2q = 5ab \tag{1.25}$$

$$3ol + q = 4hy \tag{1.26}$$

Positions of the points *hy* and *ab* on Figure 1.12a become apparent when Equations 1.25 and 1.26 are expressed as:

$$0.6ne + 0.4q = ab$$

$$0.75ol + 0.25q = hy$$

Another technique, involving plotting data on scalene, nongridded triangles, can be used to plot points on Figure 1.12a. This involves (1) calculating ratios among the three variables, (2) plotting points representing these ratios on the appropriate sidelines of the scalene triangle, and (3) drawing lines between each point and its corresponding apex on the opposite side of the triangle. The intersection of these lines is the correct position of the point to be plotted. Figure 1.12b is an expansion of the scalene triangle *ol-hy-ab* showing how this method is used in plotting point Z. Points on the sidelines were determined from the following data:

$$\% \, ol = 23$$
$$\% \, hy = 65$$
$$\% \, ab = 12$$

$$ol \, / \, (ol + hy) = .261$$
$$ol \, / \, (ol + ab) = .657$$
$$hy \, / \, (hy + ab) = .844$$

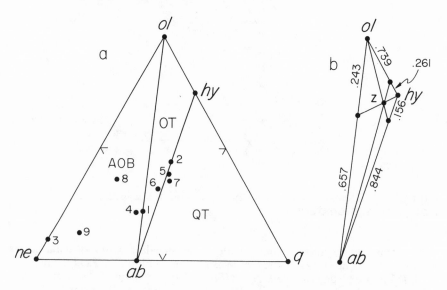

Figure 1.12 *a.* The molecular normative variation diagram *ne-ol-q. QT,* quartz tholeiites; *OT,* olivine tholeiites; *AOB,* alkali-olivine basalts. Numbered points are for same rocks as those on Figure 1.10. *b.* Graphics for plotting point in the *OT* field as explained in the text.

Because point Z can be located by the intersection of any two of the three lines, only two of the above three ratios need be calculated. The third line, however, is useful as a check, as all three lines should intersect at point Z. The same data are recalculated to *ne-ol-q* with the algorithm from Irvine and Baragar (1971):

$$q = q + 0.4ab + 0.25hy = 0 + 0.4 \times 12 + 0.25 \times 65 = 21.05 \, \%$$
$$ne = ne + 0.6ab = 0 + 0.6 \times 12 = 7.2 \, \%$$
$$ol = ol + 0.75hy = 23 + 0.75 \times 65 = 71.75 \, \%$$

Because the three values originally added to 100 percent, recalculating to 100 percent is not necessary. Note that $q21$-$ne7$-$ol72$ corresponds to the intersection of the three lines at point Z on Figure 1.12b.

1.3.2.4 Avoiding Scalene Triangles

Some petrologists find data bothersome to plot on a triangular diagram such as Figure 1.12a. Likewise, data plotted on the two-dimensional projection of a three-dimensional tetrahedron, such as Figure 1.14, are difficult to interpret. As a result, a plot such as that shown on Figure 1.13 is widely used. As for the scalene triangles of Figure 1.12a, two or more equilateral triangles are linked together in such a fashion that incompatible normative minerals (Table 1.15) never occur together at the apices of adjacent triangles. Each equilateral triangle in the example of Figure 1.13 corresponds to its equivalent scalene triangle in Figure 1.12a. The same points are plotted on both figures. One need only recalculate the three appropriate variables to 100 percent to plot a point in any one equilateral triangle.

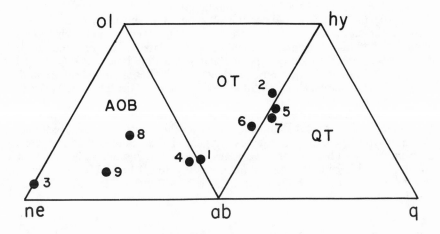

Figure 1.13 Molecular normative variation diagram *ne-ol-q* expressed as three connecting equilateral triangles. *QT,* quartz tholeiites; *OT,* olivine tholeiites; *AOB,* alkali-olivine basalts. Numbered points are from same samples as those on Figures 1.10 and 1.12a.

1.3.3 Silica Saturation and Igneous Rock Series

Igneous rock series are reviewed in Section 5.2.4, but because they can be tied in so well to topics covered in Chapters 3 and 4 through the concept of saturation, they are introduced here. One of the main ways by which we distinguish one series from another is based on SiO_2 saturation.

1.3.3.1 Two Principal Series

Most, but certainly not all, igneous rocks can be subdivided into two very broad series, which are roughly based on their relative degree of SiO_2 saturation. These are referred to as the *alkaline* (or *alkalic*) and *subalkaline* (or *subalkalic*) series. Rocks in the alkaline series generally have higher alkali (Na, K, Rb) contents and are more SiO_2 undersaturated than those in the subalkaline series. The calcalkaline and tholeiitic series can be considered as subseries within the subalkaline series. A third major series, the *peralkaline* series, is based primarily on alumina saturation (Section 1.3.4.1). Rocks in this third series are critically alumina undersaturated, but are rare compared with rocks in the other major series, so this discussion concentrates on the alkaline and subalkaline series. Feldspathoid-bearing rocks are almost exclusively in the alkaline series and quartz-bearing rocks are far more abundant in the subalkaline.

As explained previously, rocks with higher alkali contents tend to be more SiO_2 undersaturated and thus are more likely to be in the alkaline series. The content of SiO_2 and other oxides, however, obviously plays a role as well; therefore, a one-to-one correspondence between alkali content and SiO_2 saturation does not exist. Do we delineate between the alkaline and subalkaline series on the basis of alkalies or SiO_2 saturation?

The problem with delineating alkaline from subalkaline series can be seen by comparing Figures 1.10 and 1.12a. Data from the same analyses are plotted on both diagrams and ambiguity exists in classification of some samples. The dividing line between the alkaline and subalkaline fields on Figure 1.10 is more nearly horizontal than vertical. Hence the alkalies are more important than SiO_2 as discriminators, although SiO_2 is still a factor (if the line were horizontal, SiO_2 would be no factor at all).

Most *ne*-normative rocks are in the alkaline series, whereas most *hy*-normative rocks are subalkaline. Because the degree of SiO_2 saturation increases from left to right across Figure 1.12a, SiO_2 saturation clearly discriminates well between the two series. The question remains, however, as to which method is better for discrimination -- Figure 1.10 or 1.12a. No clear-cut answer exits to this question, but it is discussed further in Sections 5.2.4.2 and 5.2.4.3.

1.3.3.2 Silica Saturation and Basaltic Rocks

Figure 1.12a is directly related to the classification of basaltic rocks, which petrologists frequently believe to be parental to many other igneous rock types in the earth's crust and thus of fundamental concern. Gabbros and diabases (dolerites) are also considered to be basaltic rocks. Whether basaltic rocks are widely parental to other rocks may be an arguable point, but if one includes the oceanic crust, they certainly are the most volumetrically abundant in the crust.

The classification is based on normative minerals and SiO_2 saturation. It was originally shown as a compositional tetrahedron with clinopyroxene (represented by *di), q, ol,* and *ne* at its apices (Yoder and Tilley, 1962; modified by Green and Ringwood, 1967). The "basalt tetrahedron" is shown in Figure 1.14. It is actually part of the larger tetrahedron *q-ol-ne-la* (Schairer and Yoder, 1964). *La* refers to the mineral larnite. The smaller tetrahedron *q-ol-ne-di* suffices for our purposes.

Two planes separate this tetrahedron into three volumes. The right-hand plane *(pl-di-hy)* is known as the *plane of saturation* and the left-hand plane *(pl-di-ol)*, the *critical plane of undersaturation. Quartz tholeiites* in the right-hand volume are *q* normative, whereas *olivine tholeiites* in the central volume are *ol* normative. *Basanites* and *alkali-olivine basalts* in the left-hand volume are *ne* normative. Thus a simple definition of a *tholeiite* is a basaltic rock that is *hy* (orthopyroxene) normative. Note that *pl* and *di* are common to all three types; this is true for modal plagioclase and clinopyroxene as well. The triangle in Figure 1.12a forms the base of the tetrahedron in Figure 1.14, except that *ab* is plotted on Figure 1.12a, whereas total plagioclase *(ab + an)* is plotted on Figure 1.14.

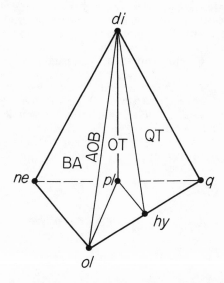

Figure 1.14 The basalt tetrahedron *q-ol-ne-di* (modified after Green and Ringwood, 1967). *QT,* quartz tholeiites; *OT,* olivine tholeiites; *AOB,* alkali-olivine basalts; *BA,* basanites.

If different types of basaltic rocks are parental to different rock series, as many petrologists believe them to be, *ne*-normative basaltic rocks are parental to the alkaline rock series. Tholeiites would be parental to the subalkaline series. Does this mean that these series can be formed in no other way? Definitely not, especially in the case of the subalkaline series. An olivine tholeiite parental to many rocks of the tholeiitic series is reasonable, but not so for the calcalkaline series.

One final point should be made about using norms to classify basaltic rocks, or for that matter, any igneous rocks. Table 1.16 is illustrative. The norm in Column A indicates that the rock is an olivine tholeiite, as it contains *ol* and *hy*. Because the norm in Column B contains *q* and *hy,* it represents a quartz tholeiite. The rock whose analysis is in Column C would be classified an alkali olivine basalt, because it contains *ne* rather than *hy* in its norm. These three rocks have extremely similar overall compositions. The Column B rock is simply an altered, oxidized version of the Column A rock, in which the only difference between the two analyses is $Fe^{2+}/(Fe^{2+} + Fe^{3+})$. The 20 percent higher alkali content of the Column C rock is the only difference between it and the Column A rock. Because alkalis can be quite mobile, their contents may have nothing to do with original igneous processes.

Error analysis (Section 1.4.2.2) would suggest that the very small values of *q* in Column B and *ne* in Column C are probably not significant, and both analyses also could be representative of olivine tholeiites. The point is that one must be careful when using norms for classification purposes.

1.3.3.3 Silica Saturation and Alkaline Rocks

Subalkaline rocks cannot be conveniently subdivided on the basis of SiO_2 saturation alone. For instance, tholeiitic basalts are subalkaline, yet they can be SiO_2 oversaturated or undersaturated. Likewise, the two main subdivisions of subalkaline rocks, tholeiitic and calcalkaline, cannot be distinguished from one another solely on the basis of SiO_2 saturation. In contrast, the more felsic (leucocratic) alkaline rocks can be (Sood, 1981).

An "alkali rock tetrahedron," analogous to the basalt tetrahedron, can be used as a basis for chemical classification of alkaline rocks. The tetrahedron has *di*, *ne*, *q*, and *kp* (kalsilite or kaliophilite) at its apices and is divided into four volumes (Fig. 1.15). The plane *di-ab-or* is the "plane of SiO_2 saturation," whereas *di-ab-lc* is the "critical plane of undersaturation," comparable with the two planes in the basalt tetrahedron.

Most rocks in the volume *ab-or-q-di* are actually subalkaline, so the remaining three volumes subdivide the alkaline series. Furthermore, rocks whose compositions fall in the volume *ne-lc-kp-di* are quite rare. So, practically speaking, most alkaline rock compositions fall in either *ne-lc-ab-di* or *lc-or-ab-di*. Many alkaline rocks, such as syenites and trachytes, are approximately SiO_2 saturated and thus their compositions fall near and on either side of the plane *di-ab-or*.

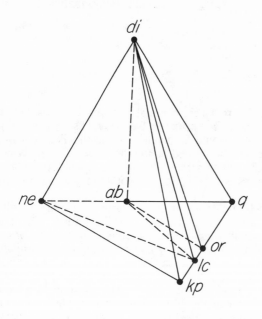

Figure 1.15 The alkali rock tetrahedron *di-ne-q-kp* (modified after Sood, 1981).

The alkali rock tetrahedron is not used as widely to classify those rocks as the basalt tetrahedron is for basalts. Perhaps the main reason is that the alkaline rock tetrahedron does not conform to a standard CIPW or molecular norm. In a standard norm *lc* and *ab* cannot coexist, whereas *ne* and *or* can. The opposite situation is true for the alkaline rock tetrahedron (Fig. 1.15). Sood (1981 p. 85) explains why this is so. Essentially, constructing the interior planes as shown on Figure 1.15 is in better agreement with the observed modal mineralogy. Compositions of leucitites, for example, commonly fall in the volume *lc-ab-or-di*, whereas those of nephelinites generally plot within *lc-ab-ne-di*.

    As a result, if we intend to use the alkali rock tetrahedron, we must either modify it to conform to standard norms or change the norm. The first can be done easily by changing the plane *lc-ab-di* to *ne-or-di*. This is shown in Figures 1.16a-b; the first is in agreement with standard norms and the second with Sood's variation. Both these figures represent the triangle *kp-ne-q*, which is the base of the tetrahedron. This triangle is the alkaline rock analog to Fig. 1.12a. Although the two relevant assemblages are *lc-ab-ne-di* and *lc-ab-or-di*, the assemblages in agreement with standard norms are *ne-lc-or-di* and *ne-ab-or-di*.

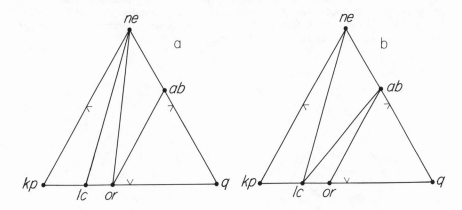

Figure 1.16 *a.* Base of the alkaline rock tetrahedron assuming standard CIPW or molecular norms were being plotted. *b.* Base of the alkaline rock tetrahedron according to Sood (1981).

    A standard molecular norm can be changed most easily by exchanging steps 21-22 with 23-24 in Appendix A. Essentially, this allows the SiO_2 deficiency to be satisfied by making *lc* from *or* before making *ne* from *ab*; a standard norm does these calculations in reverse order. One could argue that making *lc* before *ne* is actually more logical, because nepheline is a more SiO_2-undersaturated mineral than leucite. As an example of these calculations, assume an alkalic rock that still has a Si deficiency of -2.73 percent after all *hy* has been converted to *ol*. In a standard norm (Appendix A, step 21) the procedure can be summarized as follows:

$$5ab = 3ne + 2Si$$

$$\therefore\ 6.82\%\ ab = 4.10\%\ ne + 2.73\%\ Si$$

The Si deficiency is multiplied by 1.5 to obtain 4.10 percent *ne* and by 2.5 to obtain 6.82 percent *ab*. Thus the Si deficiency is satisfied by making 4.10 percent *ne*, and the provisional amount of *ab* in the norm is decreased by 6.82 percent. If *lc* is made first, this same Si deficiency is handled by:

$$5or = 4lc + Si$$

$$\therefore\ 13.6\%\ or = 10.9\%\ lc + 2.73\%\ Si$$

In this case 10.9 percent *lc* is made at the expense of 13.6 percent provisional *or* and, assuming the provisional *or* content is greater than 13.6 percent, the *ab* is unaffected. Otherwise, the Si deficiency is still not satisfied and some provisional *ab* must be converted to $3ne + 2Si$.

1.3.4 Alumina Saturation and Granitoids

Almost all mafic rocks contain diopside in the norm and thus are Al_2O_3 undersaturated. This situation does not exist for felsic rocks, where a wide range in Al_2O_3 saturation is common. *Granitoids,* quartz-bearing plutonic felsic rocks, can vary from *c* normative (oversaturated) to *ac* normative (critically undersaturated). The result is that Al_2O_3 saturation is an important criterion for classification of these rocks.

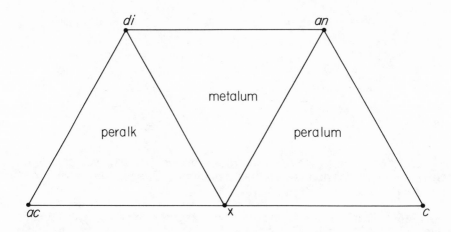

Figure 1.17 A variation diagram involving normative minerals based on Al_2O_3 saturation. Any normative mineral common to all three fields can be plotted at apex *X,* such as *ab* or *or.*

1.3.4.1 Shand's Classification

One of the earliest attempts to use any concept of saturation to characterize igneous rock suites was that of Shand (1927). He divided all igneous rocks into four suites based upon Al_2O_3 saturation:

peraluminous	$Al_2O_3 > Na_2O + K_2O + CaO$
metaluminous	$Al_2O_3 < Na_2O + K_2O + CaO$
	and $> Na_2O + K_2O$
subaluminous	$Al_2O_3 \approx Na_2O + K_2O$
peralkaline	$Al_2O_3 < Na_2O + K_2O$

All concentrations must be in mol, rather than weight or equivalent, proportions. Because his definition of subaluminous rocks is ambiguous, it is seldom used now. The molar ratio A/CNK [$Al_2O_3/(CaO + Na_2O + K_2O)$; Section 1.2.3.4] is quite useful in determining the rock series according to Al_2O_3 saturation. Note the use of the term *peralkaline,* not to be confused with *alkaline, subalkaline,* or *calcalkaline,* which are primarily based on SiO_2 rather than Al_2O_3 saturation.

1.3.4.2 Alumina Saturation and Norms

Peraluminous rocks are generally *c* normative; metaluminous, *di* normative; and peralkaline, *ac* normative. Peraluminous granitoids commonly contain primary muscovite as well as biotite *(two-mica granitoids).* Metaluminous granitoids are generally biotite-bearing, hornblende-bearing, or both, whereas sodic amphiboles are common in peralkaline granitoids.

Shand's definitions of rock series based on Al_2O_3 saturation and definitions based on norms only approximately coincide. Take the normative definition of a metaluminous rock to be one that is *di* normative, as opposed to a peraluminous rock that is *c* normative. Shand's definitions calculate some *c*-normative rocks as metaluminous. This is because some CaO is used up in making *cc, ap* (and rarely *sp*) before *an* and *di* are made; the net result is a larger available alkali/alumina ratio by Shand's calculations. This discrepancy typically is not large and is not considered further.

No tetrahedron corresponding to Figures 1.14 and 1.15 is available for Al_2O_3 saturation. It is possible, however, to construct a diagram similar to Figure 1.13 that takes into account the incompatibility of *c-di* and *an-ac* (Fig. 1.17). Rocks whose compositions fall into the right-hand triangle are peraluminous; the central triangle, metaluminous; left-hand, peralkaline. Alumina saturation increases from left to right. Any of several minerals, such as *or* or *ab,* can be plotted at point X common to all three triangles.

Figure 1.17 is only superficially similar to Figure 1.13, which can be illustrated by comparing *q-hy-ol* on Figure 1.13 with *c-an-di* on Figure 1.17. An equation can be written for *q* and *ol* reacting to form *hy* (Equation 1.24). No comparable equation can be written for *c-an-di.* A reaction can also be written for *ne-ab-q* (Equation 1.25), but not for *c-ab-ac,* or any other combination of minerals across the bottom of the figure. Even so, the figure is useful for illustrating the degree of Al_2O_3 saturation.

1.4 LINEAR ERROR ANALYSIS

1.4.1 Formulation

To the nonscientist an *error* is a mistake. Frequently this same meaning is implied in science as well, but just as frequently the term *error* has another connotation. In this section error refers to *precision* or *reproducibility.* The uncertainty with which we can measure an individual value is involved in the error of that value. Evaluation of error is not included in most petrology books, although McBirney (1984) and Le Maitre (1982) are exceptions. Taylor (1982) provides an comprehensive discussion of error analysis in thermodynamics. Some measure of dispersion about the mean value, most commonly standard deviation *(s),* coefficient of variation *(V),* or standard error *(e_s),* is used to quantify this precision (Section 1.1.3.2). Error limits can be attached to each measurement, as in the example of 125 ± 8 ppm Rb, where $s = 8$ ppm and $V = 6.4$ percent.

These measurements are frequently combined with others and are used in almost an infinite number of ways to determine new variables. Calculating a norm (Section 1.2.5) or fractionation index (see Section 5.2.2.2) from a chemical analysis is a good example. The new variables are linear combinations of the old. By *linear combination* we mean the equation expressing the relation between the new and old variables contains no squared or higher order terms, and cannot contain logarithms, trigonometic functions, and so on. Nonlinear functions are used, but we will not consider them.

1.4.1.1 A Simple Approach to Error Analysis

As usual, the best way to introduce error analysis is by an example. Petrologists calculate ratios of one element to another and use them to constrain their petrogenetic models. A widely used ratio is K/Rb, a ratio of two elements that are commonly related during crystallization of basaltic magmas. Assume the analytical precision *(V)* for both elements is 3 percent. Given the data:

$$K = 1.25 \% = 12500 \pm 375 \text{ ppm}$$
$$Rb = 50 \pm 1.5 \text{ ppm}$$

What is the error on the K/Rb ratio? The standard deviations 375 and 1.5 ppm are based on the 3 percent analytical precision.

A conservative approach would be to combine the upper limit for K with the lower limit for Rb to calculate a maximum ratio, and vice versa for a minimum ratio. Using UL and LL for upper and lower limits, respectively:

$$UL(K) = 12500 + 375 = 12875$$
$$LL(Rb) = 50 - 1.5 = 48.5$$
$$UL(Rb) = 50 + 1.5 = 51.5$$
$$LL(K) = 12500 - 375 = 12125$$

$$\text{maximum K/Rb} = 12875 / 48.5 = 265$$
$$\text{"average" K/Rb} = 12500 / 50 = 250$$
$$\text{minimum K/Rb} = 12125 / 51.5 = 235$$

1.4.1.2 A Better Approach to Error Analysis

The previous method is not totally inappropriate but it is more conservative than it needs to be. An alternative technique, which is more realistic, is to use the *square root of the summed squares* expression:

$$(\Sigma X^2)^{1/2} = (X_1^2 + X_2^2 + \ldots)^{1/2}$$

This relationship is based on the fact that errors expressed as variances are additive, for example:

$$A = B + C + D \quad \text{and} \quad a^2 = b^2 + c^2 + d^2$$

where $a^2 \ldots d^2$ are variances on $A \ldots D$. Standard deviation *(s)* or standard error (e_s) is used in the equations below rather than coefficient of variation *(V)* or variance (s^2) as the measure of error. This is because s and e_s have the same units as the original measurement, whereas V and s^2 do not. Coefficients of variation can be used for a, b, and c in Equation 1.28 but not for a and b in Equation 1.27.

The four mathematical manipulations necessary to work with linear equations are addition, subtraction, multiplication, and division. Their associated error equations are are all based on the premise that variances are additive. One equation can be used for addition and subtraction and another for division and multiplication. A capital letter (other than X, Y, or Z, which are constants) signifies a variable and the equivalent small letter is the standard deviation (or other appropriate measure of dispersion) for that variable. For division or multiplication:

$$q = Q \, [(a/A)^2 + (b/B)^2]^{1/2} \qquad\qquad (1.27)$$

where A and B are variables with attached errors a and b, respectively; X is a constant and $Q = XA/B$ or $Q = XAB$ In other words:

$$Q \pm q = [XA \pm a] \, / \, [B \pm b] \quad \text{or}$$

$$Q \pm q = [XA \pm a] \cdot [B \pm b]$$

For addition and subtraction, where X, Y, and Z are constants:

$$T = XA + YB - ZC$$

$$T \pm t = [XA \pm a] + [YB \pm b] - [ZC \pm c]$$

$$t = [(Xa)^2 + (Yb)^2 + (Zc)^2]^{1/2} \qquad\qquad (1.28)$$

Using Equation 1.27, error on the K/Rb ratio in the example above is:

$$q = 250 \times [(375/12500)^2 + (1.5/50)^2]^{1/2} = 11$$

So the K/Rb ratio can be reported as 250 ± 11. The relative error *(V)* is 4.4 percent, which is greater than the original error on each element (3 percent), but less than that determined from the more conservative approach above (6 percent).

What is meant by the statement that the K/Rb ratio in a particular rock sample is 250 ± 11? Recall our discussion in Section 1.1.3.1 about the normal (Gaussian) curve and the significance of the standard deviation. Approximately 68 percent of the area under the normal curve falls within $\pm 1s$ of the mean, whereas about 95 percent falls within $\pm 2s$. Assuming that this ratio is drawn from a population of ratios that is approximately normally distributed, we can say that roughly a 68 percent probability exists that the true ratio is between 239 and 261. It follows that about a 95 percent probability exists that the true value falls between 228 and 272. Had this ratio been the mean of a number of individual measurements, we could have based the calculations on the standard error e_s.

Some more simple numerical examples are:

addition:

$51 \pm t = [27 \pm 5] + [15 \pm 2] + [9 \pm 3]$
$t = (5^2 + 2^2 + 3^2)^{1/2} = 6$
$V = (100 \times 6) / 51 = 12\,\%$

subtraction:

$3 \pm t = [27 \pm 5] - [15 \pm 2] - [9 \pm 3]$
$t = (5^2 + 2^2 + 3^2)^{1/2} = 6$
$V = (100 \times 6) / 3 = 200\,\%$

multiplication: $3645 \pm q = [27 \pm 5] \times [15 \pm 2] \times [9 \pm 3]$
$q = 3645 \times [(5/27)^2 + (2/15)^2 + (3/9)^2]^{1/2} = 1472$
$V = (100 \times 1472) / 3645 = 40\%$

1.4.2 Some Practical Examples

Three examples illustrate the utility of these equations. The first deals with the assumed quality of major-oxide analyses, the second with norms, and the third with a fractionation index (see Section 5.2.2.2).

1.4.2.1 Quality of a Major Oxide Analysis

Perhaps the most common measure of the quality of a weight percent oxide analysis is the summation of all the major and minor oxides, including volatile species. If the summation is between 99 and 101 percent, this is considered a *superior analysis*. If between 98 and 102 percent, the analysis is *acceptable*. The quality of an analysis is a function of the combined effect of precision and accuracy.

For sake of simplicity, let us assume that our accuracy is quite good and our precision is 3 percent for each oxide except SiO_2. Because SiO_2 is almost invariably the most abundant oxide, the precision and accuracy of its analysis is most critical to this estimate of quality. What must be the precision on the SiO_2 analysis to assure a superior or an acceptable analysis? Given the following analysis in cation percent:

	cation %	or	ab	an	q
Si	65	21	18	6	20
Al	19	7	6	6	
Ca	3			3	
Na	6		6		
K	7	7			
Σ	100	35	30	15	20

The norm is used in the next example of error analysis (see Section 1.2.5.4 for calculation of this norm). Summing the variances, or squaring both sides of Equation 1.28:

$$s^2(\text{total}) = s^2(\text{Si}) + s^2(\text{Al}) + s^2(\text{Ca}) + s^2(\text{Na}) + s^2(\text{K})$$

For a "superior" analysis:

$$1^2 = s^2(\text{Si}) + 0.57^2 + 0.09^2 + 0.18^2 + 0.21^2$$

$\therefore$ $s(\text{Si}) = .77$ and Si $= 65.0 \pm 0.77\%$
$\therefore$ $V = 0.77 \times 100 / 65.0 = 1.2\%$

Precision of this quality is attainable but requires careful work. An acceptable analysis would require 1.9 percent precision on the SiO_2 analysis, which is easily obtained.

1.4.2.2 A Norm

When we calculate the norm of a quartz-normative rock, as in the earlier example, q is the last mineral determined. Many more calculations, with their attendant errors, have gone into the final normative q percentage than into those of *ab, or,* and *an.* The more mafic the rock, typically the lower is its normative quartz, and the larger the relative error (V) will be. We will take the haplogranite composition above and calculate the error on quartz relative to that for the feldspars. For an analysis in which:

$$\% \; or = 5 \times \% \; K$$
$$\% \; ab = 5 \times \% \; Na$$
$$\% \; an = 5 \times \% \; Ca$$

we assume that the 3 percent error on the original analyses also applies to these three normative minerals. As each equivalent is simply multiplied by the constant 5, this seems like a reasonable assumption. Based on Equation 1.28, the calculation is:

$$q = 100 - or - ab - an$$

$$s(q) = (1^2 + 1.05^2 + 0.90^2 + 0.45^2)^{1/2} = 1.76$$

$$\therefore \; q = 20 \pm 1.76 \, \% \quad \text{and} \quad V = 8.8\%$$

The 1^2 value is based on a superior analysis that totals 100 ± 1 percent. Consequently, the relative error on percent normative quartz in this haplogranite is almost three times as large as that on the feldspars.

Typically the absolute error (s) on q or c will be at least 1-2 percent. Values for q and c determine the degree of SiO_2 and Al_2O_3 saturation, which are important in characterizing igneous rock suites. Clearly, one is taking a considerable risk by classifying a particular rock into a certain suite on the basis of c or q values that are less than 2 percent (as an example, refer to Fig. 1.13). Within limits of error, any basalt whose norm plots just on the quartz-normative side of the plane of SiO_2 saturation (and thus is classified as a quartz tholeiite) is no different from an olivine tholeiite whose norm plots just on the olivine-normative side of the same plane. Similarly, a granitoid with only 1 percent c in the norm, considering the limits of error on c, may not actually be peraluminous.

1.4.2.3 A Fractionation Index

The last example deals with the *mafic index* (abbreviated MI), which for our purposes is defined as $MI = \Sigma FeO / \Sigma FeO + MgO$. Depending on how total Fe is defined (ΣFeO, ΣFe_2O_3, or $FeO + Fe_2O_3$; Section 1.1.1.3), the actual numerical value for MI can be different. All three versions of MI are equally valid. The mafic index is a fractionation index that generally, but not always, numerically increases with increasing crystal fractionation in mafic igneous rocks (see Section 5.2.2.2). Given a 3 percent precision on the ΣFeO and MgO analyses, what is the error on MI? The analyses are: $\Sigma FeO = 10.0 \pm 0.30$ percent and $MgO = 8.00 \pm 0.24$ percent. From Equation 1.28:

$$s(\text{denominator}) = (0.30^2 + 0.24^2)^{1/2} = 0.384$$

From Equation 1.27:

$$s(\text{MI}) \quad = 0.555 \times [(0.240/8)^2 + (0.384/18)^2]^{1/2}$$
$$= 0.020$$

$$\therefore \ \text{MI} = 0.555 \pm 0.020 \ \text{ and } \ V = 3.6\%$$

The relative error, therefore, is slightly larger on MI than on the original analyses. For this and other reasons, some petrologists prefer to use only percent MgO as a measure of fractionation in mafic rocks. Advantages and disadvantages of both are discussed in Section 5.2.2.2. Numerous additional examples of the use of linear error analysis are found in the remainder of this book.

1.5 REVIEW OF ROCK-FORMING MINERALS

A rock, by definition, is an aggregate of one or more minerals. We cannot know very much about an igneous rock unless we are quite familiar with its mineralogical constituents. This book presumes that the reader has had a basic mineralogy course but may have forgotten some details of the manner in which major rock-forming silicates and oxides are classified. The chemical classification system presented is a simplified and shortened version. More details can be found in any one of several modern mineralogy textbooks.

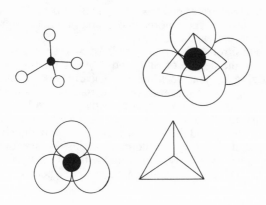

Figure 1.18 Various graphical depictions of a silica (or alumina) tetrahedron. The central Si^{4+} or Al^{3+} ion is surrounded by four O^{2-} ions in tetrahedral coordination.

1.5.1 The WXYZ System

The WXYZ system is a simple means by which the major igneous rock-forming minerals, dominantly silicates and oxides, can be classified chemically. It is one of several systems and was chosen because it is most easily learned. The only anions with which we have to deal are O^{2-}, OH^-, and rarely the trace anions F^- and Cl^-. Thus we are primarily concerned with cations; the major cations are summarized in Table 1.17. A large number of trace cations can also occupy any of the four sites, especially the W, X, and Y sites, but we do not deal with them.

1.5.1.1 Network Formers and Modifiers

The cations are divided into two large groups, *network formers* and *network modifiers*. Network formers are those cations at the centers of the tetrahedra, surrounded by four anions, which form the basic building blocks of the silicate structures. Graphically, these tetrahedra can be depicted in various ways (Fig. 1.18). These cations are said to be in tetrahedral or fourfold *coordination*. In the WXYZ system, they occupy Z positions. The term *coordination* simply refers to the number of anions surrounding a cation in a mineral structure. Network formers are most commonly Si^{4+} or Al^{3+} (small quantities of Fe^{3+} can also be network formers); hence the building blocks are referred to as *silica tetrahedra* or *alumina tetrahedra*. All of these tetrahedra are complex anions with the *formulae*:

silica tetrahedron	$(SiO_4)^{4-}$
alumina tetrahedron	$(AlO_4)^{5-}$

In hydrous minerals part of the oxygen ions are replaced by OH^- ions, or less commonly by F^- or Cl^-. In pure silicate minerals all the building blocks are silica tetrahedra; in *aluminosilicate* minerals part are alumina tetrahedra, although silica tetrahedra are normally more abundant.

TABLE 1.17 The WXYZ System for Cations in Rock-forming Minerals

|——————— Network modifiers ——————— | Network formers			
Coordination numbers			
8 12 (micas)	6 8 (garnet) 4 (mellilite)	6 4 (spinel)	4
W	X	Y	Z
Ca^{2+} Na^{1+} K^{1+}	Mg^{2+} Fe^{2+} Ca^{2+} Mn^{2+}	Fe^{3+} Ti^{4+} Al^{3+}	Si^{4+} Al^{3+}

Network modifiers are the remaining cations -- those not in tetrahedral coordination. Those occupying W positions typically have relatively low valences and relatively large ionic radii. Except in sheet silicate structures, they are in cubic or eightfold coordination (i.e., they are surrounded by eight anions). In sheet silicates, such as micas, they are in 12-fold ("cubic close-pack") coordination. X-position cations are in octahedral or sixfold coordination and, on average, have higher valences and are smaller than W-position cations. In garnets, however, they exhibit cubic coordination. The other exception for these cations is in the mellilite group, where they are in tetrahedral co-ordination. Y-position cations tend to be still smaller and more highly charged than those in the X position, although they also exhibit octahedral coordination. An exception is their coordination in spinels, which is tetrahedral. The Z-position ions are smallest and most highly charged of all cations. Average ionic size and coordination number increase from right to left across Table 1.17, whereas valence generally increases from left to right.

1.5.1.2 Coordination Numbers

These coordination numbers are determined by simple geometric considerations. This is because the cation-anion bonds are quite ionic, which indicates considerable electron transfer from the cations to the anions and minimal penetration of neighboring electron clouds. If electron clouds do not penetrate one another, we can consider the ions as simple rigid spheres in which the bond length (distance from nucleus of anion to nucleus of cation) equals radius of cation plus radius of anion. The bond length, therefore, is the distance between certain critical lattice planes and can be determined by X-ray diffraction.

Intuitively, it is apparent that the larger the value for the ratio r_c / r_a, the greater is the coordination number; i.e., more anions can be packed around a large cation than a small one (or more small anions can be packed around a cation than large anions). Assuming that the ions behave as rigid spheres, we can calculate theoretical limits of r_c / r_a for each coordination. These limits are:

r_c / r_a	coordination numbers	geometry
<0.155	2	linear
0.155 - 0.225	3	equilateral triangle
0.225 - 0.414	4	tetrahedron
0.414 - 0.732	6	octahedron
0.732 - 1.00	8	cube
>1.00	12	cube close-pack

The critical limits, then, are 0.155, 0.225, 0.414, and 0.732. These are determined through simple geometry. Recall that the lengths of sides for a 30° - 60° right triangle are 1 and $3^{1/2}$, while the length of the hypotenuse is 2. The lower limits are calculated by:

coordination
numbers

3	$r_c / r_a = (2 / 3^{1/2}) - 1 = 0.155$
4	$r_c / r_a = (3^{1/2} / 2^{1/2}) - 1 = 0.225$
6	$r_c / r_a = 2^{1/2} - 1 = 0.414$
8	$r_c / r_a = 3^{1/2} - 1 = 0.732$

A useful exercise is to confirm these calculations by geometry. Observed coordination numbers based on X-ray diffraction studies can be compared with predicted numbers based on these limits for r_c / r_a. The agreement is generally good, confirming the validity of the "rigid sphere model."

1.5.2 Silicate Structures

Silica and alumina tetrahedra play the same role in the inorganic world that the carbon atom plays in the organic world. Permutations and combinations are vastly more numerous and complex among organic hydrocarbons than among inorganic silicates. One of the main reasons for this is that carbon can be bonded to its neighboring atoms in a variety of different ways and the silicon ion cannot. Still, tetrahedra can be linked together in various ways analogous to hydrocarbons, forming such diverse structures as rings, chains, and sheets. This linking-together is referred to as *polymerization*.

TABLE 1.18 Basic Silicate Structures

Formal name	Common name	N	O/Z	Complex anion
orthosilicates	island	0	4	(ZO_4)
sorosilicates	pair	1	3.5	(Z_2O_7)
cyclosilicates	ring	2	3	(ZO_3)
inosilicates	single chain	2	3	(ZO_3)
	double chain	2,3	2.75	(Z_4O_{11})
phyllosilicates	sheet	3	2.5	(Z_2O_5)
tectosilicates	network	4	2	(ZO_2)

N: Number of oxygens shared

1.5.2.1 Seven Basic Groups

Silicate (including aluminosilicate) structures are subdivided into seven groups (Table 1.18). These groups are based on their O/Z ratios, which in turn are a function of the number of O^{2-} ions an individual Z-position cation is sharing with its neighbors. Some anthropomorphic examples illustrate this point. In the island structure a Z cation is not sharing oxygens with its neighboring Z cations, so it "owns all four oxygens unto itself." Thus the O/Z ratio is 4 and the basic complex anion is (ZO_4). In the single chain and ring structures each Z cation is sharing two oxygens with its neighbors (Fig. 1.19), so it can be considered to "own" $1 + 1 + 1/2 + 1/2$ anions, for a total of 3. Hence the O/Z ratio is 3 and the basic complex anion for these groups is (ZO_3). Similarly, in the sheet structure three oxygens are being shared (Fig. 1.19), so each Z cation "owns" $1 + 1/2 + 1/2 + 1/2$ or 2.5 oxygens; the basic complex anion is therefore (Z_2O_5). Z cations in network structures share all four oxygens and thus the basic formula for this group is (ZO_2).

The only group for which this simple approach must be expanded somewhat is the double chain structure. One double chain can be considered as having four rows, two "outside" and two "inside" rows (Fig. 1.19). Z-cations in tetrahedra of the outside rows share two oxygens (O/Z = 3), whereas Z-cations of the inside rows share three (O/Z = 2.5). Considering a chain to be infinitely long, an equal number of tetrahedra exist in the inside rows as in the outside. Thus the O/Z ratio for the entire double chain structure is 2.75 and the basic complex anion is (Z_4O_{11}).

The remaining group, the sorosilicates, is simple to explain. Each Z-cation is sharing one oxygen (Fig. 1.19), so the O/Z ratio is 3.5 and the complex anion is (Z_2O_7). Minerals that exhibit the sorosilicate and cyclosilicate (ring) structures are the most uncommon.

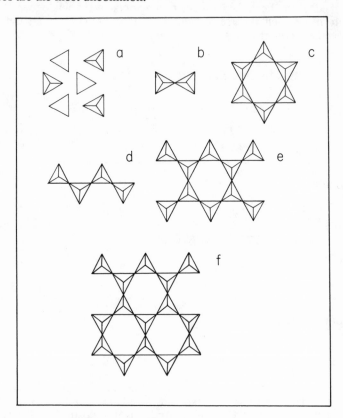

Figure 1.19 Schematic representations of six basic silicate structures: *a.* island; *b.* pair; *c.* ring; *d.* single chain; *e.* double chain; *f.* sheet. Network structure is not shown.

1.5.2.2 Polymerization

Let us return to the concept of polymerization. These polymerized rings, chains, sheets, and other structures are all negatively charged and are electrostatically bonded to one another by network modifiers. The degree of polymerization increases as the number of anions shared increases (and the O/Z ratio decreases), regularly from the island to the network structures. Moreover, the ratio of network formers to network modifiers (Z / W+X+Y) typically increases from the island to network structures. These polymerized structures are not only found in silicate minerals but also in magmas from which these minerals crystallize.

Different degrees of polymerization profoundly affect physical properties of magmas such as viscosity. The more viscous a substance is the less easily it flows and magmas with higher SiO_2 contents are relatively viscous. In general, higher SiO_2 contents are accompanied by higher Z / W+X+Y ratios and more anion sharing, hence lower O/Z ratios and more polymerization. Intuitively, a greater degree of polymerization should lead to a higher viscosity in the magma; this, in fact, is the case. For these reasons, mafic magmas, with lower silica contents, usually are far more fluid than felsic magmas.

TABLE 1.19 Basic Formulae for Principal Mineral Groups

Group	O/Z	N	S	Formula
spinel	--	--	--	XY_2O_4
garnet	4	0	I	$X_3Y_2Z_3O_{12}$
olivine	4	0	I	X_2ZO_4
epidote	3.67	0,1	I,P	$W_2(X,Y)_3(Z_3O_{11})(OH)$
melilite	3.5	1	P	$W_2XZ_2O_7$
orthopyroxene	3	2	SC	$(X,Y)ZO_3$
clinopyroxene	3	2	SC	$W(X,Y)(ZO_3)_2$
orthoamphibole	2.75	2,3	DC	$(X,Y)_7(Z_4O_{11})_2(OH)_2$
clinoamphibole	2.75	2,3	DC	$W_2(X,Y)_5(Z_4O_{11})_2(OH)_2$
mica	2.5	3	S	$W(X,Y)_{2-3}Z_4O_{10}(OH)_2$
chlorite	2.5	3	S	$(X,Y)_6Z_4O_{10}(OH)_8$
serpentine	2.5	3	S	$X_3Z_2O_5(OH)_4$
feldspar	2	4	N	WZ_4O_8
feldspathoid	2	4	N	WZ_3O_6 or WZ_2O_4
silica minerals	2	4	N	SiO_2

N: Number of oxygens shared
S: Structure:
   I   island   DC double chain
   P pair   S   sheet
   SC single chain   N   network

Oxygen ions that are shared by adjacent tetrahedra and link tetrahedra together in the melt are termed *bridging oxygens;* the remainder are referred to as *nonbridging oxygens.* Bridging oxygens are bonded to network formers, whereas non-bridging oxygens are bonded to both network modifiers and formers. A measure of depolymerization is the ratio of nonbridging oxygens to tetrahedrally co-ordinated cations *(NBO/T);* the higher this ratio, the less viscous is the melt. A number of factors cause the NBO/T ratio to change, thereby affecting depolymerization, which in turn affects melt viscosity, density, and other properties. For instance, an increase in depolymerization and decrease in viscosity can be caused by (1) greater F^- content, or H_2O, and thus available OH^- ions (Mysen and Virgo, 1985a); or (2) greater pressure, causing an increase in Fe network modifiers relative to formers (Mysen and Virgo, 1985b).

TABLE 1.20 Examples of Common Rock-forming Minerals

mineral	W	X	Y	Z	anion
SPINEL					
spinel	--	Mg	Al_2	--	O_4
magnetite	--	Fe^{2+}	Fe^{3+}_2	--	O_4
chromite	--	Fe^{2+}	Cr_2	--	O_4
GARNET					
almandine	--	Fe^{2+}_3	Al_2	Si_3	O_{12}
pyrope	--	Mg_3	Al_2	Si_3	O_{12}
grossularite	--	Ca_3	Al_2	Si_3	O_{12}
spessartine	--	Mn_3	Al_2	Si_3	O_{12}
OLIVINE					
forsterite	--	Mg_2	--	Si	O_4
fayalite	--	Fe^{2+}_2	--	Si	O_4
ORTHOPYROXENE					
enstatite	--	Mg	--	Si	O_3
hypersthene	--	MF	--	Si	O_3
CLINOPYROXENE					
diopside (di)	Ca	Mg	--	Si_2	O_6
hedenburgite (hd)	Ca	Fe^{2+}	--	Si_2	O_6
augite (di+hd+Al)					
aegirine-acmite	Na	--	Fe^{3+}	Si_2	O_6
jadeite	Na	--	Al	Si_2	O_6
CLINOAMPHIBOLE					
tremolite (tr)	Ca_2	MF_5	--	Si_8	$O_{22}(OH)_2$
hornblende (tr+Al)					
riebeckite	Na_2	Fe^{2+}_3	Fe^{3+}_2	Si_8	$O_{22}(OH)_2$
MICA					
phlogopite	K	Mg_3	--	$AlSi_3$	$O_{10}(OH)_2$
biotite	K	MF_3	--	$AlSi_3$	$O_{10}(OH)_2$
muscovite	K	--	Al_2	$AlSi_3$	$O_{10}(OH)_2$
paragonite	Na	--	Al_2	$AlSi_3$	$O_{10}(OH)_2$
FELDSPATHOID					
leucite	K	--	--	$AlSi_2$	O_6
nepheline	Na	--	--	AlSi	O_4
FELDSPAR					
albite	Na	--	--	$AlSi_3$	O_8
anorthite	Ca	--	--	Al_2Si_2	O_8
K-feldspar	K	--	--	$AlSi_3$	O_8
SILICA	--	--	--	Si	O_2

MF: (Mg, Fe^{2+})

1.5.2.3 Mineral Groups

The major igneous rock-forming minerals are all silicates (or aluminosilicates),
except for the spinels, which are oxides. Their basic formulae, using the WXYZ

system, are given in Table 1.19. The table is arranged in decreasing O/Z ratio (or increasing number of anions shared). Mineral groups are distinguished from one another primarily on the basis of their complex anions. For instance, the olivines and pyroxenes are chemically distinguished from one another by (ZO_4) versus (ZO_3). The symbolism (X,Y) in Table 1.19 indicates a variable ratio of X-cations to Y-cations in the lattice. Those silicate minerals that contain hydroxyl (OH⁻) ions are referred to as *hydrous* silicates, although most of them do not literally contain water molecules. Another misnomer is that these hydroxyl ions are commonly referred to as *structural water* (Section 1.1.1.3). Individual minerals are expressed in terms of the WXYZ system in Table 1.20.

The only mineral group in Table 1.19 that requires special explanation is the epidote group. Its O/Z ratio of 3.67 does not seem familiar in light of the discussion of basic silicate structures. This apparent discrepancy comes about because the epidote structure is a combination of the orthosilicate and sorosilicate structures. Twice as many (Si_2O_7) as (SiO_4) tetrahedra exist in the lattice of epidote minerals, so a weighted average of the O/Z ratio would be 3.67, yielding the doubly complex anion (Z_3O_{11}).

1.6 VERY BASIC BASIC

All the problems in this book can be solved with a nonprogrammable calculator, or even by hand. They are much more easily and quickly solved, with far less opportunity for error, by means of short computer programs. BASIC is one of the programming languages most widely available on desk-top microcomputers and programmable calculators. BASIC is an acronym for "beginners' all-purpose symbolic instruction code," a high-level programming language originally developed by J.G. Kemeny and T.E. Kurtz at Dartmouth College in the late 1960s and early 1970s. BASIC is a very appropriate acronym because the language is so easily learned by a beginner. Many other more powerful programming languages are available (such as FORTRAN and Pascal), but BASIC is introduced herein because of its availability, as well as the ease with which it can be learned.

The term *high-level* refers to the fact that BASIC, in contrast with an *assembly language,* is relatively close to English, with commands easily understood by novice programmers. The experienced programmer sometimes uses an assembly language, a mnemonic form of *machine language,* which the computer actually uses to perform its duties. Any high-level or assembly programming language is a program itself; on microcomputers it is generally stored on a floppy or hard disk. On programmable calculators and some microcomputers it is *hard wired* into the *read-only memory* (referred to as *ROM,* as opposed to *random-access memory, or RAM).* The information stored in RAM is continually being changed while the computer is being used, but ROM is permanently stored in the computer and cannot be changed. Any program we write and all memory storage registers assigned by the program are temporarily stored in RAM.

All high-level programming languages can be divided into two types: they produce either *compiled* or *interpreted* programs. Many versions of BASIC produce interpreted programs. A compiled program has been entirely translated into a language the computer can use (machine language) before the program is run, whereas an interpreted program is translated into machine language one line at a time. A compiled program needs to be compiled only once, but an interpreted program is interpreted every time it runs. Compiled programs run much faster once their errors have been eliminated, but interpreted programs are generally more easily edited.

The main intent of this section is to "whet the appetite" of someone who has never attempted to write a program. Hopefully, this will be a beginning. This short discussion can only provide the briefest introduction to BASIC -- only the very basic BASIC. Computer graphics, for example, are not treated. Only programs written to solve algebraic problems are considered. The BASIC discussed in this section is as generic as possible, but each version of BASIC has somewhat different "house rules," so the appropriate instruction manual should always be consulted.

1.6.1 BASIC Basics

Excluding the term BASIC itself and a few other acronyms, throughout this section a fully capitalized word or abbreviation (known as a *keyword)* indicates a command or statement that can be used in a program and is a part of the BASIC language. Any BASIC program written to solve a problem in this book consists of three parts: (1) data input, from the operator to the computer; (2) data manipulation, in which results are obtained; and (3) output of results, from the computer to the operator. Part 1 requires writing some INPUT statements and part 2 consists of one or more algebraic or quasialgebraic equations, collectively referred to as an *algorithm*. Part 2 is actually the heart of the program because it calculates the results. The reason why some "equations" are only quasialgebraic will soon become apparent. If a printer is available and a print-out of results on paper is desired, part 3 consists of one or more LPRINT statements. If results printed on the monitor screen or display are adequate, then PRINT statements will suffice for part 3.

1.6.1.1 Variables

Three types of simple variables can be used in BASIC: *numeric, integer,* and *string*. A fourth type, an *array variable,* is actually a collection of simple variables, such as numeric variables grouped together. Integer variables pertain to whole numbers and are only used in the example programs below as counters in simple FOR... NEXT... loops (Sections 1.6.2.2 to 1.6.2.4).

Numeric variables can be considered as storage registers (or "storage bins") in the memory of the computer into which real numbers are temporarily placed. A real number is an integer followed by a decimal fraction, except in the specific case where it is a whole number. In statistical terminology, a continuous variable is generally represented by a real number (Section 1.1.3.1). For a real number, the maximum number of significant figures displayed depends on the computer or calculator; eight or ten are common. Real numbers can be expressed in either normal, everyday *fixed-point* notation (e.g., 1073.2857) or in *floating-point,* scientific notation, with only one digit left of the decimal place (e.g., 1.0732857 E3, equivalent to 1.0732857×10^3).

Each numeric variable in a BASIC program must be given a name, which can be a single letter, or a letter followed by additional letters or numbers. Some versions of BASIC only allow up to two letters (or one letter followed by one number) as a designation for a numeric variable, whereas others allow many more. We will adopt the conservative approach and never use more than two. Thus A, AQ, and A3 are allowable names for numeric variables, but 3A is not;

each name must start with a letter. A few keywords, such as IF, ON, and TO, have only two letters, so they also cannot be used as variable names. If A = 3, R2 = 4, and PZ = A + R2, then PZ = 7. The equation PZ = A + R2 is an algebraic equation, but it also instructs the computer to do the following: "Take the number in storage bin A, add it to the number in bin R2, and put the answer in bin PZ."

A simple example demonstrates why considering a numeric variable as a storage bin is an extremely useful tool in programming. It also demonstrates why some "equations" can be considered quasialgebraic. Take the expression A = A + B. Only in the special case where B = 0 is this an algebraic equality. For the general case this expression is algebraically impossible, but in the language of BASIC it is not intended to be an algebraic equation. Rather, it is a set of instructions to the computer, an algorithm. It tells the computer to take the number in storage bin A, add this number to the number stored in bin B, and replace the original number in bin A with this new summation. This example emphasizes the fact that numbers are only temporarily stored, as for the case of the original number in bin A. To stress the point that the equals sign as used in BASIC commonly does not have the normal algebraic meaning, some books recommend that the beginner consider the above algorithm as A <-- A + B, although it must be written as A = A + B in a BASIC program.

A string variable is some combination of letters, numbers, punctuation, symbols, or even blank spaces, collectively referred to as *characters,* "in a string." A word or a sentence can be a string variable. Names for string variables always end with a dollar sign ($), otherwise, the same rules apply to names of string variables as to those of numeric variables. Hence A$, AQ$, and A3$ are acceptable names for string variables, but 3A$ is not. String variables are also analogous to numeric variables in that they are storage bins for "character strings." A character string must be in double quotations in a BASIC program to be recognized by the computer for what it is. If A$ = "PLAGIO-CLASE ", R2$ = "FELDSPAR", and PZ$ = A$ + R2$, then PZ$ = "PLAGIO-CLASE FELDSPAR". Note the blank space after PLAGIOCLASE in the A$ string. If A$ = "3", R2$ = "4", and PZ$ = A$ + R2$, then PZ$ = "34", not 7. In these examples the computer is instructed to string together the contents of storage bins A$ and R2$, then place the resulting string in bin PZ$. Had the instructions read A$ = A$ + R2$, the original "3" in A$ would have been replaced with the new character string "34".

Because we are not considering integer variables, an array variable for our purposes is a grouping of either numeric or string variables, but not both. A single array variable cannot contain both string and numeric variables. An array variable can be a one-dimensional, linear array, which best can be visualized as a column (stack) or row of storage bins. A table of data containing both rows and columns is analogous to a two-dimensional array variable. Three- or even higher-dimensional array variables are possible, but linear arrays suffice for our purposes.

If we place the number 3 in bin C (i.e., C = 3), then we can define the numeric array variable A(C), where (C) is referred to as the *subscript.* The array variable A(C) can then be considered as a stack of three storage bins or *elements,* designated as A(1), A(2), and A(3), each of which is a simple numeric variable and capable of storing a number. Had we defined a string array variable as A$(C), where C = 3, then the three stacked storage bins would be designated A$(1), A$(2), and A$(3). Only numbers can be placed in A(C), but all types of character strings, including numbers (in this case considered as characters and not mathematical numbers), can be placed in A$(C). Strings must be enclosed by double quotations and double quotations cannot be within a string.

1.6.1.2 Mathematical Manipulations

To solve a mathematical problem using BASIC, numbers stored as either simple numeric variables or as numeric array variables must be mathematically manipulated. As stated earlier, the set of instructions to the computer telling it how to perform these manipulations is called an algorithm. Not surprisingly, BASIC has some very specific rules about how an algorithm must be written. These instructions superficially appear to be algebraic equations, and frequently they are, but some examples already discussed demonstrated how this is not always true. These instructions, however, all have one common characteristic -- the left side of the equals sign consists of one variable and nothing else. Take the example where PZ = A + R2. The computer is instructed to perform a certain mathematical manipulation on the contents of storage bins A and R2, namely addition, and store the results in PZ. All mathematical instructions to the computer are written in this fashion. Numerous examples of mathematical algorithms are given below, and in every case only a variable is written to the left of the equals sign.

When writing a mathematical algorithm it is important to be aware of the hierarchy in which mathematical manipulations are performed. The order, from first to last, is:

1. All expressions in parentheses
2. Exponential, log, and trigonometric functions
3. Multiplication and division
4. Addition and subtraction

An example is illustrative. Take the equation:

$$Y = AW + BX^2 + (CZ)^3$$

Writing this as an algorithm:

$$Y = A*W + B*X^2 + (C*Z)^3$$

Spaces on either side of the plus and equals signs are for clarity; the computer will interpret its instructions properly with or without them. The computer will (1) multiply the number stored in C by that in Z and cube the product, (2) square the number stored in X and multiply it by the number stored in B, (3) multiply the numbers stored in A and W, (4) perform the two additions, and (5) store the results in Y.

Parentheses could be placed accordingly:

$$Y = (A*W) + (B*X^2) + ((C*Z)^3)$$

without changing the instructions to the computer. If any question exists about the order in which mathematical calculations are to be carried out, a good rule is to add parentheses in the appropriate places to eliminate apparent ambiguity. Too many parentheses are better than too few. Be certain, however, that the number of left parentheses equals the number of right parentheses, as in algebraic equations.

Some examples of mathematical algorithms taken from equations in this book are given in Table 1.21. Names of variables are apparent when algorithms are compared with equations in the text. Throughout the book, wherever possible, equations are written in semi-algorithmic fashion so they may be easily adapted to a BASIC program. Note that every algorithm in Table 1.21 has a variable and nothing else to the left of the equals sign.

When the algorithms in Table 1.21 are compared with their respective equations, it is readily apparent that learning only a few simple rules allows any algebraic equation to be recast as a BASIC algorithm. A word of caution is required, however. Keywords for mathematical functions in different versions of BASIC can have different meanings. Take the keyword LOG as an example. For most versions of BASIC written for microcomputers, LOG refers to the natural logarithm to the base e, normally written as ln. For some versions of BASIC written for programmable scientific calculators, LOG refers to the common logarithm to the base 10 and LN denotes the natural logarithm to the base e. Moreover, the variable that follows such mathematical functions as LOG, SIN, and SQR (square root) on most microcomputers must be in parentheses, e.g., LOG(X), SIN(Y), and SQR(Z). For most programmable calculators, LOG X, SIN Y, and SQR Z will suffice. Consult the appropriate instruction manual for the particular "house rules."

TABLE 1.21 Examples of Mathematical Algorithms

Equation	Section	Algorithm
1.27	1.4.1.2	Q1 = Q*SQR((A1/A)^2+(B1/B)^2)
1.28	1.4.1.2	T = SQR((X*A)^2+(Y*B)^2+(Z*C)^2)
2.34	2.2.1.4	DG = DH-T*DS+P*DV
2.35	2.2.1.4	P = (DS/DV)*T-DH/DV
2.36	2.2.1.5	T = (-DH-P*DV)/(F*R*LOG(P)-DS)
2.39	2.2.1.7	P = ((DS/DV)-R*LOG(K))*T-(DH/DV)
3.1	3.2.4.6	LX = (DH/R)*(1/T1-1/T) ∴ X = EXP((DH/R)*(1/T1-1/T))
5.5	5.2.1.2	CL = CO*F^(D-1)
5.6	5.2.1.3	CL = (CS*(1-F)^(1/D-1))/D
5.8	5.2.1.4	CL = CS/(D*(1-F)+F)
5.10	5.2.2.2	LI = SI/3+K-(FE+MG+CA)
5.14	5.2.2.2	SI = 100*MG/(MG+FE+NA+K)
5.15	5.2.2.2	FI = (K+NA)/(K+NA+CA)
5.17	5.2.2.2	PF = 100*(1-CO/CL)
5.18	5.2.2.4	LR = (OY/OX)*F^(DY-DX)
5.21	5.2.2.5	Y = (-D-C*X)/(A+B*X)
5.22	5.2.3.1	Z = N*X+(1-N)*Y

1.6.2 Four Examples

Most of the problems in this book can be solved with one of four types of programs: (1) simple, requiring no FOR... NEXT... loops; (2) requiring summations and FOR... NEXT... loops; (3) requiring READ and DATA statements, as well as FOR... NEXT... loops; and (4) using "flexible" FOR... NEXT... loops. One exception is the norm calculation, which is intrinsically not difficult but long and tedious. We will confine our examples to shorter programs. The

FOR... NEXT... loop is one of the most useful programming techniques and must be mastered. The programs that follow were written using GWBASIC (or BASICA), used with most IBM-compatible microcomputers, but they are sufficiently generic to run with any version of BASIC. For some other version of BASIC, small changes may have to be made. For most versions of BASIC, if a print-out on paper (a "hard copy") is needed, change the PRINT statements on the appropriate lines in the following four programs to LPRINT statements. All four types of programs can occasionally benefit from the use of SUBroutines, so they are also discussed.

These four programs require numbered program lines, as do most interpreted, as opposed to compiled, versions of BASIC. A program statement (instruction to the computer) either must be on a separate numbered program line or, if more than one statement is on a single line, must be separated from other statements by colons. A good practice is to number program lines in multiples of 10, so that additional lines can be inserted later if needed without renumbering all lines. Only whole numbers (integers) can be used for line numbers. The computer will run the program in the order of increasing numbers on program lines. Although not good practice, if four lines were numbered 1, 2, 98, 99, the program would run in that order.

All four programs in this section require data INPUT from the keyboard. If a data set is to be used for a variety of purposes, such as graphics, statistics, and petrologic calculations, a more convenient method is to write data into a data file. A two-dimensional array is particularly useful for this purpose. The data file is then saved on a floppy or hard disk and used as many times as required for many different programs. This saves entering the same data from the keyboard again and again. Unfortunately, no universal commands are used for writing to and reading from data files. Every version of BASIC is somewhat different in this regard, so consult the instruction manual for the available version of BASIC. Lien (1988) correlates commands in one version of BASIC with those in another and clearly explains their functions.

1.6.2.1 A Simple Program

We will first write a program to solve the exponential equivalent of Equation 3.1 (Table 1.21; see Section 3.2.4.6 for application of this equation to a binary phase diagram). As discussed, the program is in three parts: data input, mathematical manipulations, and output of results. The first attempt will be the shortest possible version of a program required to perform the calculation, but not very "user friendly" (i.e., not particularly easy to use by someone who did not write the program). We will then add a few additional statements to make the program easier to use. The program can be written in only six lines:

```
10 INPUT "DELTA H IN JOULES";DH
20 INPUT "MELTING POINT IN DEGREES KELVIN";T1
30 INPUT "TEMPERATURE IN DEGREES KELVIN";T
40 X = EXP((DH/8.315)*(1/T1-1/T))
50 PRINT "MOL FRACTION = ";X
60 GOTO 30
```

The phrases within quotations following the INPUT commands on lines 10-30 are called *prompts* (they are also character strings). When running the program they will print out on the screen and prompt the operator to enter certain data. Similarly, the phrase inside quotations for line 50, also a character string,

labels the answer and is PRINTed out with the answer. For programmable calculators these prompts and labels can be shortened to fit within the display. Note
that the PRINT labels and INPUT prompts are separated from the variables by
semicolons. Part 1 of the program, data input, consists of lines 10-30; part 2, the
algorithm, is line 40; and part 3, output of results (the answer), is line 50. The
function EXPonent in line 40 raises the base of natural logs, e, to a power equal
to the expression inside the parentheses. DELTA H and the MELTING POINT
can be considered as temporary constants, and to construct a liquidus curve on a
binary phase diagram mol fraction is required for several temperatures (Section
3.2.4.6). Consequently, line 60 is added to create a continuous loop so as many
temperatures as required can be entered without exiting the program. Because of
this continuous loop, a separate command must be used to exit the program,
which can be done by CTRL C on IBM-compatible microcomputers.

A number of simple steps can be taken to make the program more user
friendly. First, REM (abbreviation for "remark") statements can be entered for
explanatory purposes. These REM statements are ignored by the computer when
running the program. They are solely for the purpose of explaining the program
to someone who did not write it, or to remind the writer at a later date. Two
REM statements for this program might be:

```
5 REM BINARY EUTECTIC LIQUIDUS
39 REM 8.315 IN LINE 40 IS GAS CONSTANT R
```

Because of clear INPUT prompts and PRINT labels, many variables (DH, T1, T,
and X) in the program can be easily identified without REM statements.

Second, the prompts and output of results would be formatted better on
the screen if some spaces existed between lines. This can be achieved by PRINT
statements alone -- not followed by strings, either on a separate line or on a line
with other statements and separated by a colon. Third, use of CTRL C (or its
equivalent) to exit the continuous loop works but other ways to END the
program exist. One alternative is to add lines 25, 35, and 70, as well as changing
line 60 (see program that follows). The number -9999 is chosen as a value that
would never be used as a part of normal data INPUT. Line 25 reminds the
operator to type -9999 to END the program. Line 35 is a conditional IF...
THEN... statement; if T = -9999 then the program will go to line 70 and END.
Alternatively, line 35 can be written as either (1) IF T = -9999 THEN GOTO 70,
or (2) IF T = -9999 THEN END. If the second alternative is used, then line 70
can be omitted. Consequently, a more user-friendly version of the program is:

```
5 REM BINARY EUTECTIC LIQUIDUS
10 PRINT : INPUT "DELTA H IN JOULES";DH
20 INPUT "MELTING POINT IN DEGREES KELVIN";T1
25 PRINT : PRINT "TYPE -9999 TO END"
30 PRINT : INPUT "TEMPERATURE IN DEGREES KELVIN";T
35 IF T = -9999 THEN 70
39 REM 8.315 IN LINE 40 IS GAS CONSTANT R
40 X = EXP ((DH/8.315)*(1/T1-1/T))
50 PRINT : PRINT "MOL FRACTION = ";X
60 GOTO 25
70 END
```

1.6.2.2 Summations and FOR... NEXT... Loops

A program similar to the previous one solves many problems worked in this book, but it will not solve most statistical problems (Section 1.1.3). These problems require various types of summations, which are most easily accomplished with FOR... NEXT... loops. A loop begins with a FOR... statement and ends with a NEXT... statement. An integer variable (I or J are commonly used) is called the *counter;* its purpose is to monitor the number of times a loop is completed and to command the program to exit the loop at the proper time. A typical FOR... statement is FOR I = 1 TO 10, whereas the corresponding NEXT... statement is simply NEXT I. The succinct command NEXT I tells the computer to advance the counter by one and loop back to the above FOR I = 1 TO 10 statement, which determines if the counter has exceeded 10. If it has, then the loop is exited and the program goes to the first line below the NEXT I command.

The example in this section is a program to calculate an arithmetic mean, variance, standard deviation, and coefficient of variation (Section 1.1.3.2). The utilitarian version of this program is:

```
30 INPUT "# VALUES";N
35 DIM X(N)
40 FOR I = 1 TO N
50 INPUT "X";X(I)
60 NEXT I
80 FOR I = 1 TO N
90 SX = SX+X(I)
100 SS = SS+X(I)^2
110 NEXT I
120 AM = SX/N
130 V = (SS-SX^2/N)/(N-1)
140 SD = SQR(V)
145 CV = 100*SD/AM
150 PRINT "ARITHMETIC MEAN =";AM
155 PRINT "VARIANCE =";V
160 PRINT "STANDARD DEVIATION =";SD
170 PRINT "COEFF. OF VARIATION =";CV;"%"
```

This program uses two FOR... NEXT... loops, one for data INPUT (lines 40-60) and one in the algorithm (lines 80-110). It could be written with only one FOR... NEXT loop, but reasons for using two loops will become clear. The algorithm consists of lines 80-145, and output of results is PRINTed by lines 150-170. An array variable is used, X(N), and line 35 DIMensions the array for N values; i.e., it tells the computer how many storage bins are required to hold all the numbers to be placed in X(N). If N = 20, then the computer reserves 20 bins for X(N). For most versions of BASIC, if less than 11 bins are required, a DIMension statement is not necessary.

A word is required here about the two FOR... NEXT... loops. The FOR... statements in lines 40 and 80 could also be written FOR I = 0 to N-1. In the preceding program if N = 10, the program goes through each loop 10 times; on the eleventh time N is exceeded and the program continues by executing the first line below the loop. The data INPUT loop (lines 40-60) is used to store data in the numeric array variable X(N). If N = 10, elements in the array will range from X(1) to X(10) (or X(0) to X(9), if FOR I = 0 TO N-1 is written). The second loop (lines 80-110) is to calculate the summations ΣX (line 90) and ΣX^2

(line 100). Each time through the loop a new value of X is added to the number stored in SX and a new value of X^2 is added to the number stored in SS.

Once these summations are calculated and the loop is exited, lines 120-145 calculate the values for arithmetic mean (AM, line 120), variance (V, line 130), standard deviation (SD, line 140), and coefficient of variation (CV, line 145). Because SD is required to calculate CV, and V is required to calculate SD, the order in which they are calculated must be V-SD-CV. Refer to Section 1.1.3.2 for definitions of these parameters. Their values, along with appropriate labels, are then PRINTed in lines 150-170.

This program will work but is not particularly user friendly. In addition to appropriate REM and PRINT statements, a number of changes can be made to make it far easier to use. Perhaps the most important is to provide some means for correction of mistakes during data entry. Another is to provide a prompt that informs the operator the entry number of the data value to be INPUT. A third is to allow the program to accept data entry and calculate a new set of values without exiting the program. All these are accomplished in the new version of the program below.

```
5 REM UNIVARIATE STATISTICS
10 PRINT : INPUT "MAXIMUM # ENTRIES IN ANY SET";M
20 DIM X(M)
30 INPUT "# ENTRIES FOR THIS SET";N
35 PRINT : PRINT "TYPE -9999 TO CORRECT PREVIOUS MISTAKE"
40 PRINT : FOR I = 1 TO N
50 PRINT "ENTRY #";I; : INPUT X(I)
55 IF X(I) = -9999 THEN I = I-2
60 NEXT I
70 SX = 0 : SS = 0
75 REM SX AND SS FOR SUMMATIONS OF X AND X^2
80 FOR I = 1 TO N
90 SX = SX+X(I)
100 SS = SS+X(I)^2
110 NEXT I
120 AM = SX/N
130 V = (SS-SX^2/N)/(N-1)
140 SD = SQR(V)
145 CV = 100*SD/AM
150 PRINT : PRINT "ARITHMETIC MEAN =";AM
155 PRINT "VARIANCE =";V
160 PRINT "STANDARD DEVIATION =";SD
170 PRINT "COEFF. OF VARIATION =";CV;"%"
190 PRINT : INPUT "ANOTHER DATA SET (Y/N)";YN$
200 IF YN$ = "Y" THEN 30
```

Improving the prompt is accomplished by changing line 50. The old prompt would query:

X?

for every data entry, whereas the new prompt will ask:

ENTRY # 1?

and so on for up through N values. If, for example, data entry number 9 should
have been 1.09 rather than 10.9, then the following sequence would be on the
screen:

ENTRY # 8? 1.42
ENTRY # 9? 10.9
ENTRY # 10? -9999
ENTRY # 9? 1.09
ENTRY # 10? 1.35
ENTRY # 11? 1.27

    Allowing for correction of mistakes during data entry is achieved simply
by adding lines 35 and 55. Line 35 reminds the operator to type -9999 to correct
a mistake on the previous data entry. Typing -9999 causes the next prompt to
ask for the previous data entry to correct the mistake. If -9999 is typed, the
counter is decreased by two in line 55 but then is increased by one in line 60, so
the net effect is that the counter has been decreased by one when the program
loops back to line 40.
    The improvement that requires addition of the most new steps is to
change the program so that it can accept more than one data set. This, frankly,
may not be worth the effort, but it is discussed here because it illustrates some
new programming techniques. In the original version of the program, after N
values in one data set were entered, results would be printed and the program
would end. In the new version many data sets can be entered without exiting the
program. This is accomplished by adding lines 10, 70, 190, and 200, as well as
changing the DIMension statement.
    Why not simply add lines 190 and 200 to the original program and make
no other changes? If we were to do this, a cardinal rule in BASIC would be
broken -- an array can only be DIMensioned one time. Lines 190 and 200, by
use of the string variable YN$, create a conditional loop. The abbreviation
(Y/N) is the short-hand way of asking yes or no. The response is Y for yes and
N for no. If the program were to loop back to line 30, the array X(N) would be
reDIMensioned in line 40, and this is not allowed. By adding line 10, which
prompts for the maximum number of entries in any set (M), and DIMensioning
the array by X(M) in line 20, the conditional loop created by lines 190 and 200
avoids line 20 and thus does not reDIMension the array.
    The other line that must be added is line 70, which "initializes" the
numeric variables SX and SS to zero. In the original version of the program this
is not necessary because all numeric variables are initialized to zero and all
string variables to the null string when the program is rerun for a new data set.
In the new version, after the first time through the program, the storage bins SX
and SS, because they are used for summations, must be "cleared out" by initial-
izing them to zero. It is normally not necessary to initialize a numeric variable
in this fashion unless either it is used for accumulating a summation or initially,
for some reason, it must contain a number other than zero.

1.6.2.3 SUBroutines

Notice that no provision is made in this program for rounding of numbers. Some
versions of BASIC allow for rounding by use of the PRINT USING or ROUND
command, but others do not. If no PRINT USING, ROUND, or comparable
command is available, one way rounding can be accomplished is by adding a
SUBroutine. In fact, SUBroutines are extremely powerful programming tools,

with a variety of uses other than just for rounding. If one sequence of calculations is done several times throughout a program, a SUBroutine can save many programming steps. The first step necessary for a rounding SUBroutine is to expand lines 120-145:

```
120 AM = SX/N : RO = AM : GOSUB 300 : AM = RO
130 V = (SS-SX^2/N)/(N-1) : RO = V : GOSUB 300 : V = RO
140 SD = SQR(V) : RO = SD : GOSUB 300 : SD = RO
145 CV = 100*SD/AM : RO = CV : GOSUB 300 : CV = RO
```

Next the SUBroutine is written in lines 300-310:

```
300 RO = INT(RO*1000+.5)/1000
310 RETURN
```

A SUBroutine must always end with a RETURN command. The INTeger function takes the integer of any real number by eliminating all numbers right of the decimal place. Some versions of BASIC have a CINT command, which rounds a real number either up or down to its nearest integer. Line 300 rounds all data to three decimal places. If only one decimal place is desired, use 10s rather than 1000s in line 300. Similarly, 100s are used for two decimal places, and so on. If the CINT command is available, line 300 can read RO = CINT(RO*1000)/1000. The +.5 is not required if CINT is used. The last step is to enter the line:

```
210 END
```

An END command must separate all subroutines from the main body of the program.

It can be argued that a version of the rounding equation in line 300 could be typed on each of lines 120-145, which for this program would be just as conservative of programming steps. One can also argue that this method of rounding numbers is not worth the trouble. Both arguments may be valid, but the rounding procedure is included here as an example of the use of a SUBroutine. Take rounding of the arithmetic mean (line 120) as an example. After the answer is stored in bin AM, it is then also stored in bin RO. Next the command GOSUB 300 is executed and the number in bin RO is rounded to three decimal places and the rounded number replaces the original number in bin RO. The program then RETURNs to the command immediately following the GOSUB 300 command, where the rounded number replaces the original number in bin AM. An identical procedure is followed for the remaining three statistical parameters (lines 130-145).

One lesson quickly learned in computer programming is that more than one way usually exists to accomplish a particular task. The aim is always toward finding a compromise between making the program as user friendly as possible, but as conservative of memory as practical. For the example of rounding, in the absence of a PRINT USING or ROUND command, can the same task be accomplished more simply? The answer is affirmative, with the use of an array variable. First, lines 120-145 are rewritten:

```
120 S(1) = SX/N
130 S(2) = (SS-SX^2/N)/(N-1)
140 S(3) = SQR(S(2))
145 S(4) = 100*S(3)/S(1)
```

It is not necessary to DIMension the array variable S(I) because it contains only four elements. Line 147 is added:

147 GOSUB 300

Now SUBroutine 300 becomes:

300 FOR I = 1 TO 4
310 S(I) = INT(S(I)*1000+.5)/1000
320 NEXT I
330 RETURN

The final step is to replace variables AM, V, SD, and CV in lines 150-170 with S(1), S(2), S(3), and S(4), respectively. The SUBroutine only requires four lines, and four lines were available between lines 145 and 150, but it is never good practice to number lines consecutively, as mentioned previously. Some versions of BASIC provide a RENUMbering command, whereby all lines in a program are renumbered using multiples of 10. If the RENUMbering command is used, the FOR... NEXT... loop used for rounding could be inserted in the main body of the program and the SUBroutine eliminated.
        One last example of a SUBroutine will be introduced. All well-written programs that require the INPUT of numeric data incorporate a powerful technique for changing incorrectly entered data. The technique in this program allows only the previous entry to be corrected. If a mistake is immediately identified, it is easily corrected, but if it is not recognized for some time, a serious problem arises. Another SUBroutine can be added that allows correction of any number of data entries, not just the previous entry. For instance, mistakes for the 19th and 24th entries could be corrected, even though a complete data set of 50 values had been entered. First, lines 63 and 67 are included:

63 PRINT : INPUT "ALL DATA CORRECTLY ENTERED (Y/N)";YN$
67 IF YN$ = "N" THEN GOSUB 400

These lines provide the option either to change incorrect data entries or continue with the program. Next, SUBroutine 400 is written:

400 PRINT : INPUT "ENTRY # FOR INCORRECT VALUE";E
410 PRINT "CORRECT VALUE FOR #";E; : INPUT X(E)
420 PRINT : INPUT "MORE CORRECTIONS (Y/N)";YN$
430 IF YN$ = "Y" THEN 400
440 RETURN

This SUBroutine first prompts for the entry number of the incorrect value, then prompts for the correct value, and finally asks if more corrections are required. The conditional loop allows for as many corrections as necessary, in any sequence. Because it requires so little memory and is so simple to use, the routine for changing only the previous data entry (lines 35 and 55) is left in the program; SUBroutine 400 is added as an "insurance policy." Other routines could be added to make this program even more user friendly, but we are probably rapidly approaching the point of diminishing returns.

1.6.2.4 READing DATA and FOR... NEXT... Loops

This next example of a program is particularly useful for many of the conversions tabulated in Section 1.2. Data are embedded in the program following DATA statements at the end. For the conversions in Section 1.2, data are conversion factors, such as gfws or EWs, and character strings to be used within INPUT prompts and PRINT labels. READ commands are then executed at the appropriate places in the program; these READ commands read the data following the DATA statements in sequential order. The data are then used according to instructions given in the program.

The program below calculates cation percents for a major oxide analysis, which are used for a variety of purposes, including calculation of a molecular norm (Sections 1.2.3.3 and 1.2.5). In contrast to the two previous programs, only one version of this program is given. It uses four FOR... NEXT... loops, but it could be written using only two. If we did so, it would be less user friendly and would not have the feature of being able to correct mistakes in data entry. This feature is considered so important that the program is written in its present form.

```
5 REM CATION PERCENTS
10 DIM A$(11),X(11)
20 CLS : FOR I = 1 TO 11
30 READ A$(I)
40 NEXT I
50 PRINT : PRINT "ENTER ELEMENTS AS WEIGHTT % OXIDES"
55 PRINT
60 PRINT "TYPE -9999 TO CORRECT PREVIOUS MISTAKE"
70 PRINT : FOR I = 1 TO 11
80 PRINT A$(I); : INPUT X(I)
90 IF X(I) = -9999 THEN I = I-2
100 NEXT I
110 SM = 0 : FOR I = 1 TO 11
120 READ C : X(I) = X(I)/C
130 SM = SM + X(I)
140 NEXT I
150 PRINT : FOR I = 1 TO 11
160 X(I) = X(I)*100/SM
170 X(I) = INT(X(I)*100+.5)/100
180 PRINT A$(I);" =";X(I)
190 NEXT I
200 PRINT : INPUT "ANOTHER DATA SET (Y/N)";YN$
210 IF YN$ = "N" THEN 230
220 RESTORE : GOTO 20
230 END
240 DATA "SI", "TI", "AL","FE3+","FE2+","MN"
250 DATA "MG","CA","NA","K","P",60.1,79.9,51
260 DATA 79.8,71.8,70.9,40.3,56.1,31,47.1,71
```

Most of the statements in this program should be already familiar, but a few are new and require a brief explanation. The command CLS in line 20 clears the screen, but is not available in all forms of BASIC. The READ commands in lines 30 and 120 read the DATA in lines 240-260 into the program at the appropriate places. Note the quotes around the character strings in lines 240-250; these strings are used for both the INPUT prompts (line 80) and the

PRINT labels (line 180). The numeric DATA in lines 240-250 are EWs (Section 1.2.2). Line 170 rounds the results to two decimal places. The RESTORE command in line 220 causes the READ command in line 30 to start reading at the beginning of the DATA (i.e., at "SI" in line 240).

The four FOR... NEXT... loops accomplish the following: (1) lines 20-40 create a string array variable A$(I) that contains chemical symbols for the cations; (2) lines 70-100 INPUT the data (a major oxide analysis in weight percent) into the numeric array variable X(I); (3) lines 110-140 calculate mol proportions (Section 1.2.3.3) and sum them in storage bin SM; and (4) lines 150-190 recalculate the mol proportions to cation percents (line 160), then round and PRINT the results. If a provision is made for correction of mistakes during data INPUT (line 90), it is important that nothing else be done in that loop except storage of INPUT data in the numeric array variable X(I). A conditional loop is created by lines 200 and 220, if additional major oxide analyses are to be converted to cation percent.

1.6.2.5 "Flexible" FOR... NEXT... Loops

All the examples of FOR... NEXT... loops discussed have been limited in that they are either fixed (e.g., FOR I = 1 TO 11) or the number of loops could be changed but nothing else (e.g., FOR I = 1 to N). Many of the problems in Chapter 5 can be most easily solved by using FOR... NEXT... loops that have considerably more flexibility. They require the introduction of no new keywords except for the STEP command. An example program, based on Equation 5.5 (see Section 5.2.1.2), follows. It calculates the composition of the residual magma based on Rayleigh fractional crystallization for various percentages of fractionation, given the composition of the parent magma and the bulk distribution coefficient.

```
10 REM RESIDUAL LIQUID FOR RAYLEIGH FRACTIONATION
20 CLS : INPUT "COMPOSITION OF ORIGINAL LIQUID";CO
30 INPUT "BULK DISTRIBUTION COEFFICIENT";D
40 INPUT "# DECIMAL PLACES FOR RESID. LIQUID";N
50 PRINT : PRINT "F REPRESENTS MELT FRACTION."
60 PRINT "F MUST BE < = 1.0 AND > ZERO."
70 PRINT : INPUT "MAXIMUM F";MX
80 INPUT "MINIMUM F";MI
90 INPUT "INCREMENT FOR F";IC : PRINT
100 PRINT "  F"," % FRACT."," RESID. LIQ." : PRINT
110 FOR F = MX TO MI STEP -IC
120 F = INT(F*100+.5)/100
130 CL = CO*F^(D-1) : CL = INT(CL*10^N+.5)/10^N
140 REM PF IS % FRACTIONATION
150 PF = (1-F)*100 : PF = INT(PF+.5)
160 PRINT F,PF,CL
170 NEXT F
175 PRINT
180 INPUT "MORE CALCULATIONS WITH SAME CO & D (Y/N)";YN$
190 IF YN$ = "Y" THEN 50
```

Lines 20-90 are for data INPUT and the algorithm is contained within the loop incorporated in lines 110-170. The main part of the algorithm is line 130, which contains the Rayleigh fractional crystallization equation. Lines 100 and

160 PRINT the results in tabular form. The rounding routines in lines 120, 130, and 150 are not absolutely essential but allow the table of results to be neat and orderly. Recall from above that if the CINT command is available for the rounding routines, the +.5 is not necessary. The commas in lines 100 and 160 cause the data automatically to be arranged in columns. Line 40 allows the operator to choose the number of decimal places for calculated compositions of residual melts.

The loop is "flexible" because of line 110. As seen in the PRINT and INPUT commands in lines 50-90, F is the melt fraction and must always be greater than 0, but equal to or less than 1. The variable F is also used as the counter variable in the loop. Percent fractionation is simply (1-F) x 100. The first calculation for the residual composition CL is for the maximum INPUT value of F, MX; the last is for the minimum INPUT value of F, MI. The number of calculations for CL, i.e., the number of times the loop will be completed, will also depend on the value for the INPUT increment, IC. This increment IC is used in conjunction with the STEP command to determine the size of each incremental *decrease* in F when line 170 loops the program back to line 110. Line 110 could have also been written FOR F = MI TO MX STEP IC, in which case incremental *increases* in F would occur. Line 110 is written as it is so that the table of results is PRINTed in order of increasing percent fractionation. For a simple loop in which STEP is not specified, such as FOR I = 1 to N, the increment is always implicitly 1.

An example printout of data INPUT (in italics) and results based on a parent magma of 200 ppm Rb is:

COMPOSITION OF ORIGINAL LIQUID (CO)? *200*
BULK DISTRIBUTION COEFFICIENT (D)? *.01*
DECIMAL PLACES FOR RESID. LIQUID (CL)? *0*

F REPRESENTS MELT FRACTION.
F MUST BE < = 1.0 AND > ZERO.

MAXIMUM F? *.9*
MINIMUM F? *.5*
INCREMENT FOR F? *.1*

F	% FRACT.	RESID. LIQ.
.9	10	222
.8	20	249
.7	30	285
.6	40	332
.5	50	397

After 50 percent fractionation the residual magma will have a concentration of 397 ppm Rb, given a bulk distribution coefficient of 0.01 (refer to Sections 5.2.1.1 and 5.2.1.2). All trace- and many major-element modeling calculations in Chapter 5 can be easily accomplished with programs of this type.

2

CLASSICAL THERMODYNAMICS

Chapters 3 and 4 deal primarily with the application of experimental phase equilibria to igneous petrogenesis. The fundamental basis for phase equilibria is a sub-discipline of physical chemistry, classical thermodynamics. This area is not in an arcane, esoteric nether world in which petrologists should never explore. Quite the contrary, at least an introductory knowledge of thermodynamics is necessary to appreciate much of the remaining material in this book, and a thorough knowledge is essential in becoming a modern petrologist. Some of the material in this chapter may not be explicitly considered in later chapters, but be assured that it lays the groundwork. This chapter will provide only a brief introduction to thermodynamics. More complete discussions, including the application of thermodynamics to petrology, are in a number of books (e.g., Kern and Weisbrod, 1967; Wood and Fraser, 1976; Powell, 1978; Nordstrom and Munoz, 1986).

2.1 FUNDAMENTAL LAWS AND FREE ENERGY

2.1.1 Definitions and Conventions

A *system* is a certain region isolated in our thinking from the remainder of the universe, which is referred to as the *surroundings*. A system could be a single crystal (or even part of a crystal), a magma chamber, an entire batholith, a reaction vessel in a chemical laboratory, or anything we choose. Classical thermodynamics treats the system as a whole, as opposed to statistical thermodynamics, which commonly deals with a system on a molecular level. Classical thermodynamics also considers the system in a state of chemical equilibrium, in contrast with thermodynamics of irreversible reactions. Finally, as opposed to kinetics, classical thermodynamics does not consider rates of reaction.

2.1.1.1 Basics

A state of *chemical equilibrium* exists when there is no tendency for an observable chemical reaction in the system being studied. This is usually a *dynamic equilibrium,* which means that the rate of the forward (left-to-right) chemical reaction equals that of the back (right-to-left) reaction. Stating this differently, the *reactants* are combining to form the *products* at the same rate the products are forming the reactants:

$$aA \quad + \quad bB \quad = \quad cC \quad + \quad dD$$
$$\text{reactants} \qquad\qquad \text{products}$$

By convention, reactants are on the left and products are on the right. If the rates of the forward and back reactions are different, this is an irreversible reaction; i.e., the reaction is not in equilibrium.

A system can be considered as isolated, closed, or open. In an isolated system, no exchange of matter or energy occurs with the surroundings. Isolated systems are seldom if ever realized in geology. In a closed system, energy is exchanged with the surroundings but matter is not. A magma crystallizing in a closed magma chamber that is not chemically interacting with the surrounding country rocks is in a closed system, if we define the system as being the magma chamber. An open system exchanges both matter and energy with its surroundings. If the magma is interacting with the surrounding country rocks, this is considered an open system.

A system is composed of *phases,* whose chemical constituents are referred to as *components.* A phase is a mechanically separable, restricted part of the system that ideally is homogeneous with respect to all its physical and chemical properties. Consider a magma chamber undergoing crystallization. Each mineral, whether it is essentially pure or a solid solution, is a separate phase. Occasionally one mineral can be a mixture of two phases, such as an alkali-feldspar perthite. The magma is another separate phase; if two or more immiscible magmas are present in the chamber, each is a separate phase. An H_2O- or CO_2-rich fluid, appearing as bubbles in the magma, would also be a separate phase. Components are the individually variable chemical constituents that make up the phases. Components can be individual atoms (or ions), complex ions, or molecules.

2.1.1.2 Parameters of State

Various *parameters of state* can be used to define a system. These parameters are subdivided into two groups, *extensive* and *intensive.* Extensive parameters are quantitative in that they depend on the quantity or size of a system or phase; intensive parameters do not depend on quantity or size. A simple example illustrates this difference. The number of moles of a component i, n_i, is an extensive parameter because it depends upon the size of the system or quantity of the phase that contains component i. *Mol fraction (X_i),* on the other hand, is defined as:

$$X_i = n_i / \Sigma n \tag{2.1}$$

where Σn is the summation of all components in the phase. If only two components, i and j, are present in a phase, then X_i must always equal $1 - X_j$ and vice versa. Because mol fraction does not depend on the quantity of the phase in the system, it is considered as an intensive parameter. As an example, n_i might equal 10^{11} moles in a 1 km^3 magma chamber, but if $n_i + n_j + n_k + \ldots$ equals 5 x 10^{11} moles, X_i will be 0.2. If the volume of magma changes, n_i will be different but X_i will not, assuming the overall composition of the magma remains constant.

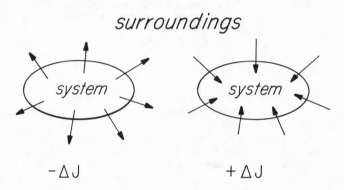

Figure 2.1 Schematic diagram showing the difference between positive and negative finite changes in a hypothetical thermodynamic parameter *J*. For a negative change, *J* is lost by the system and gained by the surroundings, and vice versa for a positive change.

Units on many thermodynamic parameters follow in the text and in Appendix F. Some additional extensive parameters are:

V   *volume;* three-dimensional size of a phase or system
E   *internal energy;* total energy of the system
S   *entropy;* a measure of disorder in the system
H   *enthalpy;* heat (*Q*; thermal energy) content of the system

Examples of intensive parameters are:

—   any of the above expressed as a molar quantity, such as molar volume (cc/mol) or molar entropy (j/mol°C; j = joules)
P   *pressure;* a measure of force applied per unit area
T   *temperature;* a measure of thermal energy (heat or *Q*); an arbitrary scale based on physical properties, such as the melting and freezing points of H_2O or the expansion of Hg
ρ   *density;* mass per unit volume
C   *specific heat* or *heat capacity;* a measure of a substance's capacity to store heat; if measured per gram, referred to as specific heat; if per mole, as heat capacity

Note that *C* measures a substance's capacity to store heat but *H* measures the amount actually stored.

2.1.1.3 Conventions

Before proceeding it is necessary to establish some sign conventions. If a system expands in volume or *gains* work (i.e., work is done on the system by the surroundings), heat, or moles, we use a positive (+) sign convention to define a change in any one of these parameters (Fig. 2.1). Conversely, if a system *loses* to the surroundings work, heat, moles, or reduces in volume, a negative (-) sign convention is used. The same convention is used for all other parameters. We refer to a chemical reaction with a $-\Delta H$ as an *exothermic* reaction (i.e., the system gives up heat to the surroundings), whereas a $+\Delta H$ indicates an *endothermic* reaction.

If a change in a parameter is large (finite), that change is signified by a Δ; if the change is infinitely small, d (δ for partial differentials) is used. This, of course, is the same convention that is used in differential calculus. For example, a finite volume increase is symbolized by $+\Delta V$, whereas an infinitely small volume decrease is represented as $-dV$. For a $+\Delta V$, $\Delta V = V_2 - V_1$ where V_2 is larger.

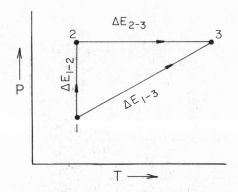

Figure 2.2 Finite changes in internal energy *E* around a closed path on a *P-T* diagram.

Chemical reactions can take place under conditions where certain parameters do not change. Specific terms are applied to these reactions, depending upon which parameter is held constant:

adiabatic	$\Delta Q = 0$
isothermal	$\Delta T = 0$
isobaric	$\Delta P = 0$
isochemical	$\Delta n = 0$
isovolumetric	$\Delta V = 0$

2.1.2 First Law

Perhaps the most straightforward definition of the First Law of Thermodynamics is *various forms of energy are equivalent and energy is indestructible.* A necessary consequence of the First Law can be seen in the "closed path" of Figure 2.2. Here a system is subjected to increasing pressure from state 1 to 2, increasing temperature from 2 to 3, and finally decreasing P and T from 3 back to 1. At each state the system has a different internal energy, but the energy change, ΔE, around the closed path will be zero. This means algebraically:

$$\Delta E_{1\text{-}2} + \Delta E_{2\text{-}3} = \Delta E_{1\text{-}3}$$

No matter what path the system takes in going from state 1 to 3, ΔE will always be the same. In thermodynamics this concept has been expressed as ΔE is a *function of state and independent of path*. Mathematically, ΔE is said to be an *exact differential*.

Changes in many other parameters are independent of path as well, such as ΔP, ΔV, and ΔH. However, changes in two parameters, ΔQ (heat) and ΔW (work), are not. In the case of Figure 2.2, a different amount of work and heat will be expended in the system's going from state 1 to 3 by path 1-2-3 than by path 1-3. To be consistent with sign conventions described previously, a change in work is defined as:

$$dW = -PdV \tag{2.2}$$

The heat change is equal to:

$$dQ = CdT \quad \therefore \quad C = dQ/dT \tag{2.3}$$

This provides another definition of heat capacity -- a change in heat with changing temperature.

A mathematical expression of the First Law defines a change in internal energy in terms of changes in heat and work (thus the term *thermodynamics*). This seems ironic because an exact differential is defined in terms of two inexact differentials:

$$dE = dQ + dW \quad \therefore \quad dE = CdT - PdV \tag{2.4}$$

The basic mathematical definition of enthalpy is:

$$H = E + PV \quad \therefore \quad dH = dE + PdV + VdP \tag{2.5}$$

Several simple equations exist that hold only under isobaric (denoted by subscript P) or isovolumetric (subscript V) conditions. These, along with many other equations in this section, can be easily proven. Any standard thermodynamics text gives their proofs. These equations are:

$$dE_V = dQ_V \quad \text{and} \quad dH_P = dQ_P \tag{2.6}$$

Combining Equations 2.3 and 2.6:

$$(\delta E/\delta T)_V = C_V \quad \text{and} \quad (\delta H/\delta T)_P = C_P \tag{2.7}$$

Equations 2.7 are called partial differential equations because a variable (in this case V or P) is held constant. For this reason δ is used rather than d to denote an infinitesimally small change.

Equations involving adiabatic processes are also of interest:

$$dQ = 0 \quad \therefore \quad dE = dW = C_V dT \tag{2.8}$$

$$dH = C_P dT \quad \therefore \quad \Delta H = \int C_P dT \tag{2.9}$$

The two equivalent Equations 2.9 are very useful and will be seen repeatedly throughout the remainder of this discussion on thermodynamics.

2.1.3 Enthalpy

Some useful outgrowths of the First Law are two kindred variations on molar enthalpy, *heat of formation* and *heat of reaction*. The heat of formation is actually a special type of heat of reaction. A common procedure is to calculate heats of reaction from heats of formation, as explained later.

2.1.3.1 Heat of Formation

The heat of formation (ΔH^o_f) is the heat released in forming a compound from its component elements in their standard states. The superscript ^{o} denotes standard pressure (1 bar). The *standard state* of an element is its stable state (solid, liquid, gas) at STP (standard T and P, 25°C and 1 bar). Oxygen is a diatomic gas (O_2) in its standard state; K is a crystalline solid; Hg is a liquid. The heat of formation of an element in its standard state is 0. Standard states can be defined for chemical constituents other than elements, or for temperatures and pressures other than STP, but this book will adopt the simplest possible definition. A more comprehensive discussion of standard states can be found in Wood and Fraser (1976). An example of a ΔH^o_f reaction is:

$$Si(\text{metal}) + O_2(\text{gas}) = SiO_2(\alpha \text{ quartz})$$

Heat of formation for this reaction is about -911 kj/mol. All heats of formation are strongly exothermic (see Appendix F). These values are usually determined experimentally by *calorimetry*.

One point of confusion about heats of formation needs to be clarified. Most authors (e.g., Kern and Weisbrod, 1967; Robie and others, 1979) use the notation that indicates that heats of formation are changes in enthalpy, i.e., ΔH^o_f. Others (Wood and Fraser, 1976, for example) simply use H^o. Both are correct because heats of formation of the elements in their standard states are zero. Consequently, using the products minus reactants expression for the example of alpha quartz:

$$\Delta H^o_f = H^o(\alpha \text{ quartz}) - H^o(Si) - H^o(O_2)$$
$$\Delta H^o_f = H^o(\alpha \text{ quartz}) - 0 - 0 = H^o(\alpha \text{ quartz})$$

This seemingly trivial point is not so trivial. Generally we will consider heats of formation as ΔH^o_f, but occasionally we will express them as H^o.

2.1.3.2 Heat of Reaction at Standard Temperature

Coupled with the concept of heat of formation is that of heat of reaction (ΔH_R). If we know ΔH^o_f values for all the products and reactants in a chemical reaction, we can calculate its ΔH_R at STP, which is written as ΔH^o_{298}. Note the use of 298K rather than 25°C; the degrees Kelvin temperature scale is always used for thermodynamic calculations. The general equation is:

$$\Delta H_R = \Sigma \Delta H_f (\text{products}) - \Sigma \Delta H_f (\text{reactants}) \qquad (2.10)$$

Analogous *products minus reactants* equations also can be used for ΔS_R, ΔV_R, and many other changes in parameters involved in chemical reactions (Sections 2.1.4 and 2.1.5). If only pure solids and/or liquids are involved, ΔH_R

at elevated T and P normally is not radically different from ΔH_R at STP. If a gas or fluid is involved, however, ΔH_R at elevated P,T conditions is quite different.

As an example of the calculation of ΔH_R, we will examine further the incompatibility of quartz and olivine:

$$Mg_2SiO_4 + SiO_2 = 2MgSiO_3 \qquad\qquad (2.11)$$
$$Fo + Q = 2Clen$$

Clen symbolizes clinoenstatite, one of three polymorphs of $MgSiO_3$. The other two are orthoenstatite (or simply enstatite) and protoenstatite. We can calculate ΔH^o_R for this reaction at standard temperature:

$$\Delta H^o_{298} = 2\Delta H^o_{Clen} - \Delta H^o_Q - \Delta H^o_{Fo} \quad \text{or}$$
$$\Delta H^o_{298} = 2H^o_{Clen} - H^o_Q - H^o_{Fo}$$

$$\Delta H^o_{298} = 2(-1547.75) + 910.70 + 2170.37$$
$$\Delta H^o_{298} = -14.43 \text{ kj/mol}$$

Data for ΔH_f and its associated error (uncertainty) are from Robie and others (1979). Clinoenstatite is the only $MgSiO_3$ polymorph listed in this reference. The error ($\pm$ 2 standard errors, $2e_s$; see Section 1.1.3.2) on ΔH^o_{298} can be calculated from Equation 1.28:

$$e = 2e_s = [(2 \times 1.215)^2 + 1^2 + 1.325^2]^{1/2} = 2.94 \text{ kj/mol}$$

Thus ΔH^o_{298} can be written as -14.43 ± 2.94 kj/mol.

2.1.3.3 Heat of Reaction at Elevated Temperature

Frequently we cannot assume that ΔH^o_R at some higher temperature is similar to that at standard temperature, so how do we deal with the general situation of calculating ΔH^o_R at elevated temperatures? We do so by including heat capacity at constant pressure (C_P) in the calculations. A number of ways exist to calculate ΔH^o_R at elevated T, but they all ultimately go back to *Hess' Law of Heat Summation*. This law perhaps should not be accorded such lofty status as it is a direct outgrowth of the First Law. It simply says that ΔH is a function of state and independent of path. Suppose we wished to calculate ΔH^o_R at 1500K. We can write the closed cycle in this fashion:

$$
\begin{array}{ccccccc}
 & & & \Delta H^o_{298} & & & \\
Fo & + & Q & = & 2Clen & & 298K \\
\downarrow\Delta H_{Fo} & & \downarrow\Delta H_Q & & \downarrow 2\Delta H_{Clen} & & \\
Fo & + & Q & = & 2Clen & & 1500K \\
 & & & \Delta H^o_{1500} & & &
\end{array}
$$

ΔH^o_{298} is the ΔH^o_R from above. Invoking Hess' Law of Heat Summation:

$$\Delta H_{Fo} + \Delta H_Q + \Delta H^o_{1500} = \Delta H^o_{298} + 2\Delta H_{Clen}$$

$$\Delta H^o_{1500} = \Delta H^o_{298} + 2\Delta H_{Clen} - \Delta H_{Fo} - \Delta H_Q$$
$$\Delta H^o_{1500} = -19.93 \text{ kj/mol}$$

This value is 5.5 kj lower than ΔH_R at 25°C, so the reaction gives off even more heat (is more exothermic) at 1500K than at standard temperature.

An explanation is required for the terms ΔH_{Fo}, ΔH_Q, and ΔH_{Clen}. They can be calculated by integrating the second Equation 2.9 ($\Delta H = \int C_P dT$) for each mineral (see Kern and Weisbrod, 1967, or Wood and Fraser, 1976). Values for C_P have been determined experimentally and empirical polynomial equations for C_P as a function of T are tabulated in Robie and others (1979). These equations commonly take the form:

$$C_P = a + bT + (c \, / \, T^{1/2}) - (d \, / \, T^2) \quad \text{or} \tag{2.12}$$

$$C_P = a + bT - (c \, / \, T^2) \tag{2.13}$$

The coefficients for C_P *(a-d)* will be different for each mineral, as no two minerals have exactly the same heat capacity. Fortunately, integrations of the second Equation 2.9 have already been done for many minerals and temperatures (Robie and others, 1979). As an example, to obtain the value for ΔH_{Fo} at 1500K from the second column (p. 363) in this reference:

$$\Delta H_{Fo} = [(H^o{}_T - H^o{}_{298}) \, / \, T] \times T$$
$$\Delta H_{Fo} = 133.594 \times 1500 = 200391 \text{ j/mol}$$

In fact, if the data available in this reference are used, a cycle does not have to be written and the reaction at 1500K can be calculated directly:

$$\Delta H^o{}_{1500} = 2(\Delta H_{Clen}) - \Delta H_Q - \Delta H_{FO}$$
$$\Delta H^o{}_{1500} = 2(-1673.00) + 901.321 + 2424.752$$
$$\Delta H^o{}_{1500} = -19.93 \text{ kj/mol}$$

2.1.4 Second Law and Entropy

A general tendency exists for chemical reactions either (1) to occur spontaneously "as written" (i.e., from left to right), (2) for the "back reaction" (right to left) to take place, or (3) for no discernable reaction to occur at all. The Second Law of Thermodynamics and the concept of *free energy* (Section 2.1.5) provide criteria by which we can predict the direction of a chemical reaction.

2.1.4.1 Basics

Perhaps the most illuminating verbal statement of the Second Law was by Clausius in 1850: *heat cannot pass of its own accord from a colder to a hotter body*. Combining this statement with the fact that dE is an exact differential whereas dQ and dW are not, one can easily derive the mathematical expression of the Second Law:

$$dS = (dQ/T)_{\text{REV}} > (dQ/T)_{\text{IRREV}} \tag{2.14}$$

where REV ("reversible") symbolizes a reaction at equilibrium and IRREV represents a spontaneous, irreversible reaction. The equation $dS = dQ/T_{\text{REV}}$ provides the basic mathematical definition of an entropy *(S)* change.

It is not immediately obvious from the above equation why entropy is a measure of disorder. The Third Law of Thermodynamics is very helpful in this regard. The Third Law states that *entropy of a perfectly ordered crystal at 0K is zero*. Hence entropy increases with increasing temperature and disorder; a value for S is always a positive quantity. Typically $S_{gas} > S_{liquid} > S_{solid}$. Likewise, a solid solution has a higher entropy than a similar pure mineral and a glass has higher entropy than a crystal.

For an adiabatic reaction:

$$\Delta S = \int dQ / T = \int 0 / T = 0 \tag{2.15}$$

For an isobaric reaction:

$$\Delta S = \int (dQ / T) = \int (dH / T) = \int (C_P dT / T) = \int C_P d\ln T \tag{2.16}$$

The "products minus reactants" equation can also be used for ΔS:

$$\Delta S = \Sigma S(\text{products}) - \Sigma S(\text{reactants}) \tag{2.17}$$

xxxxxxxxxxxxxxxxxxxxxxxxxxxxxxxxxxxxxxxxxxxxxxxxxxxxxxxxxxxxxxxxx

A brief digression into feldspar mineralogy is helpful with regard to our perceptions of entropy. The polymorph stable at relatively high temperatures will have the highest entropy, as in the polymorphs of the three K-feldspars:

$$S(sanidine) > S(orthoclase) > S(microcline)$$

In this case order-disorder relationships are primarily between the Si and Al ions of the Z position, rather than among K, Na, and Ca in the W position. In sanidine Al and Si are randomly distributed throughout the lattice in the Z site, whereas in microcline these ions are ordered with the Al occupying a site closer to the W position ions. The structure of orthoclase is intermediate between these two extremes.

Sanidine is found in volcanic and hypabyssal rocks, whereas microcline generally occurs in plutonic or metamorphic rocks. As expected, orthoclase can be found anywhere. This does not mean that volcanic rocks crystallize at higher temperatures than do plutonic rocks. Because volcanic rocks cool and crystallize so rapidly, the high entropy, disordered structure in sanidine is "locked in," and is never able to transform to a more ordered form at lower temperature. This is apparently not the case with plutonic K-feldspars, which probably are sanidine at high temperatures but slowly invert first to orthoclase and then to microcline as they cool.

xxxxxxxxxxxxxxxxxxxxxxxxxxxxxxxxxxxxxxxxxxxxxxxxxxxxxxxxxxxxxxxxx

2.1.4.2 A Numerical Example

Molar entropy data for many minerals over a wide range in temperatures can be found in Robie and others (1979). Errors are given for 298K. Because ΔS is an exact differential, a closed cycle can be written for ΔS in a manner identical to that for ΔE and ΔH, enabling S^o to be calculated at any temperature. These data are calculated on the basis of integrating Equation 2.16 using either Equation 2.12 or 2.13 for C_P. Because this has already been done by Robie and others (1979), we can calculate ΔS^o_{1500} directly. Molar entropy calculations for the reaction of Equation 2.13 at 298K and 1500K are:

$$\Delta S^o_T = 2S^o_{Clen} - S^o_{Fo} - S^o_Q$$

$$\Delta S^o_{298} = 2(67.86) - 95.19 - 41.46 = -0.93 \text{ j/mol}^\circ$$
$$\Delta S^o_{1500} = 2(243.54) - 350.80 - 144.47 = -8.19 \text{ j/mol}^\circ$$

In keeping with the Third Law, molar entropies of individual minerals at 1500K are much higher than at 298K. Error ($\pm 2e_s$) for ΔS^o_{298} using Equation 1.28 is calculated as:

$$e = 2e_s = [(2 \times 0.42)^2 + 0.20^2 + 0.84^2]^{1/2} = 1.20 \text{ j/mol}^\circ$$

The relative error for ΔS^o_{298} is about 130 percent. Assuming that a similar error applies to ΔS^o_{1500}, this reduces to about 13 percent. Because ΔS^o_{1500} is a calculated value, its error is probably larger than 1.20 j/mol°.

2.1.5 Free Energy

The concept of free energy was first articulated in the 1800s by the chemist J. Willard Gibbs, hence the term *Gibbs free energy*, symbolized in this book by G. It is one of the most useful thermodynamic parameters in that it is the *ultimate criterion for spontaneity*. It can be considered as potential energy in a chemical sense because the mineral assemblage with the lowest Gibbs free energy will be the stable assemblage.

2.1.5.1 Basics

Again consider the reaction:

$$aA + bB = cC + dD$$

If the forward reaction has a $-\Delta G$, the mineral assemblage C+D will irreversibly ("spontaneously") form at the expense of A+B (i.e., A+B are reactants and C+D are products). If the forward reaction has a $+\Delta G$, the back reaction is spontaneous and A+B will be the products. The magnitude of ΔG is the same for both reactions; only the signs are different. It follows that at chemical equilibrium $\Delta G = 0$.

The basic definition of G is:

$$G = H - TS \quad \therefore \quad dG = dH - TdS - SdT \qquad (2.18)$$

If a reaction involving finite changes takes place under isothermal conditions:

$$\Delta T = 0 \quad \therefore \quad \Delta G = \Delta H - T\Delta S \qquad (2.19)$$

2.1.5.2 A Numerical Example

Data (including error data) tabulated in Robie and others (1979) for ΔG^o_f can be treated using the products minus reactants equation in the same manner as ΔH and ΔS. In the same way that ΔH^o_f can be expressed as H^o, ΔG^o_f can be considered as G^o. Determination of ΔG^o_{298} and ΔG^o_{1500} for the reaction of Equation 2.11 results in:

$$\Delta G^o_T = 2\Delta G^o_{Clen} - \Delta G^o_{Fo} - \Delta G^o_Q \quad \text{or}$$
$$\Delta G^o_T = 2G^o_{Clen} - G^o_{Fo} - G^o_Q$$

$$\Delta G^o_{298} = 2(-1460.883) + 2051.325 + 856.288 = -14.15 \text{ kj/mol}$$
$$\Delta G^o_{1500} = 2(-1096.489) + 1542.240 + 643.096 = -7.64 \text{ kj/mol}$$

The large negative ΔG^o values for both standard temperature and 1500K confirm the incompatibility of olivine and quartz in igneous rocks (see Section 1.3.1.2). A more complete thermodynamic treatment of the incompatibility of these two minerals can be found in Kern and Weisbrod (1967). Error using Equation 1.28 for ΔG^o_{298} is calculated as:

$$e = 2e_s = [(2 \times 1.225)^2 + 1.10^2 + 1.345^2]^{1/2} = 3.00 \text{ kj/mol}$$

for a relative error of about 21 percent at 298K.
Equation 2.19 can be used to check the calculations of ΔH^o, ΔS^o, and ΔG^o for the reaction of Equation 2.11:

$$\Delta G^o_T = \Delta H^o_T - T\Delta S^o_T$$

$$\Delta G^o_{298} = -14.43 - 298(-0.00093) = -14.15 \text{ kj}$$
$$\Delta G^o_{1500} = -19.93 - 1500(-.00819) = -7.64 \text{ kj}$$

Calculations for ΔG^o, ΔS^o, and ΔH^o check internally, as ΔG^o determined in this manner agrees with ΔG^o determined directly.

2.1.5.3 Criteria for Spontaneity

Another very useful differential equation for dG is:

$$dG = VdP - SdT \qquad\qquad (2.20)$$

This equation is so important (see Sections 2.2.1.2 and 3.4.1.2) that a short proof is in order. Combining Equations 2.5 and 2.18:

$$G = E + PV - TS$$

The total differential for G must be:

$$dG = dE + PdV + VdP - TdS - SdT$$

Combining Equations 2.4 and 2.14:

$$dE = TdS - PdV$$

By substitution:

$$dG = TdS - PdV + PdV + VdP - TdS - SdT = VdP - SdT$$

Given Equation 2.20 and other equations earlier, proving why ΔG is a criterion for spontaneity is quite straightforward. Given a chemical reaction at equilibrium at constant T and P:

$$dG = VdP - SdT = 0$$

The Second Law states that:

$$\Delta S > (\int dQ / T)\text{irrev} \quad \therefore \quad T\Delta S > \Delta Q$$

for an irreversible reaction. If we consider an irreversible reaction at constant pressure and temperature:

$$\Delta H_P = \Delta Q_P \quad \therefore \quad \Delta G = \Delta Q - T\Delta S$$

$$T\Delta S > \Delta Q \quad \therefore \quad \Delta G < 0 \qquad\qquad (2.21)$$

which, as stated above, is the criterion for spontaneity.

Consider Equation 2.19 for ΔG carefully. If the $T\Delta S$ term is small relative to ΔH, or if ΔS has the opposite sign from ΔH, then ΔG and ΔH will have the same sign. Temperature is always a positive number because it is in °K. In the calculations for the reaction FO + Q = 2EN, both ΔH and ΔG have the same sign, which is a common relationship and implies that ΔH can be considered as a "secondary criterion for spontaneity." Exothermic chemical reactions commonly occur spontaneously whereas endothermic reactions do not. Exceptions take place, especially at high temperatures, when ΔS has a different sign and is large relative to ΔH.

When considering ΔH as a secondary criterion for spontaneity, we must take into account the attached errors on ΔG, ΔH, and ΔS. Equation 2.19 is rewritten for 298K as:

$$\Delta G^o{}_{298} = \Delta H^o{}_{298} - T\Delta S^o{}_{298}$$

$$[-14.15 \pm 3.00] = [-14.43 \pm 2.94] - 298[-0.00093 \pm 0.00120]$$

In this case we are safe in assuming that the negative sign is valid for both ΔG and ΔH. Had ΔS been larger or ΔH smaller, however, considering the error terms, one might not know whether the signs on ΔG and ΔH were the same or different. Similar situations have been referred to by Fyfe and others (1958) as "the plague of the small ΔG's." Generally ΔG and ΔH need to be several kj's before their algebraic signs can be accepted.

2.1.5.4 Free Energy and the Equilibrium Constant

Another topic related to ΔG concerns the manner in which ΔG is related to the chemical equilibrium constant. Consider the reaction aA+bB = cC+dD in which products and reactants are ideal gasses (Section 2.2.2.1). The equilibrium constant for this reaction is:

$$K_{eq} = \frac{(p_C)^c \cdot (p_D)^d}{(p_A)^a \cdot (p_B)^b}$$

where p_C, etc., are partial pressures of the gasses involved in the reaction. Partial pressure can be considered as mol fraction times total pressure (e.g., $p_i = X_iP$). It can be proven from basic principles that:

$$\Delta G^o_T = -RT\ln K_{eq} \qquad\qquad (2.23)$$

where R is the ideal gas constant. The equation, however, has general applicability; it can be used for many reactions other than those involving ideal gasses. It is commonly referred to as the *Van't Hoff reaction isotherm*. Robie and others (1979) have compiled data for log K_{eq} (including error terms at 298K) calculated from ΔG^o_f data using Equation 2.23. Because we are dealing with logarithms, the products minus reactants relationship applies for log K_{eq} as well. The equation is:

$$\log K_{eq} = c \log K_C + d \log K_D - a \log K_A - b \log K_B$$

As an example of the use of these relationships, during hydrothermal alteration (or metamorphism) enstatite will react with H_2O to form the hydrous sheet silicate chrysotile (a variety of serpentine) and quartz. Substituting clinoenstatite for enstatite:

$$3MgSiO_3 + 2H_2O = Mg_3Si_2O_5(OH)_4 + SiO_2 \qquad\qquad (2.24)$$
$$3Clen + 2W = Cr + Q$$

Remembering that the concentration (actually, activity; Section 2.2.2.5) of pure solids in their standard states is unity:

$$K_{eq} = 1 / p^2_{H2O} \qquad\qquad (2.25)$$

in which p_{H2O} is the partial pressure of water (assuming partial pressure equals fugacity; Section 2.2.2.3). Calculation of ΔG^o_T for 800K yields:

$$\Delta G^o_{800} = \Delta G_{Cr} + \Delta G_Q - 3\Delta G_{Clen} - 2\Delta G_W$$
$$\Delta G^o_{800} = -3485.318 - 765.287 + 3(1314.755) + 2(203.473)$$
$$\Delta G^o_{800} = 100.606 \text{ kj/mol}$$

Solving for K_{eq} at 800K using Equation 2.23:

$$\log K_{eq} = -100606 / (2.30258 \times 8.3143 \times 800) = -6.57$$

where 2.30258 is the conversion from natural to common logs and 8.3143 is the gas constant, R, in j/°mol (interestingly, R, S and C all have the same units). Dimensional analysis yields:

$$\text{j/mol} / (\text{j/mol}° \times °K)$$

indicates that log K_{eq} is dimensionless, as it must be. Alternatively, we can calculate log K_{eq} directly by:

$$\log K_{eq} = \log K_{Cr} + \log K_Q - 3\log K_{Clen} - 2\log K_W \qquad (2.26)$$
$$\log K_{eq} = 227.5696 + 49.968 - 3(85.845) - 2(13.285)$$
$$\log K_{eq} = -6.57$$

Thus K_{eq} is equal to $10^{-6.57}$ and the calculated p_{H2O} at equilibrium for this reaction at 800K is about 1900 bars.

2.1.5.5 Partition Coefficients and Geothermometry

Another variety of K_{eq} is very useful in petrology. A *partition coefficient (K_D)* is a K_{eq} for a reaction in which the phases are of variable composition. Consider the reaction:

$$NaAlSi_3O_8 + KAlSiO_4 = KAlSi_3O_8 + NaAlSiO_4 \qquad (2.27)$$
$$AB + KP = OR + NE$$

Superficially this reaction appears to be another example of a solid-state reaction between four minerals. Note, however, that the symbols for components (AB, KP, OR, NE) are used rather than symbols for phases. In fact, this reaction can be considered as an *exchange reaction* for Na and K between alkali feldspar and feldspathoid solid solutions. As temperature and pressure change Na^+ ions will replace K^+ ions on a one-for-one basis in one mineral and K^+ similarly replaces Na^+ in the other.

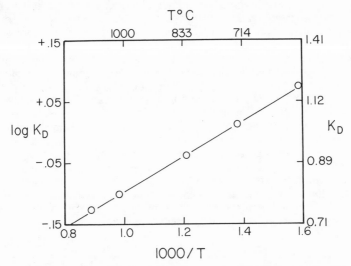

Figure 2.3   Plot of log K_D versus $1/T$ x 1000 for reaction given in Equation 2.27. Also included are T^oC and K_D scales.

An alternative form of Equation 2.27 is:

$$Na^{AF} + K^{FD} = K^{AF} + Na^{FD} \qquad (2.28)$$

where AF refers to an alkali feldspar solid solution and FD represents a feldspathoid solid solution. The equilibrium constant K_{eq}, and thus K_D, for this reaction is:

$$K_{eq} = K_D = \frac{X_K^{AF} \cdot X_{Na}^{FD}}{X_{Na}^{AF} \cdot X_K^{FD}}$$

where X refers to mol fraction.

The Van't Hoff relationship (Equation 2.23) indicates that the logarithm of K_D should be proportional to $1/T$, so we can plot these two variables to obtain a straight line. We can calculate K_Ds for several temperatures as shown in the example for Equation 2.26 and plot Figure 2.3 at a pressure of one bar. Data are from Robie and others (1979).

Figure 2.3 is the briefest of introductions to *geothermometry,* which along with *geobarometry,* are subdisciplines of petrology and geochemistry whose purpose is to estimate crystallization temperatures and pressures of rocks. Geothermometry is introduced here because it is another tie between classical thermodynamics and petrology. Figure 2.3 represents a hypothetical example of a potential geothermometer. If we have electron microprobe analyses of alkali feldspars and feldspathoids in some alkaline igneous rocks, Figure 2.3 in theory allows us to estimate the temperature of crystallization of the minerals and thus hopefully the rock.

Figure 2.3, however, is based on a number of assumptions, the most important of which are: (1) the two phases were in chemical equilibrium with each other when they crystallized; (2) the two solid solutions behaved ideally (Section 2.2.2.6); (3) errors in thermochemical data have a minimal effect on calculated temperatures; (4) chemical analyses of these minerals are of high quality; (5) minerals involved have not changed in composition after their original crystallization; and (6) changes in pressure have a relatively small effect on the equilibrium compared with changes in temperature. If pressure has a stronger effect on K_D than temperature, the reaction will be a better geobarometer than geothermometer. If both have an effect, the reaction is only useful if some independent means exists to determine one parameter or the other.

All the preceeding assumptions will most probably never be met for one set of data. Calculations such as those used to construct Figure 2.3, however, are useful in placing at least qualitative constraints on possible reactions important in geothermometry. Figure 2.3, for instance, implies that as the ratio Na^{FD}/Na^{AF} increases, the equilibrium temperature of crystallization decreases. Experimental corrobortion is needed before this K_D can be used as an accurate geothermometer (or geobarometer). Lacking experimental work, if another mineral assemblage exists in the rocks on which a geothermometer has been well calibrated, the AF-FD K_D perhaps can be calibrated from this existing geothermometer. This indirect method of calibrating a geothermometer is not as satisfactory as a more direct experimental technique, but it may be the only method available.

2.2 PREPARATION FOR PHASE EQUILIBRIA

2.2.1 Pressure-Temperature Diagrams

The preceeding discussion on classical thermodynamics has laid the groundwork for our first consideration of phase diagrams. Specifically, this section deals with *P-T* diagrams, which are among the most widely used in igneous as well as metamorphic petrology. *P-T* diagrams allow the petrologist to estimate the *P,T* conditions under which a mineral assemblage seen in the petrographic microscope or in the field might form.

2.2.1.1 A Hypothetical Example

An example is shown on Figure 2.4, a graphical representation of the hypothetical reaction aA+bB = cC+dD considered previously. Traditionally pressure
is plotted on the Y-axis, with temperature on the X-axis. Occasionally the
pressure axis will be reversed, with pressure increasing downward, to simulate
that pressure increases downward in the earth.

    The mineral assemblage A + B falls on the relatively high P - low T side
of the *phase boundary* compared to the assemblage C + D. Along the phase
boundary, also appropriately known as an *equilibrium boundary,* $\Delta G = 0$. In the
stability field where A + B are stable, $\Delta G > 0$ for the reaction as written, because
A + B are products and C + D are reactants. $\Delta G < 0$ in the C + D field. In this
example the phase boundary is linear with a positive slope ($dP/dT > 0$).
Although this is common, it is not always so. What, then, controls the position,
slope, and curvature of the phase boundary?

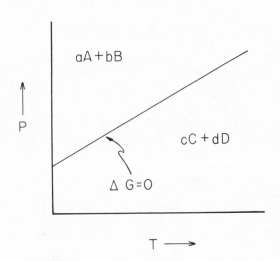

aA + bB

P

cC + dD

$\Delta\ G = 0$

T $\longrightarrow$

Figure 2.4
Schematic *P-T*
phase diagram for
the hypothetical
reaction aA + bB =
cC + dD.

2.2.1.2 The Clapeyron Equation

To determine controls on the slope and curvature of the phase boundary, we
must introduce the *Clapeyron equation,* which is so important that a short proof
is appropriate. Consider two phases, A and B, in equilibrium. If equilibrium is
maintained with changing P,T conditions, then:

$$dG_A = dG_B$$

From Equation 2.20:

$$dG_A = V_A dP - S_A dT \text{ and } dG_B = V_B dP - S_B dT$$

$$\therefore V_A dP - S_A dT = V_B dP - S_B dT$$
$$\therefore (V_A - V_B)dP = (S_A - S_B)dT$$

$$\therefore dP/dT = \Delta S\ /\ \Delta V = \Delta H\ /\ T\Delta V \tag{2.30}$$

Equation 2.30 is known as the Clapeyron equation, which states that the slope of the phase boundary dP/dT at any given P-T point is equal to the entropy change divided by the volume change of the chemical reaction. Because $\Delta S_P = \Delta Q_P / T = \Delta H_P / T$, the alternative expression involving ΔH is equally valid. The Clapeyron equation will be revisited often throughout the remainder of this book. P-T diagrams, such as Figures 2.4 and 2.5, are frequently referred to as *Clapeyron diagrams*.

Recalling differential calculus, whereas the first derivative is equal to the slope of a line, the second derivative is the curvature. The second derivative of Equation 2.30 is:

$$d^2P/dT^2 = -\Delta H / (\Delta V T^2) \tag{2.31}$$

If the curvature is positive, the boundary is concave up; if zero, a straight line; and if negative, concave down.

2.2.1.3 Use of the Clapeyron Equation

Before discussing the application of the Clapeyron and other equations to P-T diagrams, sources of data must be considered. Probably the two most widely used compilations of thermochemical data for substances of geologic interest are Robie and others (1979) and Helgeson and others (1978). As is abundantly obvious, this book mainly uses the data of Robie and others (1979), which are primarily based on calorimetric data. In contrast, data compiled in Helgeson and others (1978) are directly based on phase equilibria involving geologic materials and are internally consistent with a number of experimentally determined P-T diagrams. The data compiled by Robie and others (1979) are more convenient for the cursory treatment of thermodynamics in this book because they have already integrated the C_P equations and provided data at elevated temperatures. In addition, they include more minerals and provide error data for 298K, which are useful in placing constraints on some calculations. The data of Helgeson and others (1978) should be used if a more exacting result consistent with many studies of phase equilibria is required.

Because forsterite and quartz cannot coexist at equilibrium under any reasonable P,T conditions, the following reaction will be used to demonstrate the use of the Clapeyron equation:

$$3MgSiO_3 + Al_6Si_2O_{13} = Mg_2Al_4Si_5O_{18} + MgAl_2O_4 \tag{2.32}$$
3clinoenstatite + mullite = cordierite + spinel
3Clen + Mu = Cd + Sp

Initially we will assume that $\Delta C_P = 0$, so data for 298K will suffice. Data from Robie and others (1979) are used to calculate:

$\Delta H^o_{298} = -7.173$ kj/mol
$\Delta S^o_{298} = 14.68$ j/mol°

A value for the change in molar volume, ΔV^o, must be calculated as well. As only solids are involved in the reaction, we will ignore the fact that $\Delta V = f(P,T)$; i.e., ΔV will change with changing P and T, but we will assume that the change will be so small to be negligible. Because the molar volume of a mineral is inversely related to its density:

$$\Delta V^o = \text{gfw}[(1/\rho_2) - (1/\rho_1)] \tag{2.33}$$

Dimensional analysis shows that g/mol x cc/g = cc/mol, the normal units for ΔV. However, for dP/dT to have the proper units of bar/°K, ΔV must be in units of j/mol·bar, which only requires multiplying cc/mol by 0.1 to obtain j/mol·bar. Molar volumes are from Robie and others (1979). The products minus reactants relationship is also used to calculate ΔV:

$$\Delta V^o_{298} = V_{Cd} + V_{Sp} - 3V_{En} - V_{Mu}$$
$$\Delta V^o_{298} = 23.322 + 3.971 - 3(3.147) - 13.455$$
$$\Delta V^o_{298} = 4.397 \text{ j/mol·bar}$$

We are now prepared to estimate dP/dT. Using the $\Delta S/\Delta V$ expression:

$$(dP/dT)_{298} = 14.68 / 4.397$$
$$(dP/dT)_{298} = 3.34 \text{ bars/°K (or bars/°C)}$$

Now using the $\Delta H/T\Delta V$ expression:

$$(dP/dT)_{298} = -7173 / (298 \times 4.397) = -5.47 \text{ bars/°C}$$

Clearly a very serious discrepancy exists between the two values. Because both products and reactants are solids, the phase boundary should be approximately linear, yet one calculated slope is negative and the other is positive. In searching for a reason, the first step is to check the arithmetic. Having found no arithmetic mistake, we must look elsewhere.

Carefully observe any of the previous equations where these parameter changes were calculated. In particular for ΔH, we subtracted several large numbers to obtain a comparatively small number. Error analysis (see Section 1.4.1.2) indicates that the relative error on these parameter changes can be very large compared with the change itself. Using Equations 1.27 and 1.28, values $\pm 2e_s$ errors for ΔH^o_{298}, ΔS^o_{298}, ΔV^o_{298}, and dP/dT are:

$$\Delta H^o_{298} = -7.173 \pm 7.273 \text{ kj}$$
$$\Delta S^o_{298} = 14.68 \pm 5.80 \text{ j/mol°}$$
$$\Delta V^o_{298} = 4.397 \pm 0.023 \text{ cc/mol}$$

$$(dP/dT)_S = 3.34 \pm 0.40 \text{ bars/°C}$$
$$(dP/dT)_H = -5.47 \pm 3.43 \text{ bars/°C}$$

The subscript S refers to dP/dT calculated by $\Delta S/\Delta V$, whereas H denotes $\Delta H/T\Delta V$. Original error data for 298K are from Robie and others (1979).

The relative error on ΔS is not too large, but the relative error on ΔH is over 100 percent. For most reactions ΔH has the larger error. For this reaction, clearly we have more confidence in the slope of 3.34 bars/°C.

If we extrapolate this slope at 298K to elevated temperatures, we must make the assumption that C_P does not change appreciably with increasing temperature. As only solids are involved in the reaction, this is probably reasonable, but we can check this assumption by calculating slopes at various temperatures using the elevated temperature data from Robie and others (1979). Slopes using the $\Delta S/\Delta V$ expression are:

T^oK	dP/dT
298	3.34 bars/°C
500	3.52
700	3.39
900	3.16
1100	3.12
1300	3.37
1500	3.92

No apparent trend with increasing temperature exists and 3.34 is reasonably close to the average of 3.40 bars/°C, so the use of data at 298K is justified. Interestingly, the error for the above seven dP/dT values, as measured by $\pm 2e_s$, is $\pm$ 0.20, considerably smaller than the $\pm$ 0.40 calculated from $(dP/dT)_S$ at 298K.

2.2.1.4 Estimation of Both Slope and Intercept

The Clapeyron equation is limited in another way. The equation only allows us to estimate the slope; we cannot estimate the actual position of a phase boundary. Expressing this mathematically in terms of the familiar linear polynomial equation $Y = mX + b$, we can estimate the slope m but not the intercept b using the Clapeyron equation. An equation that will allow us to estimate both slope and intercept is:

$$\Delta G^o = -RT\ln K_{eq} = \Delta H^o - T\Delta S^o + P\Delta V^o \qquad (2.34)$$

where ΔH^o and ΔS^o are taken at some constant temperature, frequently standard temperature.

This is the simplified version of a more complex equation, the proof for which can be found in Kern and Weisbrod (1967 pp. 95-102). It assumes that ΔC_P is zero and ΔV does not change with changing temperature or pressure, so ΔH, ΔS, and ΔV in Equations 2.34 and 2.35 are at standard conditions. However, as we are dealing only with solids, this simplified version will suffice for our purposes. Remembering that $K_{eq} = 1$ for reaction 2.32 (because all phases are pure solids):

$$P = (T\Delta S^o - \Delta H^o) / \Delta V^o$$

We can rewrite this equation in the form $Y = mX - b$:

$$P = (\Delta S^o/\Delta V^o) \times T - (\Delta H^o/\Delta V^o) \qquad (2.35)$$

The m term is the Clapeyron slope $\Delta S/\Delta V$ and the intercept b is simply $\Delta H/\Delta V$. We can then solve this equation for P at different values of T and construct a phase boundary on a P-T diagram (Fig. 2.5). The temperature term in this equation is in °K, whereas the X-axis of Figure 2.5 is plotted in °C. The linear polynomial equation for the reaction of Equation 2.32 is:

$$P = [3.34 \pm 0.40] \times T + [1631 \pm 1021]$$

The error values can be used to plot the error envelope around the phase boundary; i.e., the above equation can be expressed as three linear equations for three lines:

$$P = 3.74 \text{ x } T + 2652 \qquad \text{(maximum)}$$
$$P = 3.34 \text{ x } T + 1631 \qquad \text{(average)}$$
$$P = 2.94 \text{ x } T + 610 \qquad \text{(minimum)}$$

We could link the maximum slopes with the minimum intercepts and vice versa, but this is not necessary because lines produced by such equations normally fall within the error envelope defined by these equations. For this reaction we can estimate the slope with more assurance than the intercept because the relative error on ΔS is smaller. We could turn to a more sophisticated calculation, such as one that does not assume ΔV is a constant and ΔC_P is zero, as does Equation 2.34. Ultimately we must determine the exact position of the phase boundary in the laboratory.

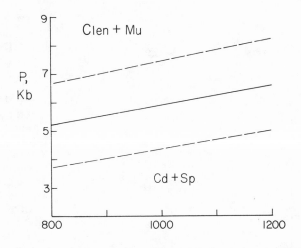

Figure 2.5 *P-T* phase diagram for the reaction given in Equation 2.32. *Solid line* represents equilibrium boundary and *dashed lines* are limits of error ($\pm 2e_s$). Calculations are given in text.

One last point needs to be made about Figure 2.5. It is not obvious whether En + Mu is the high P - low T assemblage or vice versa. On which side of the phase boundary are En + Mu stable and on which side are Cd + Sp stable? An increase in temperature will increase entropy, whereas an increase in pressure will decrease volume. So the assemblage with collectively the higher molar volume (or lower density) and higher entropy should be the high T - low P assemblage. Because ΔS and ΔV are positive for this reaction, Cd + Sp collectively have the greater volume and entropy, so they must be the high T - low P assemblage.

2.2.1.5 P-T Diagrams Involving a Volatile Phase

To this point we have only dealt with a reaction that involves solids. How do we estimate a phase boundary on a *P-T* diagram when a fluid or gas is involved? This phase boundary will not be linear so we cannot use Equation 2.35. Several possibilities exist (Kern and Weisbrod, 1967; Wood and Fraser, 1976), but as usual we will consider the simplest. A straightforward method is to construct a boundary curve using the $\Delta G = -RT\ln K$ relationship (Equation 2.23). We will assume that the fluid phase is behaving ideally, although this usually is not the case (Section 2.2.2.4).

In Section 2.1.5.4 we estimated the equilibrium pressure for the reaction of Equation 2.24 at 800K to be about 1930 bars. Using similar data for other temperatures from Robie and others (1979) allows calculation of equilibrium pressures for other temperatures. These pressures and temperatures can be plotted to construct phase boundary A for the reaction of Equation 2.24 (Fig. 2.6).

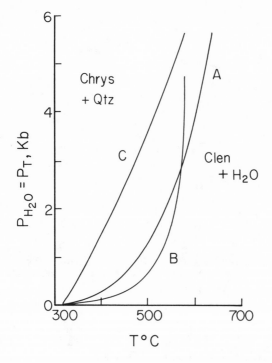

Figure 2.6 *P-T* phase diagram for the hydration/ dehydration reaction given in Equation 2.24. Curve *A* was deter- mined by using Equation 2.23 and assuming that H_2O behaves ideally. *P-T* values necessary to plot curve *B* were also calculated assuming ideal behavior of H_2O, but Equation 2.36 was used. Curve *C* was based on Equa- tion 2.23, but a correction for the nonideal behavior of H_2O was made (see discussion in Section 2.2.2.4).

An alternative for calculating the phase boundary is modified after Wood and Fraser (1976). It is based on Equation 2.34 and expressed:

$$T = (-\Delta H^o_T - P\Delta V^o_S) / (FR\ln P - \Delta S^o_T) \qquad (2.36)$$

where ΔV^o_S is the molar volume change *for solids only* and F is the coefficient (number of moles) for the fluid phase. In the case of Equation 2.24, $F = 2$. Wood and Fraser used $P - 1$ (after integration of $\int P dV$) rather than P in their equivalent of the numerator of Equation 2.36, but P approximates $P-1$ at elevated pressures. Using elevated temperature data for $\Delta H^o{}_T$ and $\Delta S^o{}_T$ from Robie and others (1979), T can be calculated at various pressures, which leads to curve B on Figure 2.6. Curve B exhibits considerably more curvature than does curve A.

Another assumption is made in the construction of Figure 2.6, which is that the water (fluid) pressure is equal to the confining (or total) pressure $(P_F = P_T)$. The following abbreviations are commonly used: P_T, total or confining pressure; P_v or P_F, vapor (gas) or fluid pressure. Within igneous systems most gasses are above their *critical points* (Section 3.1.1.4) and, therefore, behave as fluids. Consequently, we refer to fluid pressure rather than gas pressure. When $P_F = P_T$ the system is closed and saturated with respect to the fluid; i.e., fluid pressure exerted by the system on the surroundings is equal to confining pressure exerted by the surroundings on the system. The total or confining pressure can be considered as pressure exerted by the burden of overlying rocks *(lithostatic pressure)*.

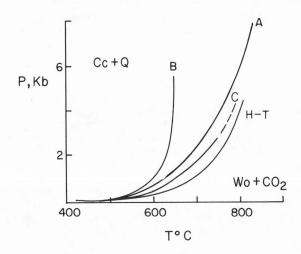

Figure 2.7 *P-T* phase diagram for the carbonation/decarbonation reaction given in Equation 2.37. Curve *H-T* is the experimentally determined curve after Harker and Tuttle (1956). Curve *A* was based on Equation 2.23, while curve *B* was determined from Equation 2.36. Ideal behavior of CO_2 was assumed for both curves *A* and *B*. Curve *C* was based on Equation 2.23 with a correction for the non-ideality of CO_2 (Section 3.2.2.4).

Regardless of the accuracy of pressure estimates for Figure 2.6, an extremely important lesson can be learned from this *P-T* diagram. The hydrous mineral chrysotile is on the low temperature - high pressure side of the phase boundary, whereas anhydrous minerals plus water are on the high temperature - low pressure side. The phase boundary is concave up. This diagram is characteristic of a group of chemical reactions involved in hydrothermal alteration and metamorphism of igneous rocks in which a fluid or gas is involved. The most likely fluid or gas will be H_2O or CO_2, so we will consider only reactions involving them.

If temperature is increasing across the boundary, these reactions are called *dehydration* or *decarbonation* reactions. The hydrous mineral or carbonate is the reactant and the other minerals plus the gas/fluid are products. It follows that decreasing temperatures and hydrous minerals or carbonates as products imply *hydration* or *carbonation* reactions. The hydrothermal alteration of enstatite to form chrysotile and quartz, then, is a hydration reaction, whereas the breakdown of chrysotile and quartz to form enstatite and water is a dehydration reaction.

Perhaps the most classic of all carbonation/decarbonation reactions is:

$$CaCO_3 + SiO_2 = CaSiO_3 + CO_2 \qquad (2.37)$$
$$\text{calcite} + \text{quartz} = \text{wollastonite} + CO_2$$
$$Cc + Q = Wo + CO_2$$

This reaction was originally studied by Goldschmidt (1912); it was determined experimentally by Harker and Tuttle (1956) and Greenwood (1967). It is also discussed at length in Wood and Fraser (1976). Using Equation 2.23, we can construct its equilibrium boundary on a *P-T* diagram (Fig. 2.7, curve A). Solution of Equation 2.36 yields curve B. Curve A is closer to the experimental curve than is curve B.

2.2.1.6 Concave-up Phase Boundaries

Why are phase boundaries distinctly concave-up for reactions involving gasses or fluids when $P_F = P_T$? The answer can be explained by the Clapeyron equation and compressibilities of fluids relative to solids or liquids. Consider Figure 2.6 or 2.7. At low pressures dP/dT is relatively small compared with the slope of the phase boundary at high pressures. In fact, with increasing pressure, some experimentally determined boundaries commonly go from vertical to negative because ΔV becomes negative. This change in dP/dT means that ΔS is small relative to ΔV at low pressures, but larger at high pressures. The change in ΔS from low to high *P-T* along the boundary is generally small compared with a similar change in ΔV; therefore, ΔV at different conditions of *P-T* exerts the most effect on dP/dT.

In turn, ΔV is most affected by changes in compressibility of the fluid, because the liquid/solid is relatively incompressible. At low pressures the fluid is quite compressible, so ΔV is large and dP/dT is small. At relatively high pressures, the reverse situation exists, and the compressibility of the fluid approaches that of a liquid. The net result is a concave-up phase boundary.

An alternative explanation can be found by consideration of the *Clausius - Clapeyron equation*:

$$dP/dT = \Delta H_v / TV_v \qquad (2.38)$$

which can be easily derived from the Clapeyron equation with the realization that volume of the vapor (V_v) must be much larger than that for either liquids or solids, so ΔV is approximately equal to V_v. We will assume that this same argument applies to V_F, which is certainly valid at relatively low pressures. This equation applies if a vapor (fluid) is involved in the reaction, as for Equations 2.24 and 2.37. Equation 2.38 is strongly affected by V_v; i.e., if V_v is large dP/dT tends to be small and vice versa. Because V_v is larger at low pressures than high, the phase boundary has a lower slope at lower pressures than at high, so it is concave up.

2.2.1.7 Application to Geothermometry and Geobarometry

In Section 2.1.5.5 the exchange reaction between Na and K in alkali feldspar and feldspathoids was introduced as a possible geothermometer (Equation 2.27). An assumption was either the crystallization pressure is low (i.e., pressures equivalent to volcanic or perhaps hypabyssal rocks) or pressure has negligible effect on K_D (K_{eq}). We can now evaluate this latter assumption by rewriting Equation 2.34, taking into account the fact that K_{eq} is not necessarily 1.0, as it was for the reaction of Equation 2.32. Rewriting Equation 2.34 for P at some temperature T, in the form $Y = mX - b$, and substituting K_D for K_{eq}:

$$P = (\Delta S°_T/\Delta V°) \times T - [(\Delta H°_T + RT\ln K_D) / \Delta V°] \qquad (2.39)$$

As expected, the slope term is the Clapeyron slope $\Delta S/\Delta V$. If the slope is near vertical, the reaction is potentially a good geothermometer, whereas a near horizontal slope is likely a good geobarometer. When ΔS is much larger than ΔV the reaction is probably a good geothermometer, and vice versa for a good geobarometer. Note that for standard temperature and $K_D = 1.0$, Equation 2.39 reduces to Equation 2.35.

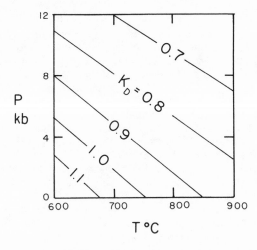

Figure 2.8 P-T diagram for exchange reaction of Equation 2.27 showing variation in constant-K_D lines.

Because all phases in the reaction of Equation 2.27 are solids the Clapeyron slopes should be approximately linear. This assumption can be examined by calculating ΔS for a reasonable temperature range over which the proposed exchange reaction might take place. Another assumption, which although untested is quite reasonable, is that ΔV changes little over this range. Because the Clapeyron slope is $\Delta S/\Delta V$, if ΔS varies little over this temperature range, slopes should also be reasonably constant. Likewise, the intercept term contains ΔH, so its variability can be tested as well. Values for ΔS and ΔH, based on Equation 2.27 and calculated from data in Robie and others (1979), are:

$T°K$	ΔS (j/mol°)	ΔH (j)
900	-9.52	-9827
1000	-10.36	-10608
1100	-10.10	-10329

Allowing for errors attached to these values (discussed later), the assumption of a constant Clapeyron slope and a constant intercept (excluding the effect of K_D) over the temperature range of interest is reasonable. For construction of Figure 2.8, averages of -10.0 j/mol° and -10250 j are respectively used for ΔS^o_T and ΔH^o_T, along with 0.289 j/mol·bar for ΔV^o.

Hence, solution of Equation 2.39 for the reaction of Equation 2.27, given different values of K_D, results in a series of parallel constant-K_D lines (Fig. 2.8). Figure 2.8 shows that the exchange reaction between Na and K in feldspathoids and alkali feldspars is dependent on both pressure and temperature, because slopes of constant-K_D lines do not approach being vertical or horizontal. Thus this reaction by itself would not be a particularly good geobarometer or geothermometer at elevated pressures.

For instance, if we had electron-microprobe data for a silica-undersaturated rock containing alkali feldspar and nepheline, we could predict the pressure of crystallization for a range in temperature, but we could obtain no unique solution based solely on this exchange reaction. An example is:

$$K_D = \frac{X_K^{AF} \cdot X_{Na}^{FD}}{X_{Na}^{AF} \cdot X_K^{FD}} = (0.32 \times 0.60) / (0.68 \times 0.40) = 0.70$$

For this K_D over a temperature range from 700 to 900°C, the predicted pressure can range from about 12 to 7 kb (Fig. 2.8). We have no way of knowing, however, which pressure or temperature, if any, is correct without an independent test. Fortunately, these constant-K_D lines have negative slopes, whereas many similar reactions produce positive slopes. If another mineral pair existed in the rock that yielded a positive slope, combination of the two equilibria allows a unique solution for both P and T -- the point of intersection of the constant-K_D lines for the two equilibria. Lacking another mineral pair, perhaps an independent estimate of temperature or pressure could be made.

Are the results shown in Figure 2.8 reasonable in the light of compositions of naturally occurring alkali feldspars and nephelines? The answer to this question is both yes and no. In volcanic and hypabyssal rocks both Na-rich and K-rich members of both solid solution series exist. The K-rich, reasonably common analog of nepheline in volcanic and hypabyssal rocks is kalsilite, which is virtually unknown in plutonic rocks. Thus shallow-level alkaline rocks provide the ranges in compositions of alkali feldspars and nepheline-kalsilites required to produce the K_D values found in Figure 2.8.

The difficulty lies with plutonic rocks. For example, a K_D for a typical nepheline syenite might be:

$$K_D = (0.70 \times 0.75) / (0.30 \times 0.25) = 7.0$$

This value, ten times greater than the K_D calculated above, is far outside the range of K_D values in Figure 2.8. Substitution of $K_D = 7.0$ into Equation 2.39, using values for ΔS, ΔH, and ΔV given above, yields impossible results. Moreover, in the earlier calculation, the predicted pressure range for crystallization is between 12 and 7 kilobars, but the mol fractions are more typical of volcanic rocks. What is the problem? Two immediately come to mind. First, an implicit assumption in the above calculations is that both solid solution series are behaving ideally (Section 2.2.2.6), but this is almost certainly not the case. Feldspars are notoriously non-ideal solid solutions (Section 3.3.1.2). Considerable substitution of Ca occurs in the lattice sites of both minerals, which can cause errors in calculated K_D values if the solutions are not ideal.

Second, error analysis (Section 1.4.1.2), based on data from Robie and others (1979), yields:

$\Delta H^o{}_T$	10250 ± 5703 j
$\Delta S^o{}_T$	10.00 ± 1.87 j/mol°
ΔV^o	0.289 ± 0.160 j/mol·bar

Relative errors are 56 percent, 19 percent, and 55 percent, respectively. Clearly, errors of this magnitude cast some doubt on the results given in Figure 2.8. Experimental verification is needed to demonstrate the true applicability of this exchange reaction. The purpose of this exercise is not to explore another geobarometer or geothermometer, but rather to learn about the methods and limitations in using thermochemical data in their construction.

To conclude this entire section on P-T diagrams, if we are satisfied with semi-quantitative constructions of phase boundaries on P-T diagrams, we can construct many types of boundaries in a variety of ways. Quantitative constructions are also possible, if data are used that are internally consistent with other phase equilibria, such as those of Helgeson and others (1978). Thermochemical data are also quite valuable learning tools, because they enable us to relate phase equilibria to physical phenomena, such as density or entropy changes. They also allow the exploration of potential geothermometers and geobarometers. When errors are calculated based on published data, however, positions of many boundaries are really only approximated and ultimately require experimental confirmation.

2.2.2 Ideality and Reality

At some point in their history most magmatic systems contain a volatile phase, in addition to a liquid solution (the magma) and solid solutions (minerals). Any of these can "behave in either an ideal or real fashion." This section will explore what is meant by this statement. Any consideration of ideal and real behavior must begin with gasses. After the behavior of gasses is examined, then we will consider solutions.

2.2.2.1 The Ideal Gas Law

A staple of every basic chemistry course is the *ideal gas law*, which states:

$$PV = nRT \qquad\qquad (2.40)$$

where n is number of moles and R is the ideal gas constant. All other terms are familiar. Equation 2.40 can be considered the *equation of state* for an ideal (or "perfect") gas. Let us examine implications of this law by considering 18 grams (1 mole) of H_2O vapor at an elevated T in a closed container, but one in which we can vary V (such as a closed cylinder with a piston at one end). Equation 2.40 implies that P on the water vapor should be proportional to T/V, as R and n are constant. If water vapor behaves ideally, as we increase T, P_v should increase proportionally. Likewise, an increase in V proportionally decreases P. If P_v remains fairly low, this ideal behavior persists; if P_v increases, this proportionality between T/V and P no longer holds. Such a gas is referred to as a *real* (or "imperfect") gas.

Pressure arises from the collective collisions of all gas molecules on the interior of the container. At low pressures each gas molecule acts independently of all others because, on average, the molecules are far enough apart so that their kinetic energies are not mutually affected by forces common to all. This is the situation necessary for ideal behavior. Why, then, do gasses deviate from ideality?

The dominant forces among gas molecules are electrostatic. The more *dipole* a gas molecule is, the stronger will be the electrostatic forces among molecules and greater the chance for deviation from ideality (a simple bar magnet is a dipole). An H_2O molecule, one of the most dipole gas molecules, has two positively charged H^+ ions on one side of the molecule and O^{2-} on the other. As a result, water vapor commonly deviates considerably from ideal behavior. Helium gas, whose atoms are not dipoles, does not deviate appreciably from ideal behavior.

TABLE 2.1 Some Van Der Waals Constants

Gas	a ($atm \cdot L^2/mol^2$)	b (L/mol)
CO_2	3.59	0.0427
CO	1.49	0.0399
H_2	0.244	0.0266
HCl	3.67	0.0408
O_2	1.36	0.0318
SO_2	6.71	0.0564
H_2O	5.46	0.0305

If the average overall electrostatic force among all gas molecules is attractive, kinetic energies will, on average, be less and the real pressure will be less than ideal. If the average force is repulsive, real pressure will be greater than ideal. This latter situation requires relatively high real pressures. An increase in temperature enhances an approach to ideal behavior because of increased entropy and thus disorder.

From Equation 2.40 $V = RT/P$ for one mole of an ideal gas. The more V deviates from RT/P, the more nonideal is the gas's behavior. As a consequence, the following definition is useful:

$$\alpha = (RT / P) - V \tag{2.41}$$

The factor α will be zero for an ideal gas; $\alpha > 0$ when real pressure is less than ideal; $\alpha < 0$ when ideal is less than real. After introduction of the next equation of state we will examine α more carefully.

2.2.2.2 The Van Der Waals Equation

Many equations of state for real gasses have been proposed, all of which either are empirical or approximations (or both), because real behavior cannot be exactly quantified. We will consider one of the oldest and simplest, but still widely in use. This equation of state for 1 mole of a real gas was first proposed in the mid-1800s by van der Waals. It is:

$$[P + (a / V^2)] (V - b) = RT \tag{2.42}$$

where a and b are "van der Waals constants" for real gasses. Units on a are atm·L^2/mol^2 and on b, L/mol. Notice the similarity between this equation of state and that for an ideal gas (Equation 2.40). When a and b are zero, the gas is ideal. If we rearrange Equation 2.42, however, it becomes apparent that a and b have the opposite effect on the magnitude of the real pressure:

$$P_{real} = [RT / (V - b)] - (a / V^2)$$

An increase in a causes the calculated real pressure to be lower, whereas an increase in b has the opposite effect. Most introductory chemistry textbooks include a proof of the van der Waals equation. Some van der Waals constants for a few gasses of petrologic interest are included in Table 2.1.

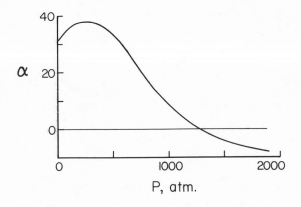

Figure 2.9
Graph showing change for α [$(RT/P_{real})-V$] with increasing pressure.

Knowing a and b for a specific gas, it is a simple matter to calculate α for any conditions of P, V, and T. For this calculation, P is normally in atmospheres and V in liters, so R is 0.08206 L·atm/mol°. As an example:

$$[P + (5.46 / 0.25^2)] (0.25 - 0.0305) = 0.08206 \times 773$$

$P_{real} = 202$ atm
$P_{ideal} = RT / V = 0.08206 \times 773 / 0.25 = 254$ atm
$\alpha = (0.08206 \times 773 / 202) - 0.25 = .064$ L

If a gas has not radically departed from ideality, Equation 2.42 approximates its real pressure reasonably well. A plot of α versus P for one mole of water vapor at constant T is shown in Figure 2.9. Because real pressure is greater than ideal at relatively high pressures, α is negative.

2.2.2.3 Fugacity

An additional factor must be considered. Not only must we be concerned about the difference between real and ideal pressure, but also we must deal with "effective" or "thermodynamic" pressure, which is referred to as *fugacity*. The relation between fugacity and chemical potential is covered in Section 2.2.3.1. The need for another concept of pressure arose when it was discovered that frequently neither real nor ideal pressure yielded correct results for other parameters when entered into some thermodynamic calculations. If the gas truly behaves ideally, then P_{ideal} can be used; if it does not, sometimes we can use P_{real} but often we must use f (fugacity). An example is Equation 2.25. At extremely low pressures or very high temperatures, using P_{real} or perhaps even P_{ideal} in the equation is probably satisfactory. For other conditions, however, f is required.

TABLE 2.2 Data for Water from Van Der Waals Equation

a	b	c	d	e	f	g
500	50	1269	1069	497	0.465	9.3
	150	423	288	217	0.755	70.1
	250	254	202	167	0.827	64.6
	350	181	154	134	0.867	62.0
600	50	1433	1490	832	0.559	-1.9
	150	478	357	280	0.785	50.8
	250	287	239	204	0.853	49.7
	350	205	180	160	0.888	48.8
700	50	1597	1911	1269	0.664	-8.2
	150	532	425	347	0.814	37.7
	250	319	276	242	0.875	38.9
	350	228	205	186	0.906	38.9

a: T^oC
b: V in cc
c: P_{ideal} in atm
d: P_{real} in atm

e: f (fugacity) in atm
f: γ_f (fugacity coeff.)
g: $\alpha = (RT/P_{real})-V$

The relation between real pressure and fugacity is expressed as a simple proportionality constant called the *fugacity coefficient:*

$$\gamma_f = f / P_{real} \tag{2.43}$$

Clearly, $\gamma_f = 1.0$ for an ideal gas. If we use the van der Waals equation of state, f can be estimated by:

$$\ln f = \ln [RT / (V - b)] + [b / (V - b)] - (2a / RTV) \tag{2.44}$$

(Kern and Weisbrod, 1967). All terms are defined as above. As in the case for real pressure, the constants a and b have an opposite effect on the magnitude of the calculated fugacity. An increase in a decreases fugacity, whereas an increase in b causes fugacity to increase. Using Equation 2.42 to calculate P_{real} and Equation 2.44 to determine f, we can then estimate the fugacity coefficient from Equation 2.43 for different values of T and V. Solution of Equation 2.44 for the example above yields:

$$\ln f = \ln [(0.08206 \times 773) / (0.25 - 0.0305)] + [0.0305 / (0.25 - 0.0305)] - [2 \times 5.46 / (0.08206 \times 773 \times 0.25)]$$

$$f = 167 \text{ atm} \quad \therefore \quad \gamma = 167 / 202 = .827$$

In summary, for this example we have calculated three values:

$$P_{ideal} = 254 \text{ atm}$$
$$P_{real} = 202 \text{ atm}$$
$$f = 167 \text{ atm}$$

We use α to quantify the difference between P_{ideal} and P_{real}, whereas γ_f is a ratio of f/P_{real}. Table 2.2 shows the results of calculations for P_{ideal}, P_{real}, f, γ, and α, for different values of T and V. The previous example is in the third row. Note the negative values for α, as discussed earlier; P_{real} is higher than P_{ideal} for these conditions of higher pressure.

A more sophisticated calculation of f_{H2O} for various values of P_{H2O} and $T°C$ has been made by Burnham and others (1969) and graphed in Carmichael and others (1974 p. 325). Considering the inaccuracies in the van der Waals approximation, agreement between Table 2.2 and Burnham and others (1969) is good for low to moderate pressures and temperatures. The main lesson to be learned from Table 2.2 is a reenforcement of a point already made -- fugacity and real pressure approach one another at either high temperature or low pressure. An increase in temperature also lowers α, and Table 2.2 confirms the pressure effect on α shown in Figure 2.9.

2.2.2.4 Effect on P-T Diagrams

We are now in a position to correct curve A on Figure 2.6, which was constructed on the basis of $\gamma_f = 1.0$ and $P = f$. Curve B cannot be corrected by this method (Wood and Fraser, 1976). The data of Burnham and others (1969) are more reliable over a wider range of temperatures and pressures than the van der Waals equation, so they have been used to construct curve C on Figure 2.6. For example, we originally calculated the "pressure" at 800K for Equation 2.22 to be about 1930 bars. We now know that in fact we were calculating fugacity, so we can convert $f = 1930$ to $P = 4500$ using the tables of Burnham and others (1969). Multiple calculations of this type yield curve C on Figure 2.6, which should be closer to "ultimate truth."

Curve C on Figure 2.7 was constructed in a similar manner from curve A using the γ_f data for CO_2 in the Appendix of Wood and Fraser (1976). Curve C agrees fairly well with the experimentally determined curve (curve H-T). On Figure 2.6 the corrected curve C is to the low temperature side of the original curve A, and the opposite situation exists for Figure 2.7. This is because most γ_f values are below 1.0 for H_2O, but considerably above 1.0 for CO_2 (Wood and Fraser, 1976, Tables A.1-A.2).

If confining pressures are less than about 3 kb (about 9 km depth), igneous temperatures are high enough such that real pressure and fugacity for water should not deviate too greatly. In geologic systems, real pressures of pore fluids associated with *burial* metamorphism or deep sedimentary *diagenesis* may deviate most from fugacities, because these processes take place under relatively high confining pressures and low temperatures. Fluid pressures associated with *contact* metamorphism may not deviate too widely from fugacities, as relatively high temperatures and low pressures are common for this type of metamorphism.

2.2.2.5 Solutions and Activity

If gasses can be either ideal or real, it should be no surprise that both liquid and solid solutions also can be ideal or real. We can consider either the concentration or *activity (a_i)* of chemical species i dissolved in a solution. For solutions, concentration is to activity as pressure is to fugacity for gasses; i.e., the activity of an ion in a solution is equivalent to its "effective" or "thermodynamic" concentration. The relationship between activity and chemical potential is covered in Section 2.2.3.1. For an ideal solution, concentration and activity are equal.

We can easily measure many activities in aqueous solutions, whereas fugacity cannot be directly measured. We measure the H^+ activity when using a *pH meter*. A pH meter has one of many types of "specific ion electrodes" that allow quantitative determinations of many ion activities in aqueous solution by electrochemical means. We could use a Na specific ion electrode and measure Na^+ activity in an aqueous solution, but we would be measuring the Na^+ *concentration* when using atomic absorption analysis. Analogous to a fugacity coefficient, an activity coefficient can be defined as:

$$\gamma_a = a_i / X_i \qquad\qquad (2.45)$$

where X_i is mol fraction (concentration) of chemical species i and γ_a is the activity coefficient, which is equal to 1.0 for an ideal solution.

For a gas dissolved in a liquid and existing in a fluid in contact and in chemical equilibrium with the liquid, the fugacity and activity coefficients are equal. An example might be H_2O bubbles in a magma in equilibrium with H_2O dissolved in the magma. The activity of H_2O dissolved in the magma will be proportional to fugacity of H_2O in the bubbles.

2.2.2.6 Raoult's and Henry's Laws

As P nears zero γ_f approaches 1.0, so it is reasonable that as X_i nears zero, γ_a also approaches 1.0. The practical result of this statement is that trace elements are more likely to behave as ideal solutions than are major elements. Excepting quartz, most igneous rock-forming minerals exhibit extensive solid solution between two or more "end-member" components. Unless one component is in a dilute solution in the others (i.e., approaches trace compositions), ideality is seldom reached. Exceptions are olivine and orthopyroxene, which behave approximately ideally throughout their entire compositional ranges. These minerals, and any others that behave ideally throughout their entire ranges, obey

Raoult's Law -- γ_a for a component *i* remains at 1.0 from $X_i = 0$ to $X_i = 1.0$. When Raoult's Law is in effect, activity and mol fraction are identical throughout the compositional range (Fig. 2.10). This figure is an example of a *binary* (two-component; see Section 3.3.1) solid solution.

For some dilute solutions where Raoult's Law does not apply, *Henry's Law* does apply. Henry's Law states that γ_a is not 1.0, but a constant, which from a practical standpoint is just as useful. This concept is expressed graphically in Figure 2.10, where Henry's Law only applies for a limited compositional range (dilute solution), whereas Raoult's Law applies throughout. If neither law is in effect, γ_a will vary in some unpredictable manner with changing X_i. If a solution is following Henry's Law, it will behave in a *nonideal but predictable manner*. By "predictable" we mean that a_i and X_i are not equal, but they will differ only by a proportionality constant, γ_a. Because chemical equilibrium constants in general (and partition coefficients in particular) are based on ratios of concentrations rather than absolute concentrations, it is not necessary for activity coefficients to be 1.0, provided they are constant over the concentration range of the data set.

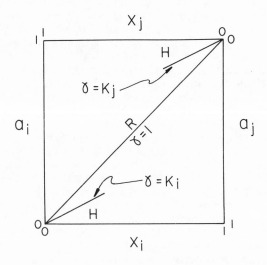

Figure 2.10 Plot of activities *(a_i* and *a_j)* versus mol fractions *(X_i* and *X_j)* for hypothetical components *i* and *j* in a binary solution. Line *R* is the line for which Raoult's Law holds, whereas the two lines labeled *H* represent Henry's Law. Along line *R*, $\gamma_a = 1$, but γ_a is equal to some constant other than 1 (either *K_i* or *K_j*) along lines *H*.

The more dilute the solution, the greater are the chances that either Henry's or Raoult's Law will be followed. For this reason, trace-element, and occasionally minor-element, analyses are very useful for petrogenetic modeling (see Section 5.2.1). Major elements are useful as well, but in a different context (see Sections 5.2.2.3 and 5.2.3).

2.2.3 Phase Rule

Of all J. Willard Gibbs' contributions to physical chemistry, perhaps none is more valuable to igneous petrologists than *Gibbs Phase Rule*. Interestingly, he originally published the phase rule in 1876, but its importance was not

discovered by petrologists until the early 1900s. Two geologists who pioneered
use of the phase rule in the early 1900s were N. L. Bowen in igneous and V. M.
Goldschmidt in metamorphic petrology.

2.2.3.1 Chemical Potential

Mathematical proof of the phase rule is dependent on understanding the concept
of *chemical potential*. If all phases involved in igneous systems were pure, we
would not need to deal with chemical potential. Up to this point, with the
exception of Equations 2.27 and 2.28, we have only considered equilibria
involving pure phases, either fluids or solids. Even the alkali feldspar and
feldspathoid solid solutions of Equations 2.27 and 2.28 were treated as pure
phases. Simple thermodynamic parameter changes, such as ΔH, ΔS, ΔV, and
ΔG were only required to characterize adequately equilibria involving pure
phases. Fluids, magmas, and most minerals are in fact not pure phases but
solutions, which require the introduction of a new concept, *partial molar
quantities*. Chemical potential can be defined as *partial molar Gibbs free
energy,* generally referred to simply as partial molar free energy.
    The most easily understood partial molar quantity is partial molar
volume. Imagine two completely immiscible liquids, such as oil and water. If
we add 1 mole of H_2O to a flask containing oil, the total volume of the oil plus
water will be increased by 18 cc (or ml), because 1 mole H_2O = 18 gm and the
density of liquid water is about 1 g/cc. However, if we add 1 mole of water
to about 25 cc of acetone (CH_3COCH_3), the total volume of water plus acetone
only will be increased by about 16.5 cc, to approximately 41.5 cc. This is
because acetone and water are completely miscible in one another and the
volume change is different from the volume of the original 18 cc of H_2O. This
volume change, 16.5 cc, is the partial molar volume of water in the solution.
    Partial molar volume, therefore, is the "effective" volume of 1 mole
solute in a solution, or change in the total volume of a solution brought about by
addition of 1 mole solute. If the net electrostatic forces between molecules (or
ions) of solvent and solute are repulsive, the partial molar volume will be greater
than the original volume of the solute. If the net electrostatic forces are
attractive, as in the case of water and acetone, the opposite situation exists.
Partial molar enthalpy, entropy, internal energy, and free energy can be similarly
envisioned. For instance, partial molar free energy (chemical potential) is the
effective free energy of a solute in a solution, or the change brought about in the
total free energy of the solution by addition of 1 mole solute.
    In the following discussion we consider the fact that chemical potential is
the criterion for spontaneity in a system in which phases are of variable composition in the same way that Gibbs free energy is the criterion in systems with
pure phases. First, we must examine how free energy varies with X_i in a solid
solution. The relevant equation is:

$$\mu^o_i = G^o_i + RT\ln a_i \qquad (2.46)$$

where μ^o_i is chemical potential of component i in the solution, G^o_i is free
energy of pure i, and a_i is activity of i in the solution. For an ideal solution:

$$\mu^o_i = G^o_i + RT\ln X_i$$

Similar equations exist for real and ideal gasses:

$$\mu^{o}_{i} = G^{o}_{i} + RT\ln f_{i} \text{ and } \mu^{o}_{i} = G^{o}_{i} + RT\ln p_{i}$$

where f is fugacity and p is partial pressure, which has only been briefly discussed previously, but a simple example is illustrative. Given an ideal gas mixture at 1000 bars P_{T}, where $X_{CO2} = 0.25$ and $X_{H2O} = 0.75$, $p_{CO2} = 250$ bars and $p_{H2O} = 750$ bars.

Because $G = H - TS$, the free energy of mixing of components in the solution must be:

$$G^{o}_{MIX} = H^{o}_{MIX} - TS^{o}_{MIX} \tag{2.47}$$

For an ideal solution, no heat is given off or absorbed by the system when the solute and solvent are mixed together, so $H^{o}_{MIX} = 0$. Consequently, G^{o}_{MIX} must be the negative quantity $-TS^{o}_{MIX}$. Because temperature in °K and entropy are always positive quantities, the free energy of mixing for an ideal solution is always a negative quantity. The result is that a plot of G^{o}_{i} versus X_{i} for an ideal binary solid solution will produce a curve that is concave up (Fig. 2.11). In contrast, as mixing of two or more components creates more disorder, S^{o}_{MIX} is always a positive quantity and a plot of S^{o}_{MIX} against X_{i} will produce a concave-down curve (not shown).

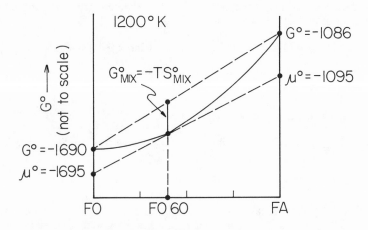

Figure 2.11 Plot of G^{o} for the olivine solid solution at 1200K. Units on G^{o} are kj/mol. See text for calculation of chemical potentials and free energy of mixing for a mixed crystal of composition FO60.

Figure 2.11 demonstrates that μ^{o}_{i} is the effective free energy of component i in the solid solution. As a consequence, the following equation must hold for an ideal binary solid solution:

$$G^{o} = X_{i}\mu^{o}_{i} + X_{j}\mu^{o}_{j} \tag{2.48}$$

Hence, the total free energy G^{o} must simply be the weighted average of the chemical potentials for components i and j. Equation 2.48 implies that free energy and chemical potential are identical for a pure phase, because it contains only one component. During our previous discussions of free energy in systems involving pure phases, we were also considering chemical potential, although it was never mentioned.

We can illustrate this relationship between G^o and X_i by using the olivine solid solution as an example. Recall the fact that $G^o = \Delta G^o_f$. Given an olivine solid solution of composition FO60 FA40 at 1200K:

$$\mu^o_{FO} = G^o_{FO} + RT\ln X_{FO}$$
$$= -1689674 + (8.315 \times 1200 \times \ln 0.6)$$
$$= -1694771 \text{ j/mol}$$

$$\mu^o_{FA} = G^o_{FA} + RT\ln X_{FA}$$
$$= -1086235 + (8.315 \times 1200 \times \ln 0.4)$$
$$= -1095378 \text{ j/mol}$$

Total G^o for the solid solution must be:

$$G^o = (0.6 \times -1694771) + (0.4 \times -1095378)$$
$$= -1455014 \text{ j/mol}$$

Multiple calculations of this type for various compositions of the solid solution allow construction of the curve on Figure 2.11. Free energy of mixing can then be estimated graphically from Figure 2.11 by:

$$G^o_{MIX} = -1455 - [(-1690 \times 0.6) + (-1086 \times 0.4)] = -6.6 \text{ kj/mol}$$

How, then, is chemical potential a criterion for spontaneity? Using the general term n_i rather than X_i for concentration:

$$G = \Sigma n_i \mu_i \quad \therefore \quad dG = \Sigma \mu_i dn_i = 0$$

for a reaction at chemical equilibrium. From the proof in Section 2.1.5.3 it follows that for a spontaneous, irreversible reaction:

$$dG = \Sigma \mu_i dn_i < 0$$

Consider a chemical reaction at equilibrium in which Sr^{2+} ions diffuse back and forth across the interface between a plagioclase and an alkali feldspar crystal. We will designate subscript A as Sr in alkali feldspar and subscript P as Sr in plagioclase. If equilibrium is maintained:

$$dn_A = -dn_P$$

Remembering that:

$$\Sigma \mu_i dn_i = 0 \quad \therefore \quad \mu_A dn_A = -\mu_P dn_P \quad \therefore \quad \mu_A = \mu_P \qquad (2.49)$$

This last equation is the criterion for equilibrium in terms of chemical potential: *at equilibrium, μ for each component is the same in all phases.* This statement is critical to the proof of the Gibbs phase rule in the next section. The chemical potential of Sr, therefore, must be the same in alkali feldspar as in plagioclase at equilibrium.

A similar proof can be made for a spontaneous, irreversible process, which leads to:

$$\mu_A > \mu_P \tag{2.50}$$

For a spontaneous process, the reaction tends to go from high to low μ and the reaction will continue to go until chemical potentials are equal. This statement, then, is the criterion for spontaneity in terms of chemical potential. Equation 2.50 implies that more Sr will diffuse from alkali feldspar into plagioclase than from plagioclase into alkali feldspar because Sr in alkali feldspar has a higher μ. This is an example and is not meant to imply that Sr will always diffuse in this direction.   For these reasons, chemical potential is frequently referred to as *escaping tendency,* because a component will diffuse from a region in which it has a high chemical potential into one in which its potential is low.   We commonly say that components diffuse "down chemical potential gradients." This diffusion takes place within a phase as well as from one phase to another.

2.2.3.2 Proof of the Phase Rule

Now that chemical potential has been introduced, proof of the phase rule is simple. Consider a system that has four components, 1-4, and three phases, a-c. At equilibrium, the chemical potential of each component must be the same in every phase, so we can write an array (matrix) of equations of the form:

$$\mu_1{}^a = \mu_1{}^b = \mu_1{}^c$$
$$\mu_2{}^a = \mu_2{}^b = \mu_2{}^c$$
$$\mu_3{}^a = \mu_3{}^b = \mu_3{}^c$$
$$\mu_4{}^a = \mu_4{}^b = \mu_4{}^c$$

For any array of this kind we can define P columns (phases) and C rows (components). In this case, P = 3 and C = 4. For the general case, there must be CP - C or C(P - 1) equations for the entire array. For this example, $4(3 - 1) = 8$ equations are present (count the equals signs). These equations are known as "relations" among the variables.

We know that $\mu = f(T, P, X_i)$, so there must be $2 + CP$ possible variables. These variables are:

$$T, P, \text{ and } X_1{}^a \ldots . X_4{}^c$$

In our example this adds to $2 + 12 = 14$ variables. However, the mol fractions are not mutually independent, because in each phase $\Sigma X_i = 1.0$. Because we have P phases, we must have P additional equations or "relations." Thus for this array there are $C(P - 1) + P$ relations among $2 + CP$ variables. For our example this would be 11 relations among 14 variables.

The number of degrees of freedom (F) in this system is defined as the number of variables minus the number of relations, so $F = 3$ for the example. This definition is remarkably similar to that for degrees of freedom *(df)* as defined in statistics (see Section 1.1.3.2). Given this definition:

$$F = 2 + CP - [C(P - 1) + P] = C + 2 - P$$

Rearranging:

$$P + F = C + 2 \tag{2.51}$$

This deceptively simple equation is Gibbs Phase Rule. The phase rule balances for the example above because $3 + 3 = 4 + 2$. This rule states that for a system in chemical equilibrium the number of phases plus the number of degrees of freedom must always equal the number of components plus two. Systems with one, two, three, or four components are termed:

$C = 1$	unary	$C = 3$	ternary
$C = 2$	binary	$C = 4$	quaternary

Thus this example is a quaternary system.

The next next step is to define carefully P, C, and F. Systems, phases, and components have already been defined adequately. Key words here are *maximum* and *minimum*. Definitions are:

C   minimum number of components
P   maximum number of phases
F   minimum number of degrees of freedom

2.2.3.3 An Example: The Haplogranite System

As an illustration let us deal with a specific quaternary system, the haplogranite system. This system consists only of quartz and feldspars. Some natural rocks, such as alaskites or leucogranites, are very close in composition to the haplogranite system. Why is it a quaternary ($C = 4$) system? The chemical elements that make up its constituent minerals are O, Si, Al, Na, K, and Ca. This would suggest that $C = 6$, but remember we are searching for the *minimum* number of components. As O^{2-} is the only anion, we could rewrite the elements as oxides: SiO_2, Al_2O_3, Na_2O, K_2O, and CaO. This would decrease C from 6 to 5, but we still have not reached $C = 4$. The solution is in the realization that quartz is a fairly pure phase of composition SiO_2, whereas the feldspars are solid solutions of the end-member compositions AB ($NaAlSi_3O_8$), OR ($KAlSi_3O_8$), and AN ($CaAl_2Si_2O_8$). So the components of the quaternary haplogranite system are Q-AB-OR-AN.

Water could exist as a separate component in the magma, even though it does not enter into any of the minerals present. If the magma is oversaturated with respect to H_2O, a separate aqueous fluid phase, appearing as bubbles in the magma, would be present. If H_2O is present as a component, the magma chamber must be an open system with respect to H_2O, because H_2O is not a component of quartz or feldspars and thus must ultimately escape the system. A more realistic situation would be that under the right conditions of temperature and pressure H_2O enters into a phase present in the rock, such as muscovite, in the form of the hydroxyl ion (OH^-). For the moment, however, let us assume that only quartz and feldspars are stable.

Assuming the system contains no H_2O, because $C = 4$, according to the phase rule $F + P$ must equal 6. The maximum number of minerals will most likely be three: plagioclase, alkali feldspar, and quartz. The alkali feldspar may have exsolved into two phases (perthite), but that is unlikely if magma is present and H_2O is not (see Section 3.3.2.5). We will make another assumption that only two minerals are presently crystallizing and some magma still remains. As in the previous example, if P equals 3, F must equal 3.

xxxxxxxxxxxxxxxxxxxxxxxxxxxxxxxxxxxxxxxxxxxxxxxxxxxxxxxxxxxxxxxx

Some conventions must be established at this point. All three letters representing terms in the phase rule, P, F, and C, are also used to designate other phenomena, such as thermodynamic parameters. To avoid ambiguity, throughout this book the following designations are used:

P	*pressure*
F	*fluid or number of moles of fluid (also used for force as in F = MA; some books use F as Gibbs free energy)*
C	*heat capacity*
P	*number of phases as defined in phase rule*
F	*number of degrees of freedom as defined in phase rule*
C	*number of components as defined in phase rule*

xxxxxxxxxxxxxxxxxxxxxxxxxxxxxxxxxxxxxxxxxxxxxxxxxxxxxxxxxxxxxxxx

What do we mean by the statement $F = 3$? We now must deal with exactly what degrees of freedom are in terms of physical reality. A number of definitions exist but the simplest is *the minimum number of intensive parameters $(T, P, X_i, X_j, ...)$ whose values must be specified to define the state of the system.* As for many verbal definitions, the exact meaning of this statement is obscure. Its meaning will become much clearer when we start dealing with specific phase diagrams in Chapter 3.

For now let us consider a magma chamber, consisting of the dry haplo-granite quaternary system, in which pressure is constant but temperature is decreasing as the magma cools. Because $F = 3$ and pressure is already fixed, only two parameters remain to be specified. Those available are T, X_{AB}, X_{OR}, X_{AN}, and X_Q. If we know temperature and the composition of any one component, or the compositions of any two components, we can determine the other three parameters, and the state of the system will be specified. This, of course, is only true for the case in which $P = 3$. As stated above, shortly all this will become much clearer.

3

UNARY AND BINARY
SYSTEMS

Processes affecting chemical and mineralogical compositions of igneous rocks can be divided into three categories on the basis of the sequence in which they occur. First, those that involve *origin* of the magma are primarily concerned with partial melting of a source rock deep in the earth's crust or upper mantle. Second, many processes are related to *modification* of the magma during its ascent and while it crystallizes. Examples are fractional crystallization, magma mixing, and assimilation (contamination) of country rocks. Third, post-crystallization *alteration* of the igneous rock commonly takes place. This alteration typically involves chemical interaction between the solid rock and an aqueous solution, at either high or low temperatures. Chapters 3-5 of this book introduce graphical and numerical methods for quantitatively dealing with processes in all three categories, but emphasis is placed on the first two.

This chapter includes many examples of unary and binary systems determined by experimental means and attempts to tie these systems to some of the thermodynamic principles learned in Chapter 2. This procedure has already been initiated in Section 2.2.1. Practical examples of igneous rocks as observed in the field and petrographic microscope, as well as variations in their chemical compositions, will also be tied to these experimental systems. Of necessity, this latter aim requires many simplifying assumptions, because igneous rocks are multicomponent rather than unary or binary systems, but some valuable lessons still can be learned.

3.1 UNARY SYSTEMS

3.1.1 Water

Water is not only the basic substance of life but, in one way or another, it participates in most geologic processes, including igneous processes. When involved in igneous activity, it is seldom in the form of a liquid, or even a gas (vapor). Rather, it is a *supercritical fluid,* the meaning of which will be explained later. Despite this fact, we will consider the unary system H_2O at all reasonable conditions of temperature and pressure, where it can be in the form of solid (ice), liquid, vapor (steam), or fluid. We will do so because it is a particularly good system in which to introduce application of the phase rule and to explain the concept of supercritical behavior. Because unary phase diagrams are, by necessity, P-T diagrams (X_i cannot be a variable in a unary system), the system H_2O also enables us to apply further some of the principles learned in Section 2.2.1.

3.1.1.1 A P-T Diagram for H_2O

The P-T phase diagram for the unary system H_2O is shown in Figure 3.1. As mentioned earlier, P is normally plotted on the vertical axis and T on the horizontal. The diagram is divided into four stability fields: ice, liquid H_2O, water vapor or steam, and the fluid field. In fact, many more fields exist, because the stability field for ice can be separated into several fields, one for each polymorph of ice. We will not be concerned with these polymorphs. Notice that the diagram is not to scale.

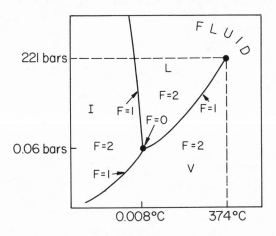

Figure 3.1 A unary phase diagram for H_2O (modified after Kennedy and Holser, 1966; Geol. Soc. America Mem. 97, p. 371-384). Pressure is on the vertical axis in bars and temperature is on the horizontal axis in °C. I, ice; L, liquid; V, vapor; F, degrees of freedom. The point on the L-V curve at 221 bars and 374°C is the critical point for H_2O.

The stability fields for ice (I), liquid H_2O (L), and vapor (V) are separated from one another by an equilibrium or phase boundary. All three of these phase boundaries meet at a common point called, appropriately enough, the *triple point*. For the system H_2O this triple point occurs at 0.008°C and 0.06 bars pressure. Another point appears on this diagram, at 221 bars pressure and 374°C. This second point is the *critical point* for H_2O, above which we cannot distinguish between liquid H_2O and steam, so we refer to the substance as a *supercritical fluid*. If either temperature is above 374°C or pressure is above 221 bars, water will behave as a fluid. We will return to supercritical fluids in Section 3.1.1.4.

3.1.1.2 Application of the Phase Rule

Figure 3.1 provides an excellent opportunity for our first practical application of the phase rule. Because we are dealing with a unary system, C = 1 and P + F = 3. Relationships are summarized in Table 3.1. Composition (X_i) cannot be a variable in a unary system, thus pressure and temperature are the only intensive parameters represented on the diagram that can be variables. Within any of the three stability fields only one phase can be in equilibrium, therefore P = 1 and F = 2. In keeping with the practical definition of degrees of freedom (F; Section 2.2.4.2), both temperature and pressure must be specified to define the state of the system at any point. Because F = 2, we refer to this situation as *divariant equilibrium*. These stability fields are two dimensional so their geometry must be an area.

Along any one of the three phase boundaries two phases are in chemical equilibrium with one another. Consequently, P = 2 and F = 1. Along any one boundary if we specify either pressure or temperature, then the other intensive parameter is automatically fixed. If we specify temperature, pressure must be a certain value, and vice versa. Hence only one variable is required to define the state of the system for a point on any one of the three boundaries. Because F = 1, this is referred to as *univariant equilibrum* and these boundaries are frequently called univariant curves.

TABLE 3.1 Application of Phase Rule to Unary Systems

C	P	F	Equilibria	Variables	Geometry
1	1	2	divariant	P and T	area
1	2	1	univariant	P or T	line/curve
1	3	0	invariant	none	point

At the triple point three phases are in equilibrium, so P = 3 and F = 0. We are not required to specify either pressure or temperature, because for the system H_2O the triple point must always be at a certain point on the diagram. Predictably, a triple point is also called an *invariant point*. This diagram only shows one invariant point, but had we included the fields for all the polymorphs of ice, there would have been several. Moreover, invariant points on *P-T* diagrams are only triple points in unary systems. In binary systems invariant points on *P-T* diagrams involve four phases at equilibrium, so they might be termed *quadruple points*. On *P-T* diagrams of ternary systems, five phases can coexist at an invariant point, and so on.

3.1.1.3 Application of the Clapeyron Equation

The Clapeyron equation may have limitations with regard to exact location of phase boundaries on a *P-T* diagram (see Section 2.2.1.4), but it is invaluable in explaining slopes and curvatures of these univariant boundaries in a qualitative sense. Three boundaries exist on Figure 3.1. The boundary between ice and liquid H_2O is very nearly a straight line with a negative slope, whereas the ice-vapor boundary is slightly concave-up with a positive slope. The third boundary, liquid-vapor, has a pronounced positive (concave-up) curvature and a positive slope.

The difference between slopes and curvatures of the ice-vapor and liquid-vapor univariant curves is distorted. This is because the P axis is distorted, because the triple point is at only .06 bars and the critical point is at 221 bars. Negative pressure is not possible, so the P scale below the triple point is expanded compared to that above. We can explain all of these observations about the three boundaries using the Clapeyron equation.

Anyone who has gone ice skating on a lake or pond knows that ice is less dense than liquid H_2O. This phenomenon, in fact, is rare in nature. For most substances the liquid form is less dense than the solid form. This is certainly true for rocks and minerals; a magma is invariably less dense than a rock of the same chemical composition. Many discussions, bordering on philosophical or even religious (some have claimed divine intervention), have taken place over the nature of our world if water were similar to most other substances and ice were more dense than water. One fact is certain -- if ice were more dense than liquid water, the ice-liquid phase boundary on Figure 3.1 would have a positive rather than negative slope.

We use the first of two expressions of the Clapeyron equation here -- $dP/dT = \Delta S/\Delta V$. For consistency, in this and subsequent examples of P-T diagrams, we assume that a phase boundary is being crossed at constant P and increasing T. If so, this slope of the ice-liquid univariant curve must be negative because ΔS is positive and ΔV is negative. ΔS will always be positive in crossing a solid-liquid phase boundary in the manner just described because a liquid is more disordered than a solid.

The volume change, on the other hand, can be positive or negative depending on the density contrast between solid and liquid. Recall that molar volume is inversely proportional to density. Because ice is less dense than water, $\Delta \rho > 0$ (ρ = density) and thus $\Delta V < 0$ for increasing T. Had the boundary been crossed in a reverse direction, $\Delta S < 0$ and $\Delta V > 0$, so dP/dT would still be negative. It is apparent that dP/dT must always be the same, no matter in what direction the phase boundary is crossed. The density of ice must be less than the density of liquid water in order for $dP/dT < 0$.

For the other two phase boundaries, with increasing T, both ΔS and ΔV will always be positive, as will dP/dT and d^2P/dT^2. In Section 2.2.1.5 we considered why the slope and curvature of a phase boundary on a P-T diagram with a gas or fluid on the high-T side will be positive. Recall that along a single phase boundary ΔS does not change much with increasing P or T, but if a gas or fluid is involved, ΔV changes considerably due to the increased compressibility of the gas or fluid with decreasing pressure. Along each of the boundaries ice-steam and liquid-steam dP/dT increases because ΔS stays relatively constant whereas ΔV decreases with increasing P.

3.1.1.4 The Critical Point and Supercritical Fluids

Let us return for a moment to the critical point. Table 3.2 gives temperatures, pressures, and densities of critical points for several gasses of petrologic interest. Above the critical point the terms *gas* or *vapor* and *liquid* no longer have meaning, so we call the substance a *supercritical fluid* or simply a *fluid*. Near its critical point a fluid is similar to a gas, but as pressure increases, it becomes more like a liquid.

Water is one of the most dipole molecules and departs most greatly from ideality (see Section 2.2.2.2). Water, interestingly, also has a relatively high critical temperature and pressure. At the critical point of water its density is about 1/3 that of liquid water. As pressure increases, the density of fluid water continuously increases until at very high pressures its density approaches that of liquid water. An increase in temperature, however, has the opposite effect. Other properties of the fluid, such as viscosity, also increase with increasing pressure until they are similar to liquid water. Other gasses exhibit similar behavior.

TABLE 3.2 Critical Points of Some Gasses

Gas	T°C	P (bars)	ρ (gm/cc)
CO_2	31.1	73.8	0.468
HCl	51.4	82.6	0.419
H_2S	100.4	90.0	0.348
SO_2	157.6	78.8	0.525
H_2O	374.2	221.1	0.322

As Carmichael and others (1974) point out, the critical pressure for water, 221 bars, is attained at depths in the earth of less than 1 km. Many other gasses of petrologic interest have even lower critical pressures (Table 3.2), so for all but the shallowest igneous processes, no matter what the temperature, "gasses" will in fact be fluids.

3.1.2 Silica

The unary system SiO_2 is quite significant in igneous petrology. The nine crystalline polymorphs of SiO_2 are:

    alpha (low) quartz and beta (high) quartz
    alpha (low) tridymite and beta (high) tridymite
    alpha (low) cristobalite and beta (high) cristobalite
    coesite, stishovite, and keatite

Two polymorphs of beta tridymite exist, but we are not concerned with them. Alpha cristobalite and alpha tridymite are observed in volcanic and hypabyssal rocks, whereas alpha quartz can be found in any type of igneous rock. Coesite and stishovite only occur on the earth's surface around meteorite impact craters or associated with nuclear explosions, implying relatively high densities and pressures of formation. Keatite is another high-pressure polymorph of SiO_2, but it has only been produced synthetically. All these high-pressure polymorphs may be present deep in the earth's mantle. The beta forms of quartz, tridymite, and cristobalite are not found in rocks at atmospheric conditions, the reasons for which will become apparent later.

3.1.2.1 A P-T Diagram for SiO$_2$

The *P-T* phase diagram for the unary system SiO$_2$ is shown in Figure 3.2. Only stability fields for alpha quartz, beta quartz, tridymite, cristobalite, and coesite are shown. On this figure three triple (invariant) points are present, representing equilibria between:

> liquid-cristobalite-beta quartz
> tridymite-cristobalite-beta quartz
> alpha quartz-coesite-beta quartz

Had we included all the polymorphs of SiO$_2$, several more triple points would have been shown.

It is not necessary to repeat application of the phase rule to this phase diagram. The phase rule applies here in the same manner it did for the unary system H$_2$O and Figure 3.1. Table 3.1 also applies to Figure 3.2. The phase boundaries on Figure 3.2 are all univariant curves and the stability fields are all divariant areas. As a result, each triple point is surrounded by three univariant phase boundaries, so eight univariant curves are present. Eight univariant curves rather than nine exist because the curve representing cristobalite-beta quartz divariant equilibria is shared by two invariant points.

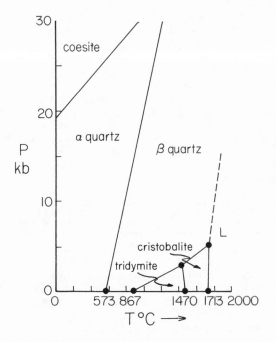

Figure 3.2 Unary phase diagram for SiO$_2$ (modified after Boyd and England, 1960).

Figure 3.2 clearly indicates why (1) tridymite and cristobalite are only found in volcanic and hypabyssal rocks, (2) coesite is only found around meteorite impact and nuclear explosion craters, and (3) alpha quartz is found in all types of igneous rocks. The more significant parameter in this case is pressure. Tridymite and cristobalite are only stable at low pressures, whereas coesite is only stable at high pressures. Alpha quartz, in contrast, is stable over a wide range of pressures. The minimum pressure at which coesite is stable is around 20 kb, which is equivalent to a depth well below the moho (about 60 km) into the earth's mantle.

The lowest temperature at which liquid SiO_2 is stable is 1713°C (Fig. 3.2). This is an extremely high temperature for metamorphic or even igneous processes, most of which seldom get above about 1400°C, yet veins of essentially pure quartz are almost ubiquitous in many igneous and metamorphic terrains. These quartz veins definitely did not form by crystallization from a pure SiO_2 magma, but rather by precipitation of SiO_2 from a hydrothermal solution. Solutions of this type, especially those subjected to high *P-T* conditions, can carry 50 or even higher weight percent SiO_2 in solution. It is a moot point whether these solutions are H_2O-rich magmas or SiO_2-rich hydrothermal solutions. The solution may be associated with a nearby true magma or may be a metamorphic fluid. Upon injection of the solution into joints or fissures in the country rocks and cooling, SiO_2 will precipitate. Temperatures need not be excessively high; they may not be greatly different from those in the surrounding country rocks.

3.1.2.2 Displacive and Reconstructive Transformations

Figure 3.2, however, does not indicate why beta quartz is never found in igneous rocks under atmospheric conditions. Granted, it is not thermodynamically stable under atmospheric conditions, but neither are tridymite, cristobalite, or coesite. The reason cannot be explained by classical thermodynamics, which never deals with reaction rates, but rather by kinetics. A similar argument can be made for the polymorphs of tridymite and cristobalite.

A close examination of the structures of all these polymorphs clearly shows that differences between alpha and beta structures are much less than differences between minerals. In other words, structural differences between alpha and beta quartz are relatively small compared with differences between quartz and tridymite, or quartz and cristobalite, etc. The chemical reaction that occurs at a phase boundary between two alpha-beta polymorphs of the same mineral is called a *displacive transformation,* whereas a reaction across a boundary between two minerals is referred to as a *reconstructive transformation.*

These terms are appropriately descriptive. During a beta-alpha transition with decreasing temperature, chemical bonds are only displaced or somewhat bent. In contrast, as temperature is lowered during a tridymite-beta quartz transition (as an example), a major reconstruction of the lattice must take place. It is reasonable that a displacive transformation is easily accomplished and takes place in a time period that is effectively geologically instantaneous. Hence, we never observe beta polymorphs in rocks under atmospheric conditions; they always convert to the alpha form on cooling across the phase boundary. The reconstructive transformation is much more sluggish, so alpha tridymite, alpha cristobalite, and coesite can persist metastably for millions of years, despite the fact that they are thermodynamically unstable. Diamond, rather than graphite, persists metastably for millions of years for much the same reason.

3.1.2.3 Application of the Clapeyron Equation

We will now revisit the Clapeyron equation. Every phase boundary on Figure 3.2 except the tridymite-cristobalite boundary has a positive slope. Again assuming that every univariant curve is crossed at constant pressure and increasing temperature, ΔS must always be positive. For all univariant curves where $dP/dT > 0$, ΔV must also be positive and $\Delta \rho$ must be negative. Because all these phases have the same chemical composition, SiO_2, molar volume is inversely proportional to density. For the tridymite-cristobalite transition, because ΔS must be positive, $\Delta V < 0$ and $\Delta \rho > 0$. Within any single stability field on Figure 3.2, ρ tends to increase with increasing pressure, but decrease with increasing temperature. These changes in ρ, and thus inverse changes in molar volumes, are so small that for our purposes they will be ignored. Let us examine the molar volumes (V^o) of the phases on Figure 3.2 to determine if they are in agreement with the predicted ΔVs across each univariant curve:

alpha quartz	22.69 cc/mol
beta quartz	22.97
tridymite	26.53
cristobalite	25.74
liquid SiO_2	27 - 28
coesite	20.64

In a qualitative sense, all these molar volumes are in agreement with the slopes of univariant curves on Figure 3.2.

Now let us turn to a more quantitative application of the Clapeyron equation to the unary system SiO_2. An informative exercise is to calculate the slope of the alpha-beta quartz phase boundary using the Clapeyron equation and compare it with the experimentally determined slope from Figure 3.2. It is informative because it illustrates the problems, and frustrations, with deciding which database to use. Initially we will use the data from Robie and others (1979). In this case the data are for 844K, temperature of the alpha - beta quartz transition at 1 bar pressure. As before, equations are written as if the phase boundary is being crossed from low to high temperature:

$$\Delta V^o = V^o(\text{beta}) - V^o(\text{alpha})$$
$$\Delta V^o = 22.97 - 22.69 = 0.28 \text{ cc/mol} = 0.028 \text{ j/bar}$$

$$\Delta S^o = S^o(\text{beta}) - S^o(\text{alpha})$$
$$\Delta S^o = 104.31 - 103.76 = 0.55 \text{ j/°mol}$$

$$\Delta H^o = H^o(\text{beta}) - H^o(\text{alpha})$$
$$\Delta H^o = -906684 - (-907160) = 476 \text{ j/mol}$$

$$dP/dT = \Delta S^o/\Delta V^o = 0.55 / 0.028 = 19.6 \text{ bars/°C}$$
$$dP/dT = \Delta H^o/T\Delta V^o = 476 / (844 \times 0.028) = 20.1 \text{ bars/°C}$$

Agreement between these two alternative expressions of the Clapeyron equation is good, yielding an average slope of about 20 bars/°C. Unfortunately, the experimentally determined slope is closer to 45 bars/°C (Fig. 3.2).

The data compilation by Robie and others (1979) is the second edition of an earlier compilation by Robie and Waldbaum (1968). One might assume that a later version may be more consistent with experimental phase equilibria, but for the alpha - beta quartz transition this is not the case. The earlier compilation, whose units are in calories rather than joules, provides the following results:

$$\Delta V^o = 0.28 \text{ cc/mol}$$
$$\Delta S^o = 0.34 \text{ cal/}^\circ\text{mol}$$
$$\Delta H^o = 290 \text{ cal/mol}$$

$$dP/dT = (0.34 \times 41.8) / 0.28 = 50.8 \text{ bars/}^\circ\text{C}$$
$$dP/dT = (290 \times 41.8) / (844 \times 0.28) = 51.3 \text{ bars/}^\circ\text{C}$$

When using thermochemical data in calories, dimensional analysis indicates that the constant 41.8 bar·cc/cal is necessary for the units on dP/dT to be bars/$^\circ$C. Robie and Waldbaum's pressure units are actually in atmospheres rather than bars, but these differences are negligible (1 atm = 1.013 bars). This slope of about 50 bars/$^\circ$C is much closer to the experimental slope of about 45 bars/$^\circ$C than the previously calculated slope of 20 bars/$^\circ$C. Herein lies an example of the problem with quantitative determination of phase boundaries (see Section 2.2.1.3 for further discussion).

3.2 BINARY SYSTEMS WITH IMMISCIBLE SOLIDS

3.2.1 The Experimental Method

Before turning to the interpretation of binary phase diagrams, a word about the experimental method is useful. Ernst (1976 Chapter 2) provides an informative summary of experimental techniques. Many of these experiments were pioneered in the early 1900s by N. L. Bowen and his colleagues at the Carnegie Institute Geophysical Laboratory in Washington, D.C.

The simplest experiments, at 1 atmosphere pressure and variable temper-ature, are done in a *quenching furnace.* Experiments in which fluid pressure is also a variable require more sophisticated equipment, a hydrothermal *reaction vessel* or "bomb." If confining pressure is a variable for "dry" experiments (to simulate mantle conditions, for example), a *piston cylinder* device is required. For some experiments, control of fugacity of the fluid phase (such as fugacity of H_2O, CO_2, or O_2) is necessary and can be done in a variety of ways (Ernst, 1976). Experiments can be conducted on synthetic materials, the results of which are the main topics of Chapters 3 and 4, or on natural rocks (for further discussion see Section 5.1).

The powdered sample (or *charge),* synthetic or natural, is placed in a chemically nonreactive container, such as a gold capsule or platinum foil, and inserted in the furnace, bomb, or whatever. For work on synthetic systems, reagent-grade chemicals may be used as starting materials. In a bomb, if P_F is to be maintained at P_T, the capsule must be sealed tightly enough so that the capsule is a closed system relative to the fluid. In some cases one capsule inside another is used. If, for example, a constant O_2 fugacity must be maintained at a certain temperature, the inner capsule, containing the charge, is placed in an outer capsule containing a *buffer* assemblage of minerals. Oxygen can diffuse through the walls of the inner capsule and reach chemical equilibrium with both the sample and the buffer assemblage, maintaining a constant O_2 fugacity for the temperature of the experiment (see Section 5.1.1.2). A typical buffer assemblage is magnetite (Fe_3O_4) and hematite (Fe_2O_3).

Temperature (and pressure, where appropriate) is maintained at desired levels until the sample hopefully reaches chemical equilibrium. This may require days, weeks, or even months, because many reactions are extremely sluggish. After all, in nature these reactions are not so concerned with time; they normally have geologic time in which to take place. It can be difficult to know before (and sometimes even after) the charge is examined whether chemical equilibrium has been reached. Trial and error are required and experience is invaluable. Two common experimental procedures are: (1) initially totally melting a number of charges of the same composition, then conducting crystallization experiments by holding each charge at a specific temperature until equilibrium is reached; or (2) conducting melting experiments by heating a number of charges of the same composition to different temperatures, then maintaining those temperatures until equilibrium is reached.

Typically, after removal from the *P-T* device, the charge is immediately *quenched*, i.e., is cooled extremely rapidly so that the phases have no chance for further reaction. A method of quenching is to drop the charge, still in its capsule or foil, into a dish of liquid. Any melt (magma) in the charge will be "frozen" as glass. Because of this, a glass is frequently referred to as a "supercooled liquid." Any fluid present either escapes when the capsule or foil is opened or exists as *fluid inclusions* (actually, gas or liquid inclusions, as they will be below their critical points). Hopefully, the minerals will be preserved exactly as they were before quenching.

The phases are then examined by any combination of a variety of techniques (e.g, polarizing microscope, X-ray diffraction, electron probe, electron microscope, fluid inclusion stage). The ultimate aim is to determine the mineralogy (i.e., which mineral phases are present in the charge), chemical composition, relative proportions, and morphology of the various phases that are stable at the *P-T* conditions of the experiment. These data, especially data for mineralogy (and presence or absence of glass) and compositions of phases are used to construct the phase diagrams.

3.2.2 Basic Principles

Binary systems (C = 2) are represented graphically by means of binary phase diagrams and their attendant equilibrium diagrams. Using the *lever rule* is absolutely essential to interpretation of the former and the *contact* principle *is* equally important in dealing with the latter.

3.2.2.1 The Lever Rule

The single most important concept that must be fully understood before one can interpret any phase diagram (excluding *P-T* diagrams) is the lever rule, also known as the *balance principle*. It is equally useful in the quantitative treatment of variation diagrams (see Section 5.2.2.3). A student who has a thorough grasp of the lever rule has little difficulty readily understanding most phase diagrams. Unfortunately, one aspect of the lever rule intuitively seems backward on initial exposure; only with experience does it seem correct. This and most other aspects of the lever rule will be dealt with in this section.

The lever rule does not resemble the phase rule or many of the laws we have considered in that it cannot be written as a single, specific formula. Although we tend to think of it in graphical terms, it can be related to a simple algebraic expression. It is the graphical expression of a "weighted average" formula learned in basic algebra. As usual, it is best illustrated by means of an example. Assume a closed magma chamber in which 30 weight percent of the original magma has crystallized as olivine. Weight percent SiO_2 of the olivine is 38 percent, whereas SiO_2 content of the remaining 70 percent residual liquid is 55 percent. What is the SiO_2 content of the original basaltic magma? This is a classic weighted average problem:

$$SiO_2 \text{ (original magma)} = (0.3 \times 38) + (0.7 \times 55) = 49.9 \%$$

Proportions of phases (in this case, 0.3 and 0.7) must be written as fractions that add to 1.0 rather than percentages that add to 100. On the other hand, compositions of phases (38 and 55 percent) can be written as percentages (or any other units, if they are internally consistent). Upon reflection this algebraic solution makes a great deal of sense. If a mineral crystallizes from a magma and if the SiO_2 content of that mineral is lower than that of the magma, then the residual magma must be relatively enriched in SiO_2 compared to the original magma.

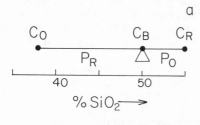

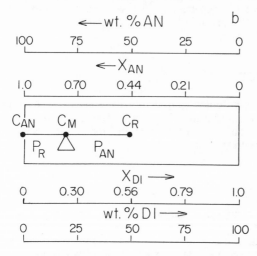

Figure 3.3 Graphical representations of the lever rule: *a.* percent SiO_2 as an example; *b.* the binary system AN-DI as an example.

Graphically this weighted average problem is represented by Figure 3.3a. Points C_O, C_B, and C_R represent the SiO_2 content of olivine, original basalt magma, and residual magma, respectively. C_O and C_R are on the ends of the "lever," and C_B is at its "fulcrum." P_O and P_R, respectively, represent the proportions (or percentages) of olivine and residual magma, where:

$$P_R = [(C_B - C_O) / (C_R - C_O)] \times 100 = 70 \%$$
$$P_O = [(C_R - C_B) / (C_R - C_O)] \times 100 = 30 \% = 100 - P_R$$

Chemical composition of a phase is determined from one side of the fulcrum, and percentage of that same phase in the system is determined from the other side (Fig. 3.3). For instance, we obtain C_R (55 percent) from the right side of the fulcrum and P_R (70 percent) from the left side. The most common mistake made by students while interpreting phase diagrams in introductory petrology classes is reading both composition and percentage of a phase from the same side of the lever. Initially, this might seem intuitively correct, but a careful comparison of the previous weighted average example with Figure 3.3a clearly indicates that this is not so. Original composition of the magma (in this example, C_B) is always at the fulcrum. *Always be certain that chemical composition of a phase is obtained from one side of the fulcrum and percentage (proportion) of that same phase is obtained from the other side.*

This example illustrates the general concept of the lever rule, but now we must turn specifically to binary systems. For all two-dimensional binary phase diagrams composition is plotted on the X-axis of the graph. Occasionally units will be X_i and X_j ($= 1 - X_i$), but more commonly they will be in weight percent. As an example we will consider the binary system AN - DI (anorthite - diopside; $CaAl_2Si_2O_8$ - $CaMgSi_2O_6$; Fig. 3.3b). For the moment we are not concerned with the vertical axis. For a binary phase diagram the vertical axis must be an intensive parameter, generally T.

At every composition $X_{AN} = 1 - X_{DI}$ and $\%_{AN} = 100 - \%_{DI}$. For this reason it is not necessary to specify composition of both components. A composition DI65 implies AN35 - DI65. By convention the composition of a phase is expressed in terms of the component on the right side of the diagram, so we would quote DI65 rather than AN35. Because the gfws of AN and DI are 278 and 216, respectively:

$$X_{AN} < (\text{weight \% AN}) / 100$$
$$X_{DI} > (\text{weight \% DI}) / 100$$

As an example, when weight percent DI and AN are both equal to 50, $X_{AN} = 0.44$ and $X_{DI} = 0.56$.

Again the simple weighted average calculation applies. Suppose 60 weight percent of an original magma of DI20 had crystallized as pure An (note the use of An, indicating the phase An, rather than the component AN; see Section 1.2.5.3). What is the composition of the residual magma in weight percent? This problem is only slightly different from the earlier weighted average problem. Expressing all compositions in terms of DI:

$$C_M = (P_{An} \times C_{An}) + (P_R \times C_R)$$
$$20 = (0.6 \times 0) + (0.4 \times C_R)$$
$$C_R = 20 / 0.4 = DI50$$

C_M, C_{An}, and C_R refer to compositions of original magma, anorthite, and residual magma, respectively. P_{An} and P_R, respectively, refer to proportions of anorthite and residual magma. By crystallizing the mineral An, which contains no DI, all succeeding residual magmas must been enriched in DI. Some petrologists, paralleling terminology used for radioactive decay, refer to the original magma as the *parent* magma, and the residual magma is called the *daughter*.

The graphical solution to this problem is shown in Figure 3.3b. Again the original magma is at the fulcrum of the lever and compositions of phases are determined on one side of the fulcrum, whereas proportions are read from the other. Expressing proportions as percentages rather than fractions and compositions in terms of DI:

$$P_{An} = [(C_R - C_M) / (C_R - C_{An})] \times 100$$
$$P_{An} = [(50 - 20) / (50 - 0)] \times 100 = 60 \%$$

$$P_R = 100 - P_{An} = 40 \%$$

Thus P_{An} is 60 percent of the length of the lever and P_R is the remaining 40 percent. The graphical and algebraic solutions are consistent.

3.2.2.2 The Contact Principle

The *contact principle* is very helpful in writing equations for all possible chemical reactions in any system, binary or otherwise. Consider the binary system $MgO-SiO_2$ with the following four crystalline phases:

periclase	Pe	MgO
forsterite	Fo	Mg_2SiO_4 or $2MgO\cdot SiO_2$
enstatite	En	$MgSiO_3$ or $MgO\cdot SiO_2$
cristobalite	Cr	SiO_2

These compositions can be represented as follows:

```
   Pe            Fo    En                        Cr
   |_____|_____|_____|

       MgO                                        SiO2
```

This type of diagram is frequently called an *equilibrium diagram* or *stability diagram*. Units are normally mol percentages. The composition of En falls halfway between Pe and Cr, because the composition of En is 50 mol percent SiO_2 and 50 mol percent MgO. The composition of Fo is 67 mol percent MgO and 33 percent SiO_2, so it must fall 2/3 of the way from Cr to Pe.

In compositional space a binary equilibrium diagram is represented by a one-dimensional line; a ternary system by a triangle (within a two-dimensional plane); a quaternary system by a tetrahedron (within a three-dimensional volume). If C = number of components in a system and D = number of dimensions required to represent those components, then D = C - 1. On an equilibrium diagram the composition of a phase is at a certain point. Compositions of two phases acting as products, or two as reactants, can be joined by a line; three products or reactants, by a triangle or three points in a line. Thus four phases require a polyhedron with four apices, or four points in a plane or line. For a chemical reaction to occur, the geometric figure (point, line, triangle, polyhedron) representing the products must be in contact with the geometric figure for the reactants on the equilibrium diagram. This is the contact principle.

Applying this principle to the MgO - SiO_2 system, the following balanced reactions are possible:

1.	Fo = Pe + En		5.	2En = Fo + Cr
2.	Fo = 2Pe + Cr		6.	3En = Pe + Fo + 2Cr
3.	2Fo = 3Pe + En + Cr		7.	Pe + En = Fo + Cr
4.	En = Pe + Cr		8.	3Pe + 2Cr = Fo + En

For the first six reactions, the point representing composition of the phase on the left side of the equation falls inside (in contact with) a line connecting compositions of phases on the right side. This line connecting two phases on a stability diagram is defined as a *join*. The last two reactions are different, because joins are required to connect compositions of phases on both sides of the equation. These joins must be in contact along some part of their lengths. For these last two reactions, the two joins are in contact along the join Fo-En.

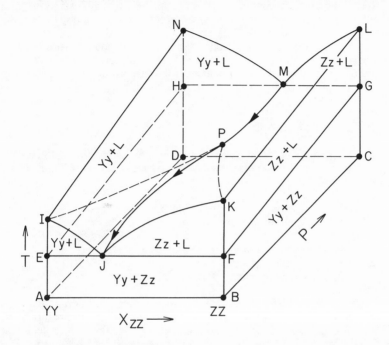

Figure 3.4 A *P-T-X* diagram for a hypothetical binary system YY-ZZ. See text for definitions of all symbols.

The following reactions, however, are not possible:

$$En = Pe + Fo$$
$$Fo = En + Cr$$
$$Pe + Fo = En + Cr$$

These latter three equations are not balanced because they cannot be balanced. The contact principle, and common sense, also dictate that any reaction with Pe or Cr on one side of an equation and any combination of other phases on the other side is impossible. These reactions are not possible because joins or points representing compositions of products and reactants are not in contact.

3.2.3 A P-T-X Diagram

Some combination of four intensive parameters is generally used to plot a standard binary phase diagram: P, T, X_i (or $\%_i$), and X_j (or $\%_j$). In fact, because $X_i = 1 - X_j$ (or $\%_i = 100 - \%_j$), only three parameters are required: P, T, and any measure of composition of either component. Three parameters graphically require three dimensions, so a two-dimensional projection of a simple three-dimensional *P-T-X* binary phase diagram for the binary system YY-ZZ is shown in Figure 3.4. Temperature is plotted on the vertical axis, composition on the horizontal axis in the plane of the page, and pressure on the horizontal axis perpendicular to the page.

The overall three-dimensional figure strongly resembles a plaster model of a stream valley as commonly introduced in introductory geology laboratories. It differs in that (1) the vertical axis of Figure 3.4 is temperature rather than elevation, and (2) the horizontal axes are composition and pressure rather than some linear measure of distance. Surface EFGH is somewhat similar to that of a subsurface water table, in that it intersects the bottom of the "stream valley," curve JM, which is sloping from back to front of the figure.

3.2.3.1 Basic Definitions

A number of points, curves, and surfaces on Figure 3.4 must be defined. Let us assume that pressure for the front face of the figure (AIJKB) is 1 bar. First, let us define several points:

I	melting point of pure phase Yy at 1 bar P
K	melting point of pure phase Zz at 1 bar P
L	melting point of pure phase Zz at P bars P
N	melting point of pure phase Yy at P bars P
J	eutectic point at 1 bar P
M	eutectic point at P bars P

The *melting point* is the temperature above which a pure compound will be completely molten. It is identical to the *freezing point,* the temperature below which a pure compound will be compleletely crystalline. A eutectic point represents a temperature, composition, and pressure at which all three phases, Yy, Zz, and melt, coexist at equilibrium. The points I-J-K-L-M-N define a surface called the *liquidus,* which is a surface above which, at any composition, the system will be entirely liquid. The curve JM, which can be referred to as the *eutectic curve,* is the locus of all eutectic points and defines the bottom of the *thermal valley.* The eutectic curve represents the lowest temperature, for a particular pressure, at which the system can be entirely liquid.

The surface EFGH is defined as the *solidus* and intersects the liquidus surface at curve JM. The solidus is a surface below which the system is completely crystalline. It follows that a *crystal mush,* a mixture of crystals and melt, is stable between the solidus and liquidus. Intersection of the solidus surface with any vertical isobaric plane must be a straight, horizontal (i.e., isothermal) line. In Figure 3.4 the solidus surface slopes from back to front, but not from side to side. For reference, plane ABCD is at constant temperature.

Below the solidus (surface EFGH) two crystalline phases, Yy and Zz, are stable. Crystalline phase Yy consists of pure component YY, whereas Zz is composed of pure ZZ. For this reason, the two solids are completely immiscible in one another; i.e., they exhibit no solid solution. Above the liquidus, however, the two components YY and ZZ are completely miscible in the melt. Two volumes exist between the liquidus and solidus surfaces -- the left-hand volume IEJNHM and the right-hand volume JFKMGL. The crystalline phase Yy and melt (a crystal mush) coexist in the left volume, and Zz and melt coexist in the right. Along the eutectic curve JM all three phases, Yy, Zz, and melt coexist at equilibrium.

3.2.3.2 Two-dimensional Sections

Working with a two-dimensional projection of a three-dimensional figure, such as Figure 3.4, in any quantitative way is not impossible but awkward. Consequently, the usual practice is to examine one of three two-dimensional sections through P-T-X diagrams:

isobaric section	T-X diagram
isothermal section	P-X diagram
isocompositional section	P-T diagram

The least common of these is the P-X diagram, and we have already been exposed to P-T diagrams in other contexts. Most of the remaining binary phase diagrams in this chapter will be T-X (or T-%) diagrams.

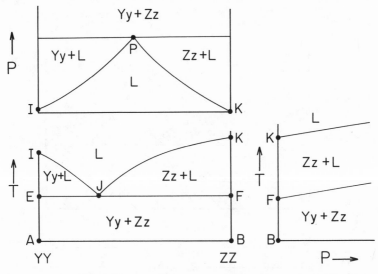

Figure 3.5 Three diagrams representing sections through the P-T-X diagram shown in Figure 3.4: P-X, T-X, and P-T. Lettered points and fields are keyed to Figure 3.4.

Figure 3.5 shows the three sections through Figure 3.4. The T-X and P-T plots are easily visualized. To return to our analogy of a stream valley, the T-X plot appears to be a cross section of a stream valley prepared for an introductory geology laboratory. The T-X section was made at $P = 1$ bar. The P-T section is taken somewhere between the eutectic and ZZ100, say at ZZ80. At ZZ100 the liquidus and solidus coincide because Zz completely melts at point K. The P-T axes are reversed from the normal arrangement so that the P-T diagram in Figure 3.5 is a true cross section of the right side of Figure 3.4. On the P-T section the liquidus and solidus appear as approximately straight lines because we are dealing only with a liquid melt and solids. As both the liquidus and solidus have positive slopes on the P-T section, according to the Clapeyron equation, ΔS and ΔV must be positive, and $\Delta \rho$ negative, with increasing T.

The P-X diagram is less easily visualized. Again think of the analogy of a stream valley in which the contour lines of constant elevation intersect the earths surface and form a "V" upstream (away from the direction of the arrows on curve JM). The intersection of an isothermal plane with the liquidus will also "V" upstream, as is shown by the dotted curve on Figure 3.4. This intersection produces the liquidus on the isothermal P-X section of Figure 3.5.

With regard to melting and crystallization, Figure 3.5 demonstrates that P and T work against one another. An increase in T causes melting, whereas an increase in P causes crystallization. The pressure referred to in Figures 3.4 and 3.5 is P_T, or confining pressure in a "dry" system (a system with no fluid phase). Had the system been saturated or supersaturated with H_2O ("excess H_2O"), $P_F = P_T$, and an increase in both T and P_F would have promoted melting. Furthermore, if $P_F = P_T$ the liquidus and solidus on the P-T section (Fig. 3.5) would be curves rather than straight lines (see Section 2.2.1.5). An increase in P_F in a "wet" system promotes melting, in contrast with an increase in P_T in a "dry" system, which inhibits melting and promotes crystallization.

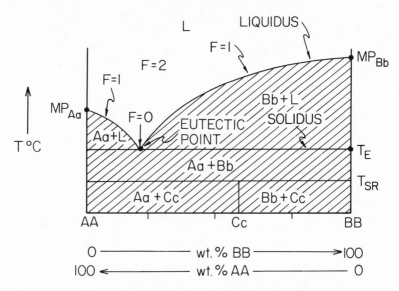

Figure 3.6 A typical T-% eutectic binary phase diagram for hypothetical components AA and BB. Diagonally ruled area is *forbidden zone.*

3.2.4 A Simple Binary Eutectic

As stated above the most common type of binary phase diagram is a two-dimensional, isobaric, T-% diagram, similar to the T-X plot in Figure 3.5. The simplest of many binary T-X diagrams is one with only a eutectic and immiscible solids (i.e., no solid solution), again similar to the T-X diagram in Figure 3.5. A typical example is shown in Figure 3.6. It represents the hypothetical binary system AA-BB, in which mineral Aa is composed of pure AA and mineral Bb is composed of pure BB. Mineral Cc is an intermediate compound, formed by the subsolidus, solid-state reaction Aa + Bb = Cc. MP_{Aa} and MP_{Bb} are the melting (or freezing) points of Aa and Bb, respectively. T_E is the eutectic temperature and T_{SR} is the temperature of the subsolidus reaction. The liquid (magma) is symbolized by L.

3.2.4.1 Application of the Phase Rule

The first step in dealing with Figure 3.6 is to reconcile it with the phase rule. This T-% diagram is an isobaric section, therefore, pressure is fixed and is no longer available as a variable. Because we have implicitly subtracted 1 from the left side of the equation $P + F = C + 2$ by reducing F by 1, we must also subtract 1 from the right side. This results in an expression of the phase rule for any phase diagram in which one variable is fixed, $P + F = C + 1$. For a binary system $C = 2$, so $P + F = 3$.

    Table 3.3 summarizes the application of the phase rule to Figure 3.6. Because pressure is fixed, three variables are involved: T, % AA, and % BB. Recall, however, that % AA = 100 - % BB. In the liquid field $P = 1$, thus both T and % BB (or % AA) are required to define the state of the system, $F = 2$, and the area above the liquidus is divariant. Along the liquidus $P = 2$ (Bb + L or Aa + L), so either T or % BB is required to define the state of the system, $F = 1$, and the liquidus is a univariant curve. As both T and % BB are fixed at the eutectic point, $F = 0$, and the point is invariant. Three phases, Aa + Bb + L, coexist at the eutectic point.

TABLE 3.3 Application of Phase Rule to a Binary System

C	P	F	Equilibria	Variables	Geometry
2	1	2	divariant	T and X	area
2	2	1	univariant	T or X	curve
2	3	0	invariant	none	point

X: composition of either AA or BB

    No mention was made of the five stability fields below the liquidus (Aa + L, Bb + L, Aa + Bb, Aa + Cc, and Bb + Cc) with regard to the phase rule. Because they are areas, they should be divariant ($F = 2$). In each field two phases coexist, so P is apparently 2, which would mean that $2 + 2 = 3$. Obviously a problem exists. In fact, the composition of neither phase actually falls in a stability field, but rather on its right or left margin. For example, within the field Bb + L, the composition of Bb falls on the right margin of the field (BB100), whereas the composition of the liquid L is on the left margin, the liquidus. For this reason, all these fields below the liquidus (the diagonally ruled area) are collectively called the *forbidden zone,* because the composition of any one phase is forbidden to fall within a field. As we cannot specify P, the phase rule really does not apply within the forbidden zone.

3.2.4.2 Qualitative Treatment

Figure 3.7 is another version of the binary system AA-BB. Let us assume that $P = 1$ atmosphere, so the experiments necessary to construct Figures 3.6 and 3.7 were probably made in a quenching furnace. Had a charge been quenched for any AA-BB composition anywhere above the liquidus, assuming chemical equilibrium, only glass would be observed in the charge. If a charge were quenched under conditions that exist between the liquidus and solidus, in the fields Bb + L or Aa + L, either Bb and glass or Aa and glass would be observed in the charge. Below the solidus, between T_E and T_{SR}, only the minerals Aa and Bb would be observed in the charge; their relative proportions would be exactly the same as the proportions of AA and BB in the original liquid. Any charge quenched below T_{SR} would contain either Aa + Cc or Bb + Cc, depending on whether the initial liquid were left (more AA-rich) or right (more BB-rich) of the composition Cc.

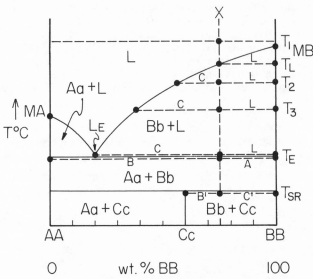

Figure 3.7 Another version of a *T*-% eutectic binary phase diagram for components AA and BB. This system is the same as that shown in Figure 3.6.

The liquidus can be considered as a locus of freezing points for liquids of intermediate composition between pure AA and pure BB. The liquidus is also an example of *freezing point depression,* a common phenomenon in chemistry. The highest freezing points, and thus melting points, on the diagram are for pure AA and pure BB. The lowest freezing point, of course, is at the eutectic point. As component AA is added to a pure BB liquid, or BB is added to a pure AA liquid, the freezing point of the solution is lowered, thus the term *freezing point depression.* This phenomenon occurs because addition of a new component, such as addition of BB to pure liquid AA, breaks electrostatic bonds in the liquid and thereby inhibits freezing. The addition of any "impurity" to a pure molten compound normally lowers the freezing point of the pure compound. This process has many practical uses in everyday life, such as adding antifreeze to radiators of our automobiles to lower the freezing point of engine coolants.

Location of the eutectic point, at AA80-BB20, tells us something about the solubility of Aa and Bb in the melt. Upon cooling any magma of composition more BB-rich than BB20 initially precipitates (crystallizes) Bb, and Aa does not precipitate until T_E. In contrast, only magmas of compositions more AA-rich than AA80 initially crystallize Aa. In comparison to solubility of Aa, solubility of Bb must be quite low, because Bb can be the first mineral to precipitate from a melt relatively impoverished in BB (>BB20). For AA to crystallize first, the melt must be fairly enriched in AA (>AA80).

3.2.4.3 Quantitative Treatment

Now that we have dispensed with some preliminaries, we can examine Figure 3.7 quantitatively. Assume a melt of original composition X (BB75); this represents the fulcrum of the lever for all subsequent determinations at this composition. Also assume that chemical equilibrium is maintained at every temperature.

If the system is quenched at temperature T_1, the charge will be entirely glass of composition BB75. Upon cooling, crystallization of Bb commences at T_L, the liquidus temperature for liquid composition BB75. If the system were quenched at T_L, a few tiny, incipient crystals should be observed in a matrix of almost 100 percent glass. If the system is quenched at T_2, a larger percentage of crystals and less glass will be observed.

At this point we can define a *tie line*, which is an imaginary line connecting the compositions of two phases in equilibrium with one another on a phase diagram. A similar line on a stability diagram is called a *join*. On all binary T-% (or P-%) diagrams this is a horizontal line connecting the composition of either (1) the melt (glass) and the crystalline phase (mineral) in equilibrium with that melt, or (2) two minerals in equilibrium with one another. Case (1) applies when the state of the system is between its liquidus and solidus, whereas case (2) only occurs below the solidus. Consequently, tie lines can only be drawn between compositions of coexisting phases when the system is within the forbidden zone. An alternate definition of a forbidden zone is any area within a binary phase diagram in which tie lines between two coexisting phases can be drawn. A tie line exactly coincides with the corresponding lever, as discussed later. At temperature T_2, the tie line, thus the lever, is the dashed line that extends from the liquidus (BB56) to BB100. Thus at T_2 the mineral Bb, of composition pure BB, is in equilibrium with a melt of composition BB56.

We can also determine the relative percentages of crystals and glass in the charge that quenched at T_2 by using the lever rule. Remember, the fulcrum of the lever is at BB75. If L represents the proportion of liquid (or glass) and C is the proportion of crystals, then at T_2:

$$\% \text{ glass} = (L \times 100) / (L + C) = 57$$

$$\% \text{ crystals of Bb} = (C \times 100) / (L + C) = 100 - \% \text{ glass} = 43$$

Hence the charge contains 43 percent crystals of Bb (composition BB100) and 57 percent glass of composition BB56. If a charge were quenched at T_3, then it would contain 60 percent crystals of Bb and 40 percent glass of composition BB38. Between T_2 and T_3 an additional 17 percent crystals of Bb precipitated and the melt correspondingly lost additional BB component. A charge quenched just above T_E would contain 69 percent crystals of Bb and 31 percent glass of composition only slightly more BB-rich than the eutectic composition, BB20.

Figure 3.7 indicates that a rather profound change occurs when the melt cools through T_E. Above T_E, the system contained 69 percent crystals of Bb and melt; below T_E, it contains 75 percent Bb and 25 percent Aa. Percentages of minerals Aa and Bb in the "rock" below T_E must be exactly the same as percentages of AA and BB in the initial melt, because the composition of Aa is AA, composition of Bb is BB, and no melt remains below T_E. The lever rule must still apply, however, so:

$$Bb = (B \times 100) / (A + B) = 75 \%$$

$$Aa = (A \times 100) / (A + B)$$
$$Aa = 100 - \% \, Bb = 25 \%$$

Because a charge quenched below T_E contains more of both Aa and Bb than a charge quenched above T_E, simultaneous precipitation of Aa and Bb must have occurred at T_E. This type of chemical reaction is called *eutectic crystallization*. Upon continued cooling at T_{SR} a subsolidus reaction occurs between Aa and Bb, in which all Aa is used up forming Cc. A charge quenched below T_{SR} contains 62 percent Cc and 38 percent Bb. These percentages are obtained from the lever rule by:

$$Cc = (C' \times 100) / (B' + C') = 62 \%$$
$$Bb = 100 - \% \, Cc = 38 \%$$

We can can also easily quantify the chemical reaction (i.e., eutectic crystallization) that takes place at T_E. Defining L_E as the composition of the eutectic liquid, then:

$$0.69Bb + 0.31L_E = 0.75Bb + 0.25Aa$$

The left side of this equation represents the state of the system just above T_E; the right side, just below T_E. Solving the equation for L_E:

$$L_E = (0.06 / 0.31)Bb + (0.25 / 0.31)Aa = 0.19Bb + 0.81Aa$$

which, within rounding error and error in reading Figure 3.7, is the eutectic composition. During eutectic crystallization the melt of composition L_E must crystallize.

A common mistake when first exposed to these concepts is to think that the first 69 percent of crystallization proceeds relatively slowly from T_L to T_E, then the remaining 31 percent precipitates in a flood of crystals. This assumes that cooling rates are approximately uniform, which is not so (Fig. 3.8). Because crystallization is an exothermic process, the cooling rate will slow down below the liquidus and the slope (dT/dt, where T is temperature and t is time) on the cooling curve (Fig. 3.8) will decrease. At T_E composition of the melt is at an invariant point, so temperature cannot drop below T_E until the eutectic melt has completely crystallized. This results in a horizontal cooling curve and thus no change in temperature at T_E as long as melt remains in the system. Once crystallization has been completed, the system cools below T_E and the "rock" continues to cool at a rate similar to the melt above T_L. The net effect of these different cooling rates is no sudden flood of crystals precipitating at T_E.

Three unique compositions exist on Figure 3.7 that are worthy of special mention. These are AA100, the eutectic composition L_E, and BB100. If a melt with any one of these three compositions cools through the liquidus, then the melt "instantaneously" crystallizes; i.e., the melt does not crystallize over a temperature interval between the liquidus and the solidus. This is apparent at the eutectic, because a melt with exactly the eutectic composition will be completely crystalline below the eutectic temperature. At the eutectic the liquidus and solidus intersect. It is less apparent at AA100 and BB100 because the solidus apparently extends all the way to AA100 and BB100. In fact, the solidus does not extend to compositions of the pure components; it applies only to the binary system AA - BB and not the two unary systems AA and BB. Stating this differently, a unary system has no separate liquidus and solidus. Points MA and MB are melting points for pure Aa and Bb, respectively, so a melt of composition AA completely crystallizes to Aa at point MA and the apparent solidus at 100AA has no effect, likewise for a melt of composition BB.

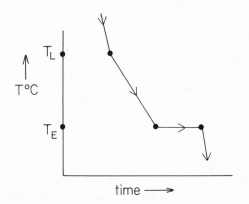

Figure 3.8 Change in temperature with time for melt *X* on Figure 3.7.

3.2.4.4 A Closed Magma Chamber

We have considered Figure 3.7 only in the context of how a particular experimental charge might appear upon quenching. A crude analogy can be made with a closed magma chamber progressively crystallizing and periodically undergoing a volcanic eruption. Such a simple process probably never takes place, and certainly a closed chamber containing only a two-component magma never existed, but this analogy can be a useful learning exercise.

Had an eruption occurred at T_1, or at any temperature above T_L, we might observe a hand specimen of obsidian (volcanic glass; composition BB75), or at least a fine-grained, *aphyric* (containing no phenocrysts) rock. Assuming the rock is not obsidian, its mineralogy could be Aa and Bb, if the chamber were shallow enough so that *P-T* conditions of the eruption and chamber are sufficiently similar. Under those conditions the same subsolidus mineral assemblage would be stable after the volcanic eruption as in the chamber. If the chamber were very deep and *P-T* conditions in the chamber and the volcano different, the mineral assemblage might be quite different from Aa-Bb. Another possibility is that the rock may have quenched as obsidian and later "devitrified," or at least partially devitrified to a new mineral assemblage. Still another possibility is that Aa and Bb initially crystallized, but upon cooling reacted to form another assemblage, such as Bb + Cc (Fig. 3.7).

Let us now make another assumption -- viscosity of the magma is high enough and density of Bb crystals and melt are similar enough so that most of the crystals stay in the chamber as they grow; i.e., they are not effectively removed from the chamber by some process such as gravity settling. If the volcanic eruption waited until the magma had cooled to T_2 to erupt, 43 percent of the rock most likely would be phenocrysts of Bb. The term *phenocryst* is purely a textural term; it refers to a crystal that is significantly larger than the surrounding crystals. The 43 percent Bb crystals could also be termed *near-liquidus phases* (also *intratelluric crystals* or *primocrysts*). These latter terms are all synonymous and have a genetic connotation; they imply crystals that grew quite early from the melt upon cooling below its liquidus. Primocrysts are typically phenocrystic, although this is not an absolute requisite. Because they grow in a liquid medium, primocrysts are commonly, but certainly not invariably, *euhedral* or *subhedral* (i.e., they have very well-formed to fairly well-formed crystal faces, as opposed to *anhedral,* with poorly developed faces).

Depending on the cooling rate after eruption, the remaining 57 percent of the rock (composition BB56) would either be a glassy or relatively fine-grained *groundmass* (or *matrix*). The groundmass mineralogy could be Aa-Bb, or any number of other assemblages, depending on the rock's subsolidus history, as described above for an aphyric rock.   In natural rocks, which are multi-component systems, the groundmass commonly has the same minerals as does the phenocryst assemblage (although mineral compositions are different), plus one or more additional minerals.   Many exceptions exist, so a phenocryst mineral may not be present in the groundmass at all.

3.2.4.5 Partial Melting

Thus far we have only been concerned with crystallization during cooling of the system. What happens during melting while the system is being progressively heated? In a simple eutectic system, such as Figure 3.7, the process of melting is essentially the reverse of crystallization. Conceptually, the process is important when we consider the origin of magmas by partial melting of solid rocks in the upper mantle or lower crust. For a variety of reasons, most petrologists believe that no more than a total of about 30-40 percent of the source rock melts during a series of partial melting events. Thus we invariably refer to partial melting, rather than total melting, as the main process by which magmas are formed.

Let us apply this constraint of 30-40 percent partial melting to a rock that contains 75 percent Bb and 25 percent Aa (Fig. 3.7). The rock is heated to T_E, at which point it begins to melt. The melt is the eutectic melt, of composition L_E (BB20). The system must stay at T_E until all Aa has melted, because L_E is at an invariant point. The first 31 percent of melting will be a homogeneous eutectic melt (L_E) of composition BB20, so an excellent chance exists that no magmas will ever be formed from this source rock except euctectic melts. If the 31 percent were extracted from the source area in a number of small increments, as is likely, each increment would have exactly the same composition, BB20. With regard to their major-element chemistry, original, pristine melts (frequently termed *primary magmas* -- magmas that have not undergone any subsequent modification) can be remarkably homogeneous over large areas. This is because many of them approximate eutectic melts and sufficient melting never took place for other melts to form.

If the source rock continued to melt beyond its eutectic temperature (up to, say, a total of 40 percent), the melt's composition would continually change by becoming successively more BB-rich, because Bb is the only mineral available to melt above T_E. For 40 percent partial melting, the temperature would be at T_3 and the final melt composition would be BB38, or 18 percent richer in BB than the eutectic melt. For the first 31 percent of melting, the melt composition would stay constant at BB20.   For the final 9 percent, its composition would change from BB20 to BB38. Small percentages of partial melting obviously favor eutectic melts. Thus a binary eutectic diagram is equally useful in explaining formation of magmas by partial melting as it is in explaining their crystallization once they formed.

3.2.4.6 A T-X Diagram Constructed from Thermochemical Data

An important consideration regarding simple binary eutectic diagrams has to do with their relationship to classical thermodynamics. If the melt is assumed to behave as an ideal solution, either T-X or P-X diagrams are easily constructed from thermochemical data. The following equations can be readily proven from basic principles covered in Chapter 2:

$$\ln X_i = (\Delta H^o_i / R) \ [(1/T_i) - (1/T)] \tag{3.1}$$

$$\ln X_i = (\Delta V^o_i / RT) \ (P_i - P) \tag{3.2}$$

For real solutions, a_i is substituted for X_i, but an ideal solution will suffice for our purposes. Equation 3.1 is used to construct an isobaric T-X diagram and Equation 3.2 is used for an isothermal P-X diagram. Terms are:

R	ideal gas constant
X_i	mol fraction of component i
ΔH^o_i	heat of fusion of pure component i at $P = 1$ bar
ΔV^o_i	molar volume change of fusion of pure component i at $P = 1$ bar
T_i	melting (freezing) point of pure component i at pressure P
P_i	pressure at which pure i melts for temperature T

Data for $P = 1$ bar can be used because only a liquid and solid are involved, so ΔH and ΔV do not change much with increasing pressure. Bars are used as pressure units for thermochemical calculations, but atmospheres are used for quenching furnace experiments. Differences are almost negligible.

For a given diagram, all terms are constant except T and $\ln X_i$, or P and $\ln X_i$, so one simply solves for X_i for different values of T or P. This will construct the liquidus for "half" the diagram, for one component, so the process is simply repeated for the other component. The point at which the two liquidi intersect is, by definition, the eutectic point. The solidus is then constructed by drawing a horizontal (isothermal for a T-X diagram; isobaric for a P-X diagram) line through that point.

An example of the use of Equation 3.1 to construct a T-X (or T-$\%$) binary eutectic diagram is the system AN-DI ($CaAl_2Si_2O_8$-$CaMgSi_2O_6$; Fig. 3.9). This is not a true binary system because some Al from the An component can preferentially dissolve in the mineral Di to produce a mineral similar to augite in composition (Osborn, 1942, Morse, 1980). This additional Al will be in the pyroxene in the form of calcium Tschermak's molecule ($CaAlAlSiO_6$ or CaTs). If sufficient CaTs is added to diopside, the result is augite. Consequently, on an equilibrium diagram, a mineral can exist in this system that cannot fall on a line connecting the composition of pure AN and pure DI. Stating this differently, we cannot write a balanced chemical reaction with only DI and AN as reactants and only augite as a product. For a system to be truly binary, compositions of all stable phases must plot on a line connecting compositions of the two components. This means that we must be able to write a balanced equation with the two components as reactants and each stable phase as a product. As a result, we refer to the system AN-DI as *pseudobinary*. We will ignore this complication and treat the system as if it were truly binary, as this does not introduce any large errors in our calculations. For a comprehensive discussion of the system AN-DI, refer to Morse (1980).

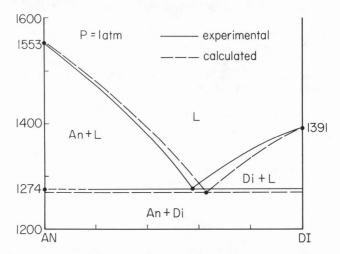

Figure 3.9 A comparison of experimentally determined and calculated binary eutectic phase diagrams for the system AN-DI. Calculations are in text. Experimental system is after Bowen (1915) and Osborn (1942).

Again, thermochemical data are from Robie and others (1979). Using Equation 3.1:

$$\ln X_{AN} = (81000/8.315) \, [(1/1830) - (1/T)]$$
$$\ln X_{DI} = (77404/8.315) \, [(1/1664) - (1/T)]$$

Rearranging both equations and solving for T:

$$T = -9741 \, / \, (\ln X_{AN} - 5.323)$$
$$T = -9309 \, / \, (\ln X_{DI} - 5.594)$$

This latter step is not absolutely necessary; one can also solve for X for different values of T. A BASIC computer program that performs these calculations is given in Section 1.6.2.1. Solving the first equation for T for different values of X yields one side of the liquidus and the second equation gives the other side. As stated, the point of intersection of the two sides is the eutectic. Figure 3.9 compares the liquidus calculated from the above two equations (for $P = 1$ bar) with the experimentally determined liquidus ($P = 1$ atm). Agreement is reasonably good. A comparison of the two eutectic points is illustrative:

	Calculated	Experimental
temperature	1270°C	1274°C
composition	DI63	DI57

Notice that the data were converted from mol fraction to weight percent. This is done by applying the principles learned in Section 1.2.3. As an example, we will convert a composition of $X_{AN} = 0.6$ and $X_{DI} = 0.4$ to weight percents. The gfws of AN and DI are 278.3 and 216.6, respectively. Calculations are:

			Weight proportion	Weight percent
AN	0.6 x 278.3	=	166.98	66
DI	0.4 x 216.6	=	86.64	34
	Total		253.64	100

A handy rule of thumb is: to convert from mol to weight units, *multiply* by the appropriate gfws (or EWs) and recalculate to 100 percent; to convert from weight to mol units, *divide* by gfws or EWs and recalculate to 100 percent. If the rule of thumb is forgotten, a check to determine if multiplication or division is the correct procedure is to compare a molar unit with its equivalent weight unit. In the example above AN has a larger gfw than DI, so its weight fraction should be larger than its mol fraction, which it is (0.66 compared with 0.60).

3.2.5 A Thermal Divide

Let us return for a moment to our comparison of a binary phase diagram with a topographic cross section. If a thermal valley is analogous to a stream valley, and topographic divides separate stream valleys, it is reasonable that thermal divides separate thermal valleys. Thermal divides are among the most useful predictive phenomena in phase equilibria because they have a direct bearing on differences between igneous rock suites (see Section 5.2.4). This section will provide one example of the relationship between thermal divides and igneous rock series.

Two major igneous rock suites are termed alkaline and subalkaline (see Section 1.3.3.1). To summarize that section, most alkaline and subalkaline rocks can be distinguished from one another based on whether they are *ne*-normative (alkaline series) or not (subalkaline series). Alkaline rocks are critically SiO_2 undersaturated, whereas subalkaline rocks are moderately SiO_2 undersaturated or oversaturated.

3.2.5.1 The System NE-Q

Our comparison binary system is the system NE-Q at 1 atmosphere P, where NE is nepheline ($NaAlSiO_4$) component and Q is SiO_2 component (Fig. 3.10). This phase diagram has been simplified from the original in two respects: (1) very minor solid solution of NE in albite occurs, which we will ignore, and (2) the system becomes quite complicated at compositions more NE-rich than about NE75, so we will only consider magmas more Q-rich than NE75. Recalling some basic principles discussed in Section 1.3.1, the presence of nepheline, a feldspathoid, in the norm and mode suggests critical SiO_2 undersaturation, whereas quartz in the norm and any SiO_2 polymorph in the mode is indicative of SiO_2 oversaturation. Any mineral assemblage to the left of composition Ab (top of the thermal divide) will contain Ne in the norm and mode, and any composition to the right will contain Q in the norm and some SiO_2 polymorph in the mode.

The thermal divide separating the two thermal valleys is quite apparent on Figure 3.10. The top of the divide is exactly at composition Ab $(NaAlSi_3O_8)$. The composition Ab separates the system NE-Q into two sub-systems, NE-AB and AB-Q, each of which is, given the simplifying assumptions above, a simple binary system with a eutectic. Another feature of the system NE-Q is the polymorphic transformation between tridymite and cristobalite. This is represented by the horizontal (isothermal) phase boundary at 1470°C in the SiO_2 + liquid field, with tridymite stable below 1470°C and cristobalite stable above. We can dispense with consideration of the phase rule or thermo-chemical calculations, because they can be applied in exactly the same manner as for AA-BB and AN-DI. If thermochemical data are used with Equation 3.1, the subsystems NE-AB and AB-Q would be constructed separately and then joined. Likewise, another demonstration of the lever rule at this point would be redundant.

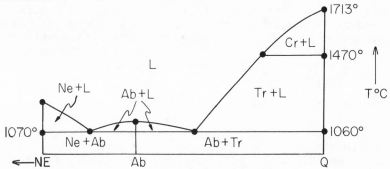

Figure 3.10 Simplified version of part of the binary system NE-Q at 1 atmosphere P (modified after Schairer and Bowen, 1956).

3.2.5.2 Practical Applications of the System NE-Q

Again let us assume that a simple binary system can tell us something about natural multicomponent rocks. Any "rocks" that formed as a result of crystal-lization on the left side of the thermal divide contain Ne and Ab, and thus most likely will be SiO_2-undersaturated and in the alkaline series. The Ab in fact is plagioclase, but in felsic rocks at least it is sodic plagioclase. In addition, some KP $(KAlSiO_4)$ component is in solid solution in Ne in natural rocks. Rocks on the right side of the divide contain Ab and Tr, are SiO_2-oversaturated, and belong to the subalkaline series. If we consider volcanic rocks that form due to magma quenching somewhere between the liquidus and the solidus, then four rock types exist on Figure 3.10, from left to right:

1. Ne-phyric, alkaline
2. Ab-phyric, alkaline
3. Ab-phyric, subalkaline
4. Tr-phyric, subalkaline

An *aphyric* volcanic rock contains no phenocrysts, whereas a *phyric* rock does. A Ne-phyric rock, then, contains Ne as a phenocrystic phase, assuming that phenocrysts are near-liquidus phases (primocrysts).

It is apparent from Figure 3.10 that any magma whose composition is just to the left of the top of the thermal divide (composition pure Ab), upon crystallization of Ab, becomes gradually more Ne-rich and changes composition down the liquidus toward the Ne-Ab eutectic. Magma compositions move away from thermal divides down liquidi toward thermal valleys just as sheet wash during a rain runs down topographic divides toward stream valleys. If the magma is not interrupted by quenching, its composition eventually arrives at the Ne-Ab eutectic and coprecipitate these two minerals. In contrast, any magma to the right of the top of the thermal divide also crystallizes Ab, but its composition move aways from Ab down the liquidus toward the Ab-Tr eutectic and becomes progressively more Q-rich. As a result, alkaline magmas commonly become more SiO_2-deficient as crystallization proceeds, whereas crystallization of the same mineral causes subalkaline magmas to become more SiO_2-enriched. This fundamental difference in these two magma types is not always observed, but it occurs frequently enough to validate this argument.

This figure also relates to the petrographic classification of igneous rocks (see Fig. 1.9; Section 1.3.1.3), the basis of which is the QAPF double triangle. To summarize, Q represents a SiO_2 mineral; A, alkali feldspar; P, plagioclase; and F, feldspathoid. The two triangles are QAP and APF, so Q and F never coexist. As nepheline is a feldspathoid and tridymite is a SiO_2 mineral, Figure 3.10 provides an experimental basis for Figure 1.9.

3.2.6 A Peritectic

Up to this point we have only considered crystallizing magmas in which once a mineral crystallizes, it never undergoes further reaction with the magma. Another situation is possible -- at lower temperatures a mineral becomes unstable and reacts with the magma to form a new, different mineral. To examine this phenomenon, we must introduce the *peritectic* or *reaction point*.

3.2.6.1 Equilibrium and Fractional Processes

Before introducing a binary system that contains a peritectic as well as eutectic, some basic definitions are in order. So far we have studiously avoided discussing the differences between equilibrium and fractional processes, because they can best be understood in the light of a peritectic. Because these processes can be either partial melting or crystallization, we must define four terms:

1. *Equilibrium crystallization.* Cooling rate of the magma is slow enough so that chemical equilibrium between magma and all crystals is maintained throughout the crystallization process. For equilibrium to be maintained, all crystals throughout the magma chamber must continuously be in contact with the magma as long as magma is present. This is obviously an idealized condition that is never perfectly achieved in nature.
2. *Fractional crystallization.* Crystals are continuously being removed from further reaction with the magma by some physical process, such as gravity settling. The only crystals in equilibrium with the magma are the last crystals to form. This is a common phenomenon during magmatic crystallization and is one of the principal ways in which magmas are modified after formation by partial melting. If fractional crystallization takes place in a closed magma chamber, it can be mathematically modeled by an idealized process referred to as *Rayleigh fractional crystallization* (RFC; see Section 5.2.1.2).

3. *Equilibrium melting*. The opposite of equilibrium crystallization. All crystals (the "source rock") stay in physical contact and in chemical equilibrium with the melt during the heating and melting process. *Equilibrium batch melting* (EBM, see Section 5.2.1.4) refers to a process whereby equilibrium between source rock and melt is maintained until a "batch" of certain magnitude has melted (e.g., 5 percent of the source rock), at which time it is separated from the source and moves upward. *Dynamic melting* (see Section 5.2.1.4) is a special type of equilibrium batch melting in which a part of each batch is not separated from the source and mixes with the next batch. Dynamic melting is probably the most common way in which magmas form.

4. *Fractional melting*. Only superficially the opposite of fractional crystallization, as will become apparent later in this section. Melt is being continuously removed from the source rock as it forms, so only the last melt to form is in equilibrium with the source rock. It is similar to equilibrium batch melting with infinitely small batches. It probably seldom if ever takes place in its ideal form because infinitely small batches cannot be physically removed from the source rock. An idealized mathematical model for fractional melting is referred to as *Rayleigh fractional melting* (RFM, see Section 5.2.1.3).

Equilibrium crystallization in its ideal form leads to chemically homogeneous igneous bodies, whether intrusive or extrusive. Phenocryst and groundmass proportions, or textures, may change, but bulk chemical compositions ideally do not. As such, equilibrium crystallization may seldom take place on the scale of a magma chamber, but rather be confined to a much smaller scale, such as that of a hand specimen. Igneous intrusions that are approximately chemically homogeneous are not unknown, however. Most often equilibrium crystallization should be considered as an ideal point of departure for more realistic models. On the other hand, equilibrium melting, especially partial equilibrium batch melting, may very well closely approximate the process most responsible for the formation of magmas.

3.2.6.2 The System LC-Q

We are now in a position to examine a binary system with a peritectic as well as a eutectic. The system we will examine is another binary system involving a feldspathoid, in this case leucite ($KAlSi_2O_6$), and SiO_2. This system, LC-Q, at 1 atmosphere pressure is shown in Figure 3.11. The left side of the liquidus has a break in slope at the peritectic (point P: also defined as the reaction point). The solidus, the heavy dark line, has a vertical offset at composition Or. As before, all areas below the liquidus can be considered as the forbidden zone. As with the system NE-Q, we need not repeat the application of the phase rule, as results are identical to those for the imaginary system AA-BB in Section 3.2.4.1. To summarize application of the phase rule, the region above the liquidus is divariant; the liquidus is univariant; eutectic and peritectic points are invariant; and the region below the liquidus is the forbidden zone.

Had we wished to construct a figure similar to the experimental system LC-Q from thermochemical data and Equation 3.1, we could do so by realizing that the system LC-Q can be considered as two overlapping systems, LC-Q (hypothetical) and OR-Q, each of which has only a binary eutectic. Point K is the melting point for pure Or (actually, high sanidine). We will use the abbreviation Or simply to indicate a K-feldspar without regard to which polymorph it is. The dashed curve K-P is the upper metastable part of the univariant Or-melt liquidus. The dashed curve P-E' is the lower metastable part of the Lc-melt liquidus. Point P, the peritectic, can be considered as the point of intersection of the two liquidi, Lc-melt and Or-melt. Point E is the eutectic for the system OR-Q (and also for the real system LC-Q), whereas point E' is the eutectic for the hypothetical system LC-Q, assuming a simple eutectic. The horizontal dashed line through E' is the metastable solidus for the hypothetical simple eutectic system LC-Q. As with the system NE-Q (Fig. 3.10), an isothermal phase boundary for the polymorphic inversion tridymite-cristobalite occurs at 1470°C.

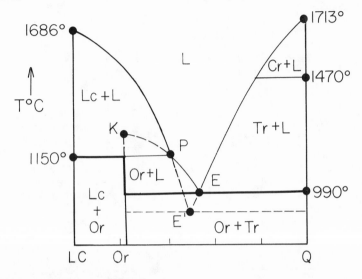

Figure 3.11 The binary system LC-Q at 1 atmosphere *P* (modified after Schairer and Bowen, 1955). See text for explanation of points *K* and *E'*. Points *P* and *E* are the peritectic and the eutectic, respectively. Solidus is shown as *heavy lines* from 1150°C to 990°C.

First let us consider equilibrium crystallization in the binary system LC-Q at 1 atmosphere *P* (Fig. 3.11). Any magma on the right (high-SiO_2) side of point P upon cooling behaves exactly as we have observed in Sections 3.2.4.2 and 3.2.4.3. Magmas with compositions between points P and E crystallize Or until the eutectic temperature at 990°C is reached, at which time the melt undergoes eutectic crystallization and coprecipitates Or and Tr in the proportions 41 percent Tr and 59 percent Or (read the lever or tie line below 990°C between Or and Q with point E as the fulcrum). Any magma with a composition between point E and 100Q precipitates a SiO_2 polymorph between the liquidus and the solidus, then also undergoes eutectic crystallization in the same eutectic proportions, Tr41 Or59. A quenched experimental charge or "volcanic rock" can be either Or phyric or Tr phyric, depending on whether the magma is on the Or + melt or Tr + melt portion of the liquidus when quenching occurs. As all these rocks contain a SiO_2 polymorph and are SiO_2 oversaturated, they would be classified as subalkaline.

3.2.6.3 Liquid and Crystal Paths

Another concept must be introduced before considering magmas whose compositions fall to the left of point P. On any phase diagram, binary or otherwise, we can define a *liquid path* (LP) and a *crystal path* (XP). These are paths that the melt and crystal compositions follow for either equilibrium or fractional crystallization or melting. For equilibrium crystallization, any point on XP marks the bulk composition of all crystals to crystallize to that point. Recall that during ideal equilibrium crystallization all crystals remain in equilibrium with the melt at all times. Morse (1980) offers a more complete discussion of liquid and crystal paths.

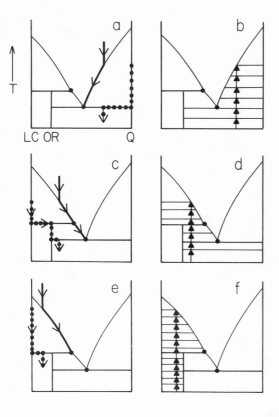

Figure 3.12 Equilibrium crystallization in the system LC-Q at 1 atmosphere for three different initial magmas: Q75 *(a-b)*, Q30 *(c-d)*, and Q10 *(e-f)*. In *a, c,* and *e, heavy solid curves* are LPs and *heavy dotted lines* are XPs. Examples of all possible tie lines, each with a lever fulcrum, are shown in *b, d,* and *f*.

Because we are initially examining equilibrium crystallization, let us monitor LP and XP for a magma of composition Q75 (Fig. 3.12a). The heavy solid line starting above the liquidus and then following the liquidus and ending at the eutectic is LP for a magma of composition Q75 undergoing equilibrium crystallization. The heavy dotted line down the Q100 line, then along the solidus to the original magma composition (final rock composition), and then down, is XP. Between the solidus and the liquidus, the crystal composition remains at Q100, but the melt is becoming progressively more LC-rich, because crystals of pure SiO_2 are crystallizing. Upon reaching the solidus, XP isothermally moves along the solidus, becoming more Or-rich while eutectic crystallization is proceeding. The final chemical composition of the rock must be the same as the initial composition of the magma, i.e., Q75, although mineralogically the rock will contain 68 percent Tr and 32 percent Or (read the lever just under 990°C between Or and 100Q with Q75 as the fulcrum).

Because the rock contains a SiO_2 polymorph, it will be subalkaline. Actually, a rock with 68 percent Tr is an extremely unlikely igneous rock, but it will serve to illustrate some basic principles. Figure 3.12b is a companion diagram to Figure 3.12a and shows all possible tie lines for equilibrium crystallization of a Q75 magma. The fulcrum of each lever is always at Q75.

3.2.6.4 Equilibrium Melting

Equilibrium melting behaves in precisely the opposite way. We need only to reverse the arrows on LP and XP to describe equilibrium melting. A rock containing 68 percent Tr and 32 percent Or would begin melting at the eutectic temperature, 990°C. The rock would continue to melt isothermally, producing a eutectic melt, thus this process can be termed *eutectic melting*. During this period LP is on the liquidus and XP is moving isothermally toward 100Q. In other words, the crystalline source rock is becoming increasingly enriched in Q because Or is preferentially melting. When XP reaches 100Q, all Or has been melted and only 45 percent of the source rock remains (read the lever just above 990°C between point E and 100Q with Q75 as the fulcrum).

This remaining source rock or crystalline residuum is frequently called the *restite*. It is very unlikely that partial melting would go beyond this point, because 55 percent of the rock has already melted. More likely, partial melting will never get to this point. If it does go beyond, only Tr is left to melt, so XP moves up the 100Q line while the LP moves up the liquidus, becoming increasingly enriched in Q. If the rock totally melts, which is very unlikely in nature, obviously the final melt composition must be the same as the original rock composition, Q75.

3.2.6.5 Reaction Crystallization

Figures 3.12c and 3.12d are companion diagrams that describe equilibrium crystallization (and, for the reverse process, equilibrium melting) for any magma whose composition falls between Or and point P on Figure 3.11. We will consider equilibrium crystallization first. A magma of composition Q30 will crystallize Lc and LP will move down the liquidus to point P, the peritectic at 1150°C. During this time XP has been moving vertically down 100LC. At this juncture a useful learning exercise is to examine the state of the system just above and just below 1150°C:

Just above 1150°C:	29% Lc and 71% melt of composition Q42 (essentially, the peritectic composition)
Just below 1150°C:	62% Or and 38% melt of composition Q42

Obviously something drastic happened at 1150°C, far different from eutectic crystallization (where, for a binary system, two minerals coprecipitate). The system above 1150°C contained 29 percent Lc, whereas the system below contains no Lc and 62 percent Or. What happened to the Lc? It reacted with the peritectic magma to form Or during a process termed *reaction crystallization*. We can represent this reaction algebraically:

Above 1150°C   Below 1150°C

$0.29Lc + 0.71L_P = 0.62Or + 0.38L_P$ (3.3)

$$L_P = 1.88OR - 0.88LC$$
$$L_P = 1.88(0.22Q + 0.78LC) - 0.88LC$$
$$L_P = 0.41Q + 0.59LC$$

This calculated composition for the peritectic liquid L_P is, allowing for rounding error and error in reading the graph, the same as Q42. Hence, the algebraic solution confirms the graphical solution. The term $(0.22Q + 0.78LC)$ is the weight composition for Or expressed as Q and LC.

While LP is at the peritectic, XP is moving isothermally along the solidus toward composition Or (Fig. 3.12c). As LP continues down the liquidus toward the eutectic, XP moves vertically (isochemically) down the line representing the composition Or toward the lower level of the solidus. When LP reaches the eutectic, coprecipitation of Or and Tr in the eutectic proportion takes place until XP reaches the original magma composition, Q30. At this point crystallization is complete, and the final rock would contain 10 percent Tr and 90 percent Or. Read the appropriate lever to confirm these percentages. Because it contains a SiO_2 polymorph, it would be classified as subalkaline.

3.2.6.6 Incongruent Melting

As before, equilibrium melting is exactly the opposite process; just mentally reverse the arrows on Figure 3.12c. A rock with 10 percent Tr and 90 percent Or would begin melting at 990°C and produce a eutectic melt until all Tr had melted. During this time XP would be moving toward Or isothermally along the lower level of the solidus. When all Tr had melted, 24 percent of the rock would be molten and the remaining 76 percent "restite" would be pure Or (again confirm these percentages). If equilibrium batch melting were operative and each batch were 6 percent of the original source, four batches of the eutectic composition would have formed and separated. As LP moves along the liquidus toward the peritectic, the partial melt is becoming enriched in Or because only pure Or is melting.

When LP reaches the peritectic, an interesting phenomenon occurs, which is the opposite of reaction crystallization. It is defined as *incongruent melting*. The forward reaction for Equation 3.3 is reaction crystallization, whereas the back reaction is incongruent melting. Examine Equation 3.3 carefully. The system contains 62 percent Or and 38 percent peritectic liquid just below 1150°C, but it contains 29 percent Lc and 71 percent peritectic liquid just above. Consequently, during incongruent melting all the Or must have melted to form Lc plus 33 percent more peritectic melt. Incongruent melting, then, is the melting of a solid to produce a liquid and another solid of different composition. This is in contrast with *congruent melting*. Had we simply melted pure Lc, Or, or Tr, we would produce a congruent melt of exactly the same composition as the mineral being melted. Incongruent melting can also be contrasted with eutectic melting, where one mineral preferentially melts but no new mineral is formed.

During incongruent melting, LP is at the peritectic and XP is moving isothermally along the upper level of the solidus toward 100LC. Incongruent melting will cease when all the Or has melted, at which time LP continues up the liquidus toward the original rock composition, Q30, and XP moves isochemically up the line representing 100Lc. When LP reaches Q30, the rock is completely molten.

3.2.6.7 Behavior of LC-rich Magmas

One final magma type can exist on Figure 3.12. This type has a composition between Or and 100LC. An example would be a magma of composition Q10 (Figs. 3.12e and 3.12f). Assuming equilibrium crystallization, upon cooling to the liquidus, Lc will begin crystallizing and will continue as LP moves down the liquidus. Meanwhile, XP is moving isochemically down the line representing Lc100. Upon reaching 1150°C, the liquid will react with Lc to form Or, but not all the Lc will react away as it did in the earlier example (for a magma of composition Q30). At this point LP is at the peritectic and XP is moving isothermally along the upper level of the solidus toward Q10.

Again a useful exercise is to examine the state of the system just above and below 1150°C:

Just above 1150°C: 76% Lc and 24% melt (Q42)
Just below 1150°C: 49% Lc and 51% Or

Because the solidus is at 1150°C in this part of the diagram, the magma must be completely solidified below 1150°C, which it is. During crystallization at this temperature the amount of Lc decreased from 76 to 49 percent, so this must be reaction crystallization. The amount of both Lc and Or would have had to *increase* for this to be eutectic crystallization. The rock is critically SiO_2 undersaturated, as it is feldspathoid-bearing, so it would be in the alkaline suite.

Again, mentally reverse the arrows on Figure 3.12e and equilibrium melting is described. Upon heating, a rock with 49 percent Lc and 51 percent Or will begin melting at 1150°C, where it will incongruently melt to produce 24 percent melt of the peritectic composition Q42. Had equilibrium batch melting taken place and each batch consisted of 4 percent of the original source, six batches would have formed and separated. All would have had the peritectic composition. During this process LP is at the peritectic and XP is moving isothermally toward 100Lc. If melting proceeds beyond the peritectic, and it may not, only Lc will be left in the source rock to melt. During this stage of melting, LP moves along the liquidus to Q10, while XP moves isochemically up the Lc100 line. Melting is completed when LP arrives at Q10.

3.2.6.8 Recognition of a Peritectic and a Eutectic

We will now examine a stability (equilibrium) diagram for the subsolidus phases and invariant points in the system LC-Q:

```
Lc        Or      P   E               Tr
|____|_____|___|___|_____|

LC                                  Q
```

The two minerals involved in the equilibria at point P (the peritectic) are Lc and Or, whereas Or and Tr participate in the equilibria at the eutectic (point E). Note that point P falls *outside* the Lc-Or join, but point E falls *inside* the Or-Tr join. This provides a simple way to determine at a glance on a binary diagram which invariant points are eutectics and which are peritectics. For a binary system, a eutectic falls inside the join of the two minerals involved in equilibria for that eutectic, whereas a peritectic falls outside. These rules of thumb are simply extensions of the contact principle (Section 3.2.2.2). Another way to distinguish between a peritectic and a eutectic, which is perhaps even simpler, is: a eutectic is always found at the bottom of a thermal valley, whereas a peritectic is located at the break in slope on the side of a liquidus.

3.2.7 Fractional Processes

Fractional crystallization (also termed *crystal fractionation*) is a common process during crystallization of igneous rocks and many physical processes have been suggested as being responsible. Perhaps the three most common are: (1) crystal settling (or, much less commonly, floating) under the influence of gravity, (2) zoned crystals, and (3) inward displacement of melt ahead of a crystallizing front. Flowage differentiation and filter pressing are mentioned as well. This section only briefly describes all these processes (Section 3.2.7.5). It will emphasize the manner in which fractional crystallization is interpreted on binary phase diagrams. Consult any standard petrology textbook for a more detailed discussion of the physical processes involved.

3.2.7.1 Liquid Paths, Crystal Paths, and Rock Hops

We continue our examination of the binary system LC-Q by considering the effect of fractional crystallization. Liquid and crystal paths for fractional crystallization of a magma of composition Q10 are shown on Figure 3.13. XP for fractional crystallization is marking a different composition than it did for equilibrium crystallization. Recall that during ideal equilibrium crystallization *all* crystals are in equilibrium with the melt and XP marked the bulk composition of all those crystals. In contrast, only the *last* crystals to form are in equilibrium with the melt during fractional crystallization. Consequently, for fractional crystallization, XP marks only the composition of the last crystals to form, not the bulk composition of all crystals that have precipitated to that point.

Two other important differences between equilibrium and fractional crystallization should be noted. For ideal equilibrium crystallization, the reacting system always includes all crystals and liquid at every point of the crystallization process between the liquidus and the solidus. This is not true for ideal fractional crystallization. Because crystals are effectively being removed from the reacting system as they form, the "reacting system" at any point in the crystallization process only includes the melt. The other difference pertains to reaction crystallization. On reflection it is apparent that reaction crystallization cannot occur during ideal fractional crystallization because the crystals must be in physical contact with the magma for reaction crystallization to occur. Reaction crystallization, therefore, is confined to equilibrium crystallization.

Carefully compare Figure 3.13 with Figure 3.12e. Figure 3.13 shows LP and XP for fractional crystallization of a magma of composition Q10, whereas Figure 3.12e represents equilibrium crystallization for the same magma. Both liquid paths and crystal paths are different. LP for equilibrium crystallization stops at the peritectic, but it continues on to the eutectic for fractional crystallization. Upon reaching 1150°C, XP for equilibrium crystallization isothermally follows the solidus to Q10. For fractional crystallization, upon reaching 1150°C, XP "hops" to the vertical line representing composition Or; it then isochemically moves down Or to the eutectic temperature, 990°C, at which point it hops again to the eutectic. Morse (1980) has referred to these "hops" by XP as *rock hops*.

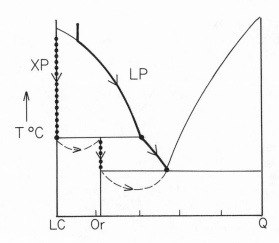

Figure 3.13 Fractional crystallization of magma with initial composition Q10 in the system LC-Q. *Dashed curves* represent rock hops. The *dotted lines* labeled *XP* and the eutectic point, all connected by rock hops, are the crystal path. The *heavy solid curves* labeled *LP* are the liquid path.

We have described the graphical changes in LP and XP for fractional crystallization. What do they mean in terms of what is physically (and chemically) happening in the magma chamber? Between the liquidus and 1150°C, as Lc precipitates, fractional and equilibrium crystallization are the same, except crystals are being effectively removed all the while during fractional crystallization. The first notable difference occurs at 1150°C. Because the 76 percent Lc crystals that have precipitated to this point are no longer available for reaction, the 24 percent remaining magma finds itself at the peritectic with no Lc crystals with which to react, so it acts as a new magma and begins crystallizing Or; XP isothermally hops from 100LC to Or. As Or crystallizes, LP moves down the liquidus toward the eutectic as the magma becomes increasingly enriched in Q component. Just above the liquidus, 38 percent of the remaining 24 percent magma (or 9.1 percent of the original magma) has crystallized as Or. This 38 percent can be determined by reading the lever between Or and point E just above 990°C with the composition of point P as the fulcrum. The remaining 14.9 percent magma crystallizes in the eutectic proportions as Or + Tr, so XP must hop again to point E.

A check can be made to determine if these calculations are correct. Compositions of these three assemblages (first 76 percent pure Lc, then 9.1 percent pure Or, last 14.9 percent Or + Tr in eutectic proportions) should, when their relative percentages are taken into account, add to the original composition of the magma, Q10. Remember that Or contains 78 percent LC and 22 percent Q by weight. Calculations are:

	LC	Q
76 % Lc	76	
9.1 % Or		
9.1 x 0.78	7.1	
9.1 x 0.22		2.0
14.9 % eutectic mix		
41 % Tr (14.9 x 0.41)		6.1
59 % Or		
14.9 x 0.59 x 0.78	6.9	
14.9 x 0.59 x 0.22		1.9
Total	90.0	10.0

The calculated composition of the original magma does add to Q10, so the compositions and proportions of the three assemblages are confirmed.

3.2.7.2 Two Hypothetical Examples

Returning to our example of fractional crystallization of a magma of composition Q10 (Fig. 3.13), it is quite unlikely that it will undergo gravity settling or floating. The viscosity of such a magma will be extremely high. Furthermore, density differences between melt and crystals are not very large. Ideal fractional crystallization on the scale of an igneous intrusion for such a magma may not take place to any great degree, but let us assume that it will and the operative process is inward melt displacement during in situ (in place) fractional crystallization of a dike. An idealized, hypothetical crosssection of the dike is shown in Figure 3.14a. The outer 76 percent (by weight; 38 percent on each side) would contain pure Lc. The next zone inward, 4.55 percent on each side, would be pure Or. The inner zone, the remaining 14.9 percent, would be composed of eutectic mix. These percentages are obtained from the previous calculations. Note that the outer zone is critically SiO_2 undersaturated, whereas the inner zone is SiO_2-oversaturated. This model, of course, is oversimplified, because it assumes that fractional crystallization was 100 percent effective and no trapped liquid was present.

Figure 3.14a is based upon a model of in situ fractionation, but an equally plausible (perhaps even more plausible) model exists. Figure 3.14b is a highly schematic sketch of a "plumbing system" for a volcano. It includes a conduit from a lower chamber, which in turn has a feeder from an even deeper chamber, or perhaps the source rock itself. The chamber is undergoing ideal fractional crystallization, with minimal trapped liquid or transport of primocrysts up the conduit to the volcano. Periodically a volcanic eruption occurs in which a volumetrically insignificant, relative to the chamber, amount of magma is removed. Rocks in the volcano should, from bottom to top, represent LP. Successively higher (younger) rocks in the chamber, assuming a model of crystal settling, trace XP (Fig. 3.14b). The chamber eventually becomes a compositionally layered intrusion.

This simplistic model is complicated in nature by a number of factors, among them (1) trapped liquid in the pore space of the cumulus crystals, (2) effect of convection currents (Marsh, 1988), (3) melt inhomogeneities owing to ionic diffusion (McBirney and Noyes, 1979; Hildreth, 1981), (4) upward transport of primocrysts in the conduit to the volcano, (5) interaction with the wall rocks (McBirney,1979), and (6) a chamber open with respect to both its feeder and conduit. In (6) the magma chamber is being periodically tapped to feed the volcano, but is also being periodically replenished from below with relatively unfractionated magma, which will mix with magma remaining in the chamber (O'Hara, 1977). Any of these factors will lead to complications in the ideal case of successive volcanic events representing LP, while successive layers in the zoned intrusion represent XP, but the basic principles are important.

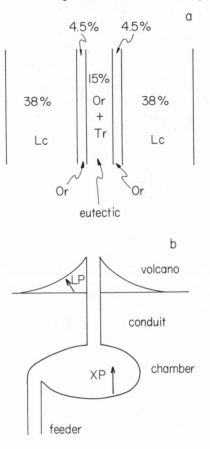

Figure 3.14 *a*. A zoned dike caused by in situ fractional crystallization due to inward displacement of melt. Initial magma composition was Q10 in the system LC-Q (Fig. 3.11). Calculations of mineral percentages are given in text. *b*. Schematic diagram of a "plumbing system" for a volcano. *XP*, crystal path; *LP*, liquid path.

3.2.7.3 Pressure Effects

To this point we have only considered systems crystallizing at 1 atmosphere pressure, so they are probably only applicable under surface or at most moderately deep conditions. The effect of pressure on these systems is very important with regard to its relation to igneous rock suites. It is clear from Figure 3.4 that the effect of increasing pressure on the hypothetical binary system YY-ZZ is to increase the eutectic temperature and cause the eutectic point to migrate toward ZZ.

Figure 3.15 shows two schematic sketches of the system FO-Q, Figure 3.15a at very low pressures and Figure 3.15b at much higher (7-10 kb). The system under low pressure has a peritectic (Fig. 3.15a). The effects of progressively increasing pressure on the system are to increase liquidus and solidus temperatures and to cause the peritectic to migrate to the left toward the FO side of the system. When the peritectic migrates to the left side of composition EN, it is no longer a peritectic, but is now a eutectic, resulting in a thermal divide (Fig. 3.15b). McBirney (1984) includes a particularly descriptive figure of this phenomenon.

Implications of Figure 3.15 on fractional crystallization are rather profound. This can be illustrated by considering two magmas of compositions X and Y. At relatively low pressures magma X, strongly SiO_2-undersaturated, will differentiate all the way to the En + Cr eutectic (heavy solid curve), so the last residual liquids will be SiO_2 oversaturated. At relatively high pressures, magma X will differentiate only to the Fo + En eutectic, so the last residual magmas will be SiO_2 undersaturated. In both cases successive residual liquids are becoming more SiO_2 enriched. At low pressures magma Y (dashed curve) behaves much like magma X, but at high pressures it initially crystallizes En and successive residual magmas become more SiO_2 deficient until the Fo + En eutectic is reached. Consider compositions to the left of En to approximate olivine tholeiites, while those to the right of En approximate quartz tholeiites (see Section 1.3.3.2). At low pressures an "olivine tholeiite" can fractionate to a "quartz tholeiite," but at relatively high pressures this is impossible (see also Section 5.1.1.2).

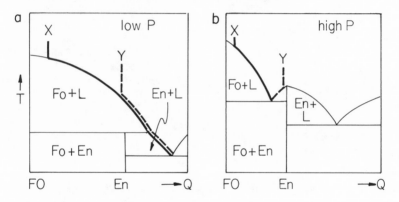

Figure 3.15 Schematic diagrams of the system FO-Q at *a.* a pressure approaching 1 atmosphere, and *b.* a pressure of 7-10 kb.

3.2.7.4 Fractional Melting

We must briefly consider fractional melting, although, as stated, it probably never occurs in its ideal form. LP and XP for fractional melting are shown in Figure 3.16 for a composition of Q10 (49 percent Lc and 51 percent Or). XP is exactly as for equilibrium melting for this composition (Fig. 3.12e), but LP is quite different. LP undergoes a melt hop and hops from the peritectic to 100LC at the melting point of pure Lc. After all Or has been used up by melting at the peritectic, only Lc is left to melt, and it cannot melt until its melting point has been reached.

An interesting feature of this type of melting is the long temperature interval during the heating process when nothing melts. Melting does not occur while the pure Lc left in the source is being heated from the peritectic temperature to the melting point of Lc. This temperature interval coincides with a time interval in which no melting occurs. Reading the lever A-B-P indicates that about 25 percent melting will take place at the peritectic. Additional melting can only take place if the rock could be heated to a sufficiently high temperature. However, natural source rocks undergoing partial melting probably seldom reach sufficiently high temperatures to melt a pure phase.

3.2.7.5 Processes Causing Fractional Crystallization

Five processes that cause fractional crystallization are: (1) crystal settling or floating, (2) zoned crystals, (3) inward displacement of melt ahead of a crystallizing front, (4) flowage differentiation, and (5) filter pressing. These latter two processes are probably less common and are only briefly mentioned.

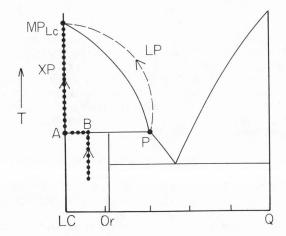

Figure 3.16 An example of fractional melting in the binary system LC-Q (Fig. 3.11). Crystal path *(XP)* follows dotted lines, while liquid path undergoes a melt hop from the peritectic *P* to the melting point of pure Lc *(MP_{Lc})*.

Crystal settling occurs much more commonly in mafic than in felsic magmas for two reasons: (1) viscosities of mafic magmas are much less than those of felsic magmas, and (2) the density contrast between primocrysts and melt, particularly if ferromagnesian silicates or oxides are involved, is generally greater for mafic magmas than felsic magmas. Various opinions exist about the effects of convection currents on gravity settling. Some petrologists believe that convection currents aid gravity settling by helping "sweep" primocrysts to the bottom of the chamber. Others feel that convection currents help keep the crystal mush "stirred" and thereby inhibit crystal settling. Very likely either effect is possible under the proper circumstances.

The main examples of crystal floating seem to be floating of plagioclase crystals in mafic magmas, although numerous cases have been reported where plagioclases actually underwent crystal settling in these magmas. Indeed, Charles Darwin himself, as a result of observations during his *H. M. S. Beagle* expedition, in 1844 reported evidence for plagioclase settling in a mafic magma and suggested that it could be explained by a process we now call crystal fractionation. Relatively iron-rich magmas are more dense than iron-poor magmas; in mafic magmas the iron content commonly, although not always, increases with increasing fractional crystallization. Likewise, the AB component in plagioclase normally increases with fractional crystallization. An increase in

AB component decreases the density of plagioclase. Thus the density of a highly differentiated mafic magma might be great enough, when combined with relatively AB-rich plagioclases, for plagioclase primocrysts to float.

Roughly equidimensional crystals can either settle or float in a magma under the influence of gravity according to Stoke's Law:

$$V = [2gr^2(\rho_S - \rho_L)] / 9\mu \tag{3.4}$$

Terms are:

V	settling velocity
g	constant for acceleration of gravity
r	radius of the crystal (assuming a sphere)
ρ_S	density of the solid (crystal)
ρ_L	density of the liquid (magma)
μ	viscosity of the magma

Ferromagnesian minerals in mafic magmas are generally more dense than plagioclase. Consequently, because of differences in magma viscosities and mineral densities, an augite primocryst might undergo gravity settling and sink more rapidly downward through the magma chamber, whereas a plagioclase primocryst would sink much more slowly and perhaps even float. The most important term controlling settling velocity is particle (crystal) size, because it is the only squared term (r^2) in Equation 3.4. As expected, this equation indicates that a large density contrast promotes crystal settling or floating, but a high viscosity is a hindrance.

The second common process by which crystal fractionation occurs is the formation of zoned crystals. Many plagioclase crystals are chemically, and thus optically, zoned in thin-section. Other minerals are zoned as well, but the zonation may not be so obvious as in plagioclase and may require electron probe analyses for confirmation. We have yet to discuss the effect of fractional crystallization on solid solutions (see Section 4.3.1.5), so the treatment here will be brief. *Normal zoning* in plagioclase will serve as an example. A plagioclase crystal that has an AN-rich core and becomes progressively more AB-rich toward the rim is said to be normally zoned. The AN-rich core forms relatively early in the crystallization history of the magma, at comparatively high temperatures. As the crystal grows, successive outer layers cover ("armor") inner layers and prohibit them from further reaction with the melt. Because only the crystal rim is in equilibrium with the melt and the interior is prevented from further reaction with the melt, this process is a type of fractional crystallization. Other types of zoning are observed, such as *reverse zoning* and *oscillatory zoning*.

Had equilibrium crystallization taken place, the final crystal, after all melt had crystallized, would be homogeneous throughout. If diffusion rates had been great enough between the interior of the crystal and the melt, as well as in the melt itself, equilibrium crystallization would have been maintained and the entire crystals would have remained in equilibrium with the melt during the complete crystallization process. It follows that extremely slow cooling rates promote equilibrium crystallization and homogeneous crystals because they allow continuous communication (diffusion) of ions between the core of the crystal and the melt. Diffusion rates are slower in magmas with high viscosities, so zoned crystals may also be promoted by high magma viscosities.

The third common cause for fractional crystallization is melt displacement inward from a crystallizing front. Anyone who has ever observed a partially frozen ice cube in a tray has seen a similar phenomenon. Crystallization is initiated on all six walls of the cube compartment because they are the coldest and provide a surface to which the first crystals can attach. Continued crystallization occurs as a front of crystals moving inward. The interior of the cube is the last to crystallize. Assume that ordinary tap water is used to make the cubes. If one were to extract carefully the unfrozen interior liquid and compare its chemical analysis for Na, K, Ca, Mg, Cl, etc., with that of the ice on the exterior of the cube, the liquid would be enriched in all these ions compared to the ice. The original tap water would have a composition for these ions intermediate between that for the ice and the "residual" interior liquid. This is because these ions do not fit well into the crystalline structure of ice. They are *incompatible* (see Section 5.2.1.1) in the ice.

Magmas in chambers can crystallize in a similar but considerably more complicated fashion. Ions that cannot enter lattice positions of early-formed, near-liquidus phases become enriched in the interior residual melt. The outer crystals nearer the chamber walls are protected from further reaction with the melt by the interior crystals. Only the crystals at the solid-liquid interface are in equilibrium with the melt. This process is a cause for fractional crystallization and leads to *normally zoned* igneous intrusions. Notice that for a normally zoned intrusion the relatively high-temperature material is on the outside, just the opposite of a normally zoned crystal. Through time the crystal front moves inward and ions excluded from the exterior crystals become progressively enriched in the interior residual magma, and thus eventually in the interior crystals, either as new minerals or in solid solutions of old minerals. Many factors can complicate this simple process and it can be operative while other processes, such as gravity settling or multiple injection, are taking place.

Flowage differentiation occurs when a crystal mush moves through a conduit. Friction between the walls of the conduit and the flowing crystal mush causes the mush in the center of the conduit to flow more rapidly than that on the margins. The net effect of these differential flow rates presumably is to cause crystals in the mush to migrate toward the center of the conduit. This creates an intrusive body, such as a dike, with a primocryst-rich center and primocryst-poor margins. In addition, relative sizes of the primocrysts should increase inward. A similar phenomenon causes logs moving down a stream to migrate toward and be concentrated in the center of the stream. The effectiveness of this process in magmatic systems is open to question.

When grapes are squeezed in a wine press the juice is separated from the remaining solid material by a process analogous to *filter pressing*. The liquid escapes and the solids are left behind. An example might be a magma chamber containing a crystal mush at reasonably high confining pressures. If a fracture develops in the surrounding wall rocks the relatively mobile liquid may be separated from the crystals and injected into the fracture, eventually leading to an intrusive body such as a dike. This process is known as filter pressing. Assuming the magma chamber eventually becomes the main intrusive body, it should be relatively primocryst-rich compared to the dike.

All five of these processes provide various means by which fractional crystallization can occur, but do all five actually take place? Certainly the first three do, but less conclusive evidence exists for the latter two. Phenocryst-rich centers of dikes relative to their margins are normally considered to be a strong indication of flowage differentiation, but in some cases these can be explained by multiple injection or other phenomena.

3.2.8 Liquid Immiscibility

In every system we have examined so far only one melt existed. In certain circumstances in nature it is possible for two, or perhaps even more, liquids to coexist. In everyday life liquid immiscibility is observed when one mixes oil and water. If only a small amount of oil is added to water, small globules or droplets ("spherules") of oil float on the water.

Although liquid immiscibility apparently exists between silicate and sulfide or oxide magmas, considerable debate exists about how common it is between two dominantly silicate magmas. Philpotts (1976, 1982), for example, is a strong advocate for liquid immiscibility. On the other hand, some studies (e.g., Watson, 1976; Roedder, 1979) have suggested that rather uncommon silicate magma compositions must exist before liquid immiscibility is possible. It is true that many laboratory experiments that produced two immiscible liquids involved melts whose compositions are comparatively rare in nature. For this reason, liquid immiscibility will be introduced in this section, but we will not dwell on it.

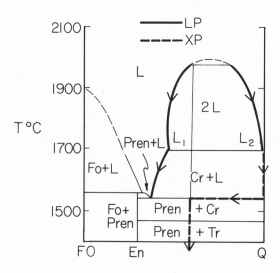

Figure 3.17 Liquid immiscibility in the binary system FO-Q at 1 atmosphere pressure. Phase diagram modified after Bowen and Andersen (1914) and Grieg (1927). *Pren*, abbreviation for protoenstatite; other abbreviations are identical to those used for other figures.

3.2.8.1 Why Liquid Immiscibility?

Perhaps the best-known binary system that is affected by liquid immiscibility is the system FO-Q at 1 atmosphere (Fig. 3.17). The familiar tridymite-cristobalite phase boundary is present, although in this system it is a subsolidus reaction (compare with Figs. 3.10 and 3.11). The parabolic curve between L_1 and L_2 encloses the two-liquid field and is the region in which liquid immiscibility exists. This curve is defined as the *solvus*. This solvus is actually above the liquidus, which for this part of the diagram is the isothermal line L_1-L_2. Some solvi are below the solidus (Section 3.3.2.1). As for other binary diagrams, the liquid field above the liquidus (and, in this case, above the solvus) is divariant; the liquidus and the solvus are univariant; points P, E, L_1, and L_2 are invariant; and all areas below the liquidus and in the two-liquid field are in the forbidden zone.

Qualitatively, the existence of a two-liquid field, and thus liquid immiscibility, can be explained by structural differences in the melt. A liquid should not be considered a random arrangement of ions, molecules, or atoms. It has network formers and network modifiers in much the same way as does a mineral (see Section 1.5.2.2). The ratio of network formers to modifiers will be higher in a Q-rich melt than a FO-rich melt (i.e., a Q-rich melt will be more polymerized). In fact, a melt of composition Q100 will have a ratio of infinity, as pure SiO_2 contains no network modifiers. In comparison, the ratio in a pure FO (Mg_2SiO_4) melt will be 1/2.

Two liquids will not intimately mix and form a true solution if their structures are not "tolerant" of one another. The structures have a better chance of being mutually tolerant if their network former/modifier ratios are similar. At high temperatures, entropy, and thus degree of disorder, is relatively high. Consequently, the two structures, in a sense, are comparatively tolerant of one another and can intimately mix, so no immiscibility occurs. Upon cooling, the structures become less tolerant as disorder decreases, until finally the situation becomes so "intolerable" that *unmixing* (or *exsolution*) starts. The terms *unmixing* and *exsolution* refer to the separation of one liquid phase into two as the two-liquid field is reached upon cooling. Actually, the same terms are used when a solid unmixes into two solids (Section 3.3.2). An everyday example of this process is the preparation of soup. Animal fat (oil) may be dissolved in boiling soup, but when the soup cools, the liquid fat becomes immiscible and will float on top of the aqueous solution comprising the soup. A very small amount of fat will appear as globules floating on the soup.

Some optimum composition in the system FO-Q, at about Q70, exists at which this intolerance of structures is optimized. A single liquid of this composition, upon cooling, will begin unmixing at the highest possible temperature in the two-phase field, about 2000°C. At this temperature and composition, at the top and center of the two-phase field, immiscibility is confined to an infinitely narrow compositional range. As temperature, and thus degree of tolerance, is lowered, this compositional range for which liquid immiscibility occurs progressively widens. Its maximum extent is from about Q47 to Q97, half of the entire compositional range, at slightly less than 1700°C. Below this temperature liquid immiscibility disappears because the system is below the solvus.

3.2.8.2 Quantitative Treatment

It is not necessary to discuss crystallization histories of any magmas with compositions from Q0 to Q47, because they will behave in an analogous fashion to those in the system LC-Q. We will only be concerned with magmas in which liquid immiscibility occurs, Q47-Q97. Melts on the left (relatively SiO_2-deficient) side of the solvus will be referred to as M_L, those on the right as M_R. If equilibrium crystallization is operative with no quenching, once the system is below the isothermal liquidus at slightly less than 1700°C, all traces of liquid immiscibility are gone and the liquid (M_L) crystallizes as described in Section 3.2.4.2.

Liquid and crystal paths for equilibrium crystallization are shown on Figure 3.17 for a melt of initial composition Q60. Upon cooling and reaching the solvus the single melt exsolves (unmixes) into two, and with continued cooling, LP for M_L moves down the left side of the solvus toward L_1 and M_L becomes progressively enriched in FO component. In contrast, LP for M_R moves down the right side toward L_2 and is continually being enriched in Q. The lever rule can be applied anywhere along here. As an example, if the melt were quenched at 1900°C, one would observe globules of M_R glass in a matrix of M_L glass:

> 21 percent M_R glass of composition Q91
> 79 percent M_L glass of composition Q51

Let us examine the state of the system just above and just below the liquidus (line L_1-L_2):

Just above L_1-L_2:	25% M_R of composition Q97
	75% M_L of composition Q47
Just below L_1-L_2:	24% Cr
	76% M_L of composition Q47

Although a first approximation, one can envision this process as follows: the Q component of liquid M_R crystallized at the liquidus, producing 24 percent Cr, while the remaining 1 percent FO component dissolved in the liquid M_L. Thus, upon continued cooling, liquid immiscibility exists no more. Below the liquidus LP and XP follow similar paths as on Figure 3.12a.

If liquid immiscibility occurred, and if the two liquids were of different density, chemical constituents could be easily fractionated. Sulfide-rich or oxide-rich magmas, considerably more dense than silicate magmas, apparently can be fractionated from silicate magmas in this manner. Some sulfide ore deposits may form by this process, where a sulfide magma sinks to the bottom of the chamber. As stated above, however, considerable debate exists as to whether this mechanism is common for two silicate magmas. Many petrologists believe that this mechanism, for two silicate magmas, is not common, although numerous cases have been documented (e.g., Wiebe, 1979; Philpotts, 1976, 1982).

3.3 BINARY SYSTEMS WITH SOLID SOLUTIONS

Most igneous rock-forming minerals exhibit at least partial solid solution between two or more components. The SiO_2 polymorphs are the major exception, and even they can have trace impurities occupying lattice sites. Mineral groups, such as the olivines and orthopyroxenes, are complete solid solutions between two "end members" that actually approach ideal behavior. The clinopyroxenes and alkali feldspars are examples of minerals that can exhibit complete solid solution at relatively high temperatures but undergo exsolution at lower temperatures. Hence, we must now consider complete and partial solid solution, as well as subsolidus exsolution (unmixing).

For true solid solution to occur the two or more ions proxying for one another *must be randomly disposed (i.e., exhibit no ordered arrangement) throughout the lattice site*. The lattice site itself, of course, is ordered with respect to any other lattice sites in the crystal structure. In an augite, for instance, the X site is ordered relative to the W, Y, and Z sites, but Fe^{2+}, Mg, Mn, etc. are randomly disposed throughout the X site.

3.3.1 A Transition Loop

We can now turn to the binary phase diagram for the system FO-FA at 1 atmosphere pressure (Fig. 3.18). The upper curves (one calculated, one experimental) represent the liquidus and the lower curves, the solidus. The liquidus plus solidus are referred to as a *transition loop,* representing a transition between two phases, both of which are binary solutions. Transition loops can also exist below the solidus (Section 3.3.2.6). The system FO-FA exhibits complete miscibility between these components below the solidus as well as above the liquidus; both these areas are divariant. The liquidus and the solidus are univariant and the area between these two curves is the forbidden zone, thus tie lines between coexisting phases can only be drawn in this area. The system contains no invariant point.

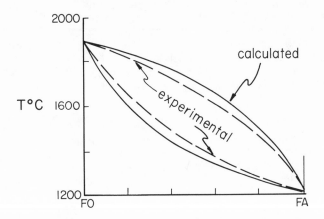

Figure 3.18 A comparison of the 1-atmosphere experimental (after Bowen and Schairer, 1935) and calculated (based on data from Robie and others, 1979) solid solution transition loops for olivine (FO-FA).

3.3.1.1 Construction from Thermochemical Data

Assuming ideal solutions for both liquids and solids, we can construct the liquidus and the solidus using the appropriate thermochemical data and a modification of Equation 3.1. If we define X_S as mol fraction FA in the crystals and X_L as mol fraction FA in the melt, then:

$$\ln(X_L / X_S) = (\Delta H^o_{FA}/R)\,(1/T_{FA} - 1/T) \tag{3.5}$$

A similar equation can be written for FO:

$$\ln[(1-X_L) / (1-X_S)] = (\Delta H^o_{FO}/R)\,(1/T_{FO} - 1/T) \tag{3.6}$$

ΔH^o_i is heat of fusion and T_i is melting point for the pure component i. In contrast with Equation 3.1, we have two unknowns, X_S and X_L, but fortunately we also have two equations. A number of ways exist to solve these two equations for X_S and X_L, but perhaps the simplest is to define two terms:

$$A = X_L / X_S \quad \text{and} \quad B = (1 - X_L) / (1 - X_S)$$

From these relations we can easily prove:

$$X_S = (B - 1) / (B - A) \quad \text{and} \quad X_L = AX_S \qquad (3.7)$$

We now need only solve Equations 3.5 and 3.6 for different values of T between T_{FO} and T_{FA}, calculate A and B for each temperature, and then determine X_S and X_L for that temperature. The locus of all T-X_S points defines the solidus and the locus of T-X_L points is the liquidus.

Mathematically this calculation is quite straightforward. The problem is deciding which data to use for ΔH of fusion. Robie and others (1979) give a value of 92173 j for Fa but no value for Fo. Bowen and Schairer (1935), who experimentally determined the system, state that heats of fusion for both pure compounds are similar. Their approximate value is equivalent to about 58600 j, quite a difference from 92173 j. It is no surprise that using 58600 j in both Equations 3.5 and 3.6 leads to a calculated transition loop that is extremely close to the experimentally determined loop (Fig. 3.18). Larger values of ΔH for FO and FA lead to greater curvatures for the liquidus and the solidus, so using 92173 j for both ΔH_{FO} and ΔH_{FA} in Equations 3.5 and 3.6 results in the outer solid curves on Figure 3.18, considerably different from the experimental loop (inner dashed curves).

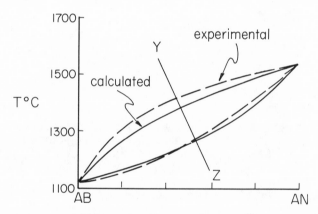

Figure 3.19 A comparison of the 1-atmosphere experimental (after Bowen, 1913; and Kushiro, 1973) and calculated solid solution transition loops for plagioclase (AB-AN). The line Y-Z divides the calculated transition loop into two mirror-image halves.

3.3.1.2 Recognition of Ideal Behavior

Because the symmetry of the experimental and theoretical curves for FO-FA agree so well (Fig. 3.18), both the liquid and solid solutions must approach ideal behavior. The plagioclase solid solution, however, does not behave ideally, although complete miscibility does exist between pure crystalline AB and AN (Fig. 3.19). The experimentally determined transition loop is shown as dashed curves, and the loop is shown as solid curves based on the equivalents of Equations 3.5 and 3.6 written for AB and AN using thermochemical data from Robie and others (1979).

Whether or not the thermochemical data are accurate is not critical in determining ideal behavior. In fact, for plagioclase they seem accurate, because the areas encompassed by the two loops are quite similar. The general expressions of Equations 3.5 and 3.6 (written for any components i and j), which are based on an assumption of ideality, will always generate a loop that is bilaterally symmetrical around a line through the middle of the loop (line Y-Z). The experimental loop for AB-AN is not symmetrical about Y-Z, but the calculated loop is. Thus the plagioclase solid solution is not ideal. The experimental FO-FA loop (Fig. 3.18) would be symmetrical about a similar line, confirming the ideal behavior of olivine.

3.3.1.3 Relation to Partition Coefficients

In Section 2.1.5.4 the equation $\Delta G^o = -RT\ln K_{eq}$ was shown to be useful in generating a plot of $\ln K_{eq}$ versus $1/T$, which could be a useful geothermometer at low pressures. For this plot K_{eq} equals K_D, the partition coefficient. Because Raoult's Law holds and ideality is generally maintained throughout the olivine solid solution, we should be able to express the olivine transition loop in terms of partition coefficients and a geothermometer. The $\Delta G^o = -RT\ln K_D$ equation is not required, because we already know mol fractions for crystals and liquids coexisting at equilibrium for various temperatures. We define K_D as X_S/X_L, where X_S is determined from the solidus and X_L is read from the liquidus for a specific temperature. The techniques learned in Section 1.2.3 must be employed to convert the transition loop in units of weight percent to a loop in mol fraction.

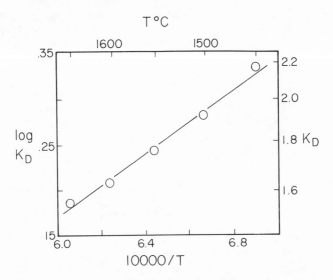

Figure 3.20 Plot of log K_D versus $1/T$ x 10000 based on the experimentally derived olivine transition loop (Fig. 3.18). Scales for $T\,°C$ and K_D are also included.

Expressing K_D in terms of Mg, an example of data (for one temperature) necessary to construct the line on Figure 3.20 is:

$T\,°C = 1600 \;\therefore\; 1/T = 6.25 \times 10^{-4}$
solid: FO72 $\;\therefore\; X_S = 0.79$
liquid: FO40 $\;\therefore\; X_L = 0.49$
$K_D = 0.79/0.49 = 1.61 \;\therefore\; \log K_D = 0.207$

Compositions for coexisting melts and crystals are determined from the
experimental curve on Figure 3.18. Similar calculations for other temperatures
allow construction of the straight line expressing the relationship between $1/T$
and $\log K_D$ on Figure 3.20. The slope of this line must be proportional to
$-\Delta G°/R$. Thus the olivine transition loop can be considered in the light of
partition coefficients and geothermometry.

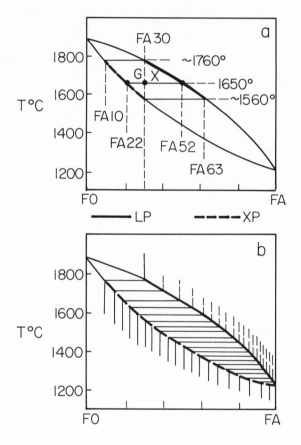

Figure 3.21 The
binary phase
diagram FO-FA as
applied to *a.*
equilibrium crys-
tallization, and *b.*
fractional crystal-
lization of an initial
magma of compo-
sition FA30.

3.3.1.4 Use of the Lever Rule

How is the lever rule used to interpret a binary solid solution diagram? First let
us consider equilibrium crystallization in the system FO-FA (Fig. 3.21a). Upon
cooling and reaching the liquidus at 1760°C, a melt of composition FA30 will

initially begin precipitating crystals of composition FA10. This composition is determined by drawing an isothermal tie line at 1760°C between the liquidus and the solidus; FA10 is determined at the point of intersection of the tie line and the solidus and is the solid composition in equilibrium with a melt of composition FA30. Upon further cooling and crystallization LP continues down the liquidus and XP down the solidus. Crystallization is complete when XP reaches the original melt composition, FA30, which is at a temperature of about 1560°C. During equilibrium crystallization in a closed system the final composition of the crystals must be equal the initial composition of the melt. The last drops of liquid had a composition of FA63, the melt composition in equilibrium with a solid composition of FA30.

Assume that during cooling the system is quenched at 1650°C. The tie line (lever) indicates that crystals of composition FA22 are in equilibrium with a melt of composition FA52. Moreover, we can use this lever, with FA30 as the fulcrum, to determine the percentages of glass and crystals in the charge. Remembering the rule of thumb that the composition of a phase is read from one side of the lever and the percentage from the other:

$$\% \text{ glass} = G \times 100 / (G + X)$$
$$\% \text{ crystals} = X \times 100 / (G + X) = 100 - G$$

For this example glass (FA52) makes up about 28 percent of the charge and crystals (FA22) the remaining 72 percent.

For the initial melt composition of FA30 an infinite number of tie lines can theoretically be drawn between 1760 and 1560°C. If each lever at each successively lower temperature were read, a progressive increase in FA content for both crystals and melt (as shown by XP and LP), along with an increase in the percentage of crystals relative to melt, would be indicated. On this diagram percent crystals is always read from right side of the lever and percent melt (glass) from the left side. Thus at 1760°C percentage of melt is 100 percent and at 1560°C percentage of crystals is 100 percent. As temperature is lowered, Fe is continuously diffusing into the crystals and is exchanging with Mg, which is diffusing into the melt. The reason why Mg content of the melt decreases rather than increases is because the percentage of crystals is also increasing, and Mg is needed in the later-formed crystals. This can be seen by reading the lever at two temperatures and writing two equations:

	Melt	Crystals
1700°C:	FA30 = (0.50 x FA43) + (0.50 x FA17)	
1650°C:	FA30 = (0.28 x FA52) + (0.72 x FA22)	

Both melt and crystals contain more FA component at the lower temperature, but because more crystals are present at 1650°C the equation still balances.

3.3.1.5 Fractional Crystallization

Liquid and crystal paths in the system FO-FA for fractional crystallization are shown in Figure 3.21b. Again using an initial magma composition of FA30, LP and XP for fractional crystallization are respectively identical to LP and XP for equilibrium crystallization from 1760 to 1560°C. LP and XP for equilibrium crystallization do not extend below 1560°C, but they both continue all the way to 1217°C (FA100) for fractional crystallization. Layering in an intrusion with FO-rich olivines at the base grading upward to FA-rich olivines, due to gravity settling, is called *cryptic layering*.

This difference in equilibrium and fractional crystallization can be explained by recalling that the reacting system for fractional crystallization only includes the residual liquid plus very last crystals to form. As a result, the composition of the "reacting system" is continually becoming more FA-rich, which is shown diagrammatically in Figure 3.21b. In equilibrium crystallization this composition stays constant at FA30. During fractional crystallization there is no reason for crystallization to cease when XP arrives at FA30 upon cooling because the residual melt (FA63) in equilibrium with those crystals is essentially a "new" crystal-free system ready to precipitate new crystals. This process theoretically continues all the way to FA100, where the very last drop of melt must have the same composition as the very last crystal.

3.3.1.6 Fe^{2+} and Mg^{2+} in Ferromagnesian Minerals

Pure Fo has a higher melting point than pure Fa, and thus FO-rich olivines crystallize earlier and at higher temperatures than do FA-rich olivines. This phenomenon is not unique to olivines. A relatively Mg-rich member of all ferromagnesian silicate solid solutions crystallizes earlier and at higher temperatures than does an Fe-rich member. This includes ortho- and clinopyroxenes, amphiboles, and biotites. For instance, a phase diagram for the binary system EN-FS, involving orthopyroxenes, appears quite similar to Figure 3.18. En ($MgSiO_3$) has the relatively high melting point and Fs ($FeSiO_3$) the low.

This phenomenon can be explained from two approaches: using classic crystal chemical arguments and from thermochemical data. To make the crystal chemistry argument first, an alternative expression of the above process is: when Mg^{2+} and Fe^{2+} ions compete for X-position lattice sites in ferromagnesian silicates crystallizing from a melt, statistically Mg^{2+} ions win the competition and become enriched in the early-formed minerals. The Mg^{2+} ion is more electropositive than the Fe^{2+} ion; i.e., the Mg^{2+} ion forms a relatively ionic bond with O^{2-}, whereas Fe^{2+} forms a relatively covalent bond. In addition, the Mg^{2+} ion is smaller than the Fe^{2+} ion. These two facts may account for the preferential entry of Mg^{2+} into X-position sites relative to Fe^{2+}.

Another consideration is that compounds whose bonds are relatively ionic generally have higher melting points than do those with more covalent bonds. This implies that, at least with regard to melting points, the Mg-O bond is "stronger" than the Fe-O bond. The greater strength of the Mg-O bond relative to the Fe-O bond is supported by the free energies of formation of their respective oxides at STP:

$$\Delta G^o_f \text{ (kcal)}$$

MgO	-136.046
FeO	-60.080

The difference is quite remarkable -- the value for MgO is over twice that for FeO. This may provide the best explanation of all for the relatively high melting points of Mg-rich ferromagnesian minerals. Whatever the explanation, this phenomenon is quite consistent and widespread among igneous rock-forming ferromagnesian silicates.

3.3.2 Subsolidus Reactions

Subsolidus (solid state) chemical reactions are primarily thought to be associated with metamorphic phenomena, but they are also associated with igneous processes. We have seen two examples of subsolidus reactions on binary phase diagrams. Figure 3.7 demonstrated a situation where the subsolidus mineral assemblage Aa + Bb undergoes a solid-state chemical reaction at temperature T_{SR} to form either Aa + Cc or Cc + Bb, depending on the bulk chemical composition of the system. This could be considered a metamorphic reaction. Subsolidus polymorphic phase transformations, such as the cristobalite-tridymite reaction on Figure 3.17, can also occur. A number of other types of subsolidus reactions, which involve subsolidus transition loops and solvi, are introduced in this section.

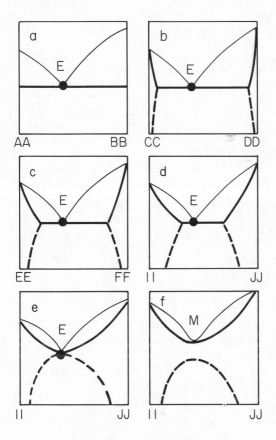

Figure 3.22 Examples of hypothetical binary systems with different degrees of partial solid solution, ranging from complete immiscibility in *a* to complete solid solution between the solidus and solvus in *f*. Extent of solid solution increases progressively from *a* through *b-e* to *f*. *Light solid curves,* liquidus; *heavy solid* curves, *solidus; heavy dashed curves,* solvus. Temperature on vertical axis for every graph.

3.3.2.1 Partial Solid Solution

We have already examined systems exhibiting complete subsolidus miscibility (solid solution; Figs. 3.18 and 3.19) and systems with complete subsolidus immiscibility (Figs. 3.9-3.11). Many igneous rock-forming minerals, however, have properties that are between these two extremes and, over certain temperature ranges, can be considered as partial solid solutions. Common examples are pyroxenes, alkali feldspars, and Fe-Ti oxides. We have also seen an example of liquid immiscibility involving a solvus (Fig. 3.17). A partial solid solution also involves a solvus, but this solvus is below the solidus. Moreover, in some systems the solvus can appear somewhat more complicated than the simple parabolic curve we saw for the solvus of liquid immiscibility (Fig. 3.17). Any treatment of subsolidus reactions must include partial solid solutions.

The difference between complete solid immiscibility and complete miscibility can best be envisioned as a continuum between these two extremes (Figs. 3.22a-f). On these figures the upper light solid curves represent the liquidus, the middle heavy solid curves mark the solidus, and the lower heavy dashed curves are the solvus. For Figures 3.22b-f, areas below the solidus and outside the solvus are stability fields of one solid solution. Two phases (solid solutions) are stable inside the solvus. One extreme is shown by Figure 3.22a, complete solid immiscibility. The other extreme can be seen in Figure 3.22f, complete solid solution, albeit over a rather limited temperature range (between the solidus and the solvus).

Hence, for complete solid solution to exist at a given temperature, the one-phase field must extend from the composition of one pure end-member to the other (Fig. 3.22f). If, for a particular temperature, this one-phase field is interrupted by a solvus and two-phase field, this indicates the existence of partial solid solution at that temperature. Some petrologists refer to the area within the solvus as an *immiscibility gap*. It is also a forbidden zone, because isothermal tie lines can be drawn connecting points on both sides of the solvus; these points represent compositions of the two crystalline phases in equilibrium.

Between these two extremes (Figs. 3.22a and 3.22f) are gradations (Figs. 3.22b-e) showing progressively increasing partial solid solution, or decreasing immiscibility. In Figure 3.22b the degree of solid solution is extremely limited. It is confined to compositions that are either extremely CC-rich or DD-rich because the solvus is very near CC100 and DD100. At any temperature only a small amount of DD can be dissolved in a mineral consisting mainly of CC, and vice versa. Hence the immiscibility gap is extremely wide. In Figures 3.22c-d the immiscibility gap becomes progressively narrower as the two-phase field narrows at the expense of the two one-phase fields on either side. From Figure 3.22b through 3.22d more and more of one component can dissolve in a crystalline phase enriched in the other component, so the degree of solid solution at any given temperature increases.

Each of Figures 3.22a-c represents a different system, the "dry" systems AA-BB, CC-DD, and EE-FF. Figures 3.22d-f represent the same system, II-JJ-H_2O. The solvus is about the same width and reaches a maximum at about the same temperature on Figures 3.22d-f; the difference in the three figures can be found in the melting points of pure Ii and Jj and thus the liquidus and solidus temperatures. Water pressure is highest for Figure 3.22d and lowest for Figure 3.22f; increasing P_{H2O} lowers the melting point (Section 3.2.2.3).

In Figure 3.22d the melting points are so low the solvus intersects the solidus, resulting in an invariant point (a binary eutectic) at point E. Similarly, Figures 3.22b-c have eutectic points. Figure 3.22a has a eutectic point but no solvus. Figure 3.22f is different; the solvus and solidus do not intersect because the melting points of Ii and Jj are considerably higher. This system has no true invariant point, but rather a *minimum point* at M. Because the solvus and the solidus do not intersect on this figure, complete solid solution exists between the solvus and the solidus. Figure 3.22e is intermediate between Figures 3.22d and 3.22f in that the solidus barely touches the top of the solvus. If they truly touch, a eutectic point is present.

3.3.2.2 Effect of Pressure

Why does the top of the solvus remain at about the same temperature while the melting points, and thus liquidus and solidus, are at different temperatures? The solvus in fact does not stay at exactly the same temperature, but is typically raised 10-20°C per 1 kb increase in water pressure. In contrast, melting points, as well as liquidi and solidi, are commonly lowered by 100-200°C per 1 kb increase in water pressure. Recall that an increase in total (load or confining) pressure in a "dry" system (a system with no H_2O) has the opposite effect -- it causes melting points, liquidi, and solidi to be raised.

Water molecules, or H^+ and OH^- ions formed by the dissociation of H_2O at high temperatures, have the ability to break bonds in silicate polyhedra in the melt, thereby lowering melting points. The solvus and solidus are more likely to intersect at high water pressures than at low because, as a result of lower melting points, the solidus is lowered while the solvus is slightly raised. Thus Figures 3.22d-f exemplify progressively lower water pressures in the same system.

3.3.2.3 A Regular Solution

The solvi on Figures 3.22b-d are drawn as being bilaterally symmetrical about a vertical line half-way between the two side lines (i.e., DD50, FF50, or JJ50). Any solid solution that exhibits a symmetrical solvus is called a *regular solution*. It is apparent that any solid solution with a solvus cannot be an ideal solution, so a regular solution can be considered as a special case of a real solution. The term *regular* refers to the fact that the solvus is shaped in a regular, simple, mathematically predictable way. As an example, for a regular solution of crystalline EE-FF at a specific temperature, exactly the same amount of EE is dissolved in an FF-rich crystal as FF dissolved in an EE-rich crystal (Fig. 3.22c).

3.3.2.4 The System AB-OR-H_2O

Many mineral groups have subsolidus solvi; among the most common are the alkali feldspars and pyroxenes. We will take the alkali feldspars as an example. Figure 3.23 is the ternary system albite-orthoclase-H_2O (AB-OR-H_2O; $NaAlSi_3O_8$-$KAlSi_3O_8$) at 5 kb water pressure. Because H_2O is not present as a component in any of the stable minerals, we will consider the system to be binary. The phase diagram is superficially similar to Figure 3.22c, but some notable differences exist. The most obvious difference is that the solvus on Figure 3.23 is strongly assymetrical; therefore, the solid solution AB-OR is decidedly not regular. The melt plus fluid field above the liquidus is divariant, as are the fields Naf_{SS} (Na-rich feldspar solid solution) plus fluid and Kf_{SS} (K-rich feldspar solid solution) plus fluid. The liquidus, solidus, and solvus are all univariant, and the eutectic point E is invariant. Areas between the liquidus and solidus, as well as below the solvus, are forbidden zones.

Let us examine equilibrium crystallization on Figure 3.23, given a number of different initial melt compositions. The assumption throughout this discussion is that cooling rates are slow enough such that equilibrium is maintained throughout. This situation is generally most easily attained in a plutonic environment. A melt of composition OR95, on cooling, will completely crystallize through the Kf_{SS} + L + F transition loop, producing a solid solution of composition OR95. Upon further cooling, this composition will never intersect the solvus, so exsolution (unmixing) never occurs. A melt of composition OR5 would have a similar history.

A melt of composition OR80 would also completely crystallize while cooling through the Kf_{SS} + L + F transition loop, yielding a solid solution of composition OR80 just below the solidus. The LP for this magma is shown on Figure 3.23 as a heavy solid curve, and the heavy dashed curve represents XP. Upon further cooling, this solid-solution composition intersects the solvus and begins exsolving an AB-rich solid solution of composition OR11. As cooling proceeds unmixing continues with the OR-rich phase becoming more OR-rich and the AB-rich phase progressively occupying a larger proportion of the two-phase mixture and becoming slightly more AB-rich. Anthropomorphically, upon cooling both phases purge themselves of the opposite, "unwanted" component (i.e., the OR-rich phase purges itself of AB and vice versa) and the guest grows in size at the expense of the host.

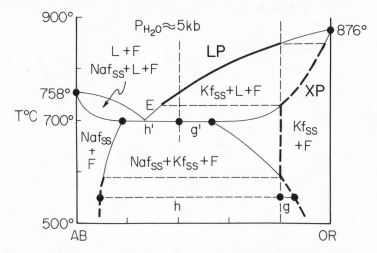

Figure 3.23 The system AB-OR-H_2O at P_{H2O} = 5 kb. Diagram modified after Morse (1968). The fluid F is stable in every field. Naf_{SS} and Kf_{SS} are appreviations for Na-rich and K-rich solid solutions, respectively.

If the system were quenched at 550°C, for example, we could read the lever rule to determine the relative amounts of the OR-rich and AB-rich phases. Because the OR-rich phase makes up the largest proportion of the mixture for this bulk composition of OR80, we refer to it as the *host,* and the AB-rich phase is called the *guest.* The lever is read accordingly:

% host = (h x 100) / (h + g) = 90
% guest = (g x 100) / (h + g) = 10

Composition of the host would be OR86, of the guest, OR10.

Any initial melt whose composition falls between about OR18 and OR52 would have a different history from those described above. After crystallizing through either the Naf_{SS} + L + F or Kf_{SS} + L + F transition loops, these compositions never enter either the Kf_{SS} + F or Naf_{SS} + F fields, but rather immediately begin exsolving into a host and a guest. For example, immediately below the solidus, at about 700°C, an alkali feldspar from an initial melt of composition OR40 would exsolve into the following:

$$\% \text{ host} = (h' \times 100) / (h' + g') = 65$$
$$\% \text{ guest} = (g' \times 100) / (h' + g') = 35$$

The host composition would be OR52 and the guest composition OR18. Upon further cooling the relatively K-rich phase becomes considerably enriched in OR component and the Na-rich phase somewhat increases in AB component.

3.3.2.5 Perthites and Antiperthites

This type of mixture of two alkali feldspars, where the host (always the most abundant phase) is relatively OR-rich and the guest is relatively AB-rich, is termed a *perthite*. An initial melt of composition OR15 would have a somewhat similar history, but would yield an AB-rich host and OR-rich guest. Such an alkali feldspar is called an *antiperthite*.

Refer again to Figure 3.23. Imagine an infinite number of levers drawn from one side of the solvus to the other with OR40 as the fulcrum of each lever. As temperature is lowered the length of g' progressively increases relative to h'. If the host is always the most abundant phase, then the host-guest relationship reverses itself at lower temperatures. Just below the solidus the relatively K-rich phase is the host, but near 500°C the host is relatively Na-rich; the reversal takes place at about 650°C. As a result, a continuum exists between perthites and antiperthites, depending on both initial melt compositions and equilibrium temperatures. They should not be looked on as separate minerals.

In felsic plutonic rocks the alkali feldspar is almost invariably a perthite, or less commonly an antiperthite. Perthites and antiperthites are described on the basis of the size and shape of the guest phase. With regard to size, they are classified accordingly:

Macroperthite:	the guest phase is coarse enough such that the two phases can be distinguished with the unaided eye.
Microperthite:	polarizing microscope is required to distinguish the two phases.
Cryptoperthite:	the guest phase is so fine that the alkali feldspar appears as one phase under the polarizing microscope, but X-ray diffraction analysis reveals that it is two phases.

Antiperthites are similarly classified. Into which category a particular perthite or antiperthite falls depends upon the degree of exsolution. Very slow cooling rates and high H_2O pressures promote extensive exsolution. With regard to the shape of the guest phase, possible shapes include (1) small, irregular, rounded "blebs," (2) long, irregular "strings," (3) relatively large, irregular "patches," and (4) lamellae parallel to some crystallographic plane. Perthite lamellae can be distinguished from twin lamellae in that boundaries between twin lamellae in thin-section appear as ruler-straight lines and boundaries between perthite lamellae are more irregular.

3.3.2.6 Subsolidus Transition Loops

The alkali feldspar solvus in Figure 3.23 is shown as an asymmetrical, smooth curve. In fact, the curve may not be smooth and may be connected to one or more subsolidus transition loops. Each transition loop marks a phase transformation between two alkali feldspar polymorphs. A possible configuration for the OR-rich side of the alkali feldspar solvus is shown in Figure 3.24. The exact shape and temperature of this solvus and transition loop are partly conjectural, as they have not been exactly duplicated in the laboratory. Subsolidus reactions of this type are commonly extremely sluggish and difficult to produce experimentally. On Figure 3.24 orthoclase solid solution (Or_{SS}; monoclinic symmetry) is the stable polymorph of K-feldspar above the transition loop and microcline solid solution (Mc_{SS}; triclinic symmetry) is stable below. See Section 2.1.4.1 for a discussion of K-feldspar polymorphs. Inside the transition loop Or_{SS} + Mc_{SS} coexist.

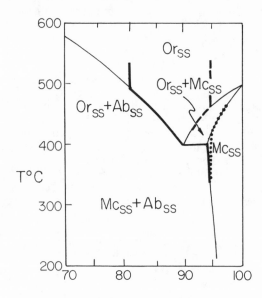

Figure 3.24 Example of a subsolidus transition loop (modified after Ragland, 1970). This loop is on the OR-rich side of the subsolidus OR-AB solvus. Horizontal axis is weight percent OR.

   This subsolidus transition loop can be interpreted in the same manner as a loop involving a liquidus and a solidus. As an example, take Or_{SS} of composition OR95. Assuming sufficiently slow cooling rates such that equilibrium is maintained throughout, upon cooling this composition will intersect the top curve of the loop at about 470°C and begin converting to Mc_{SS} of composition OR97. Upon continued cooling, Or_{SS} will move continuously down the top of the loop (dashed curve) while coexisting Mc_{SS} is moving down the bottom of the loop (dotted curve) and increasing in volume relative to Or_{SS}. An isothermal tie line, and thus lever, could be drawn at any temperature to determine the compositions and relative proportions of the two phases. Below the transition loop Mc_{SS} (OR95) upon further cooling will eventually intersect the solvus, at about 340°C, and begin exsolving an Ab_{SS} guest.
   An Or_{SS} crystal of composition OR80 will have a different history. Upon cooling it will begin exsolving Ab_{SS} at slightly less than 500°C and the composition of the Or_{SS} host will move down the solvus (heavy solid curve)

until it reaches 400°C, at which point it isothermally inverts to Mc$_{SS}$ and changes composition from about OR90 to OR94 along the base of the transition loop. The Mc$_{SS}$ host then continues down the solvus. In this case the inversion from Or$_{SS}$ to Mc$_{SS}$ is isothermal. For an initial composition of OR95, the inversion took place over a temperature range (from about 470-410°C). Keep in mind that all these temperatures are relative and their absolute values might differ by 100°C or more.

3.3.2.7 Subsolidus Reactions and a Peritectic

Not only can subsolidus reactions be involved with eutectic or minimum points, but they can also occur with a reaction point (peritectic). A good example can be found within the pyroxene group, where subsolidus exsolution is a common phenomenon, but also where orthopyroxene can react with a peritectic liquid to form a clinopyroxene. A hypothetical example of a binary system that contains both a subsolidus solvus and a peritectic in shown in Figure 3.25. Point P is the peritectic; a and b are the melting points of pure A and B, respectively. The liquidus is defined by curves a-P and P-b; the solidus is a-r-s-b; and the solvus is c-r-s-d. The solidus and solvus coincide along isothermal line r-s and the immiscibility gap along the bottom of the diagram is represented by c-d.

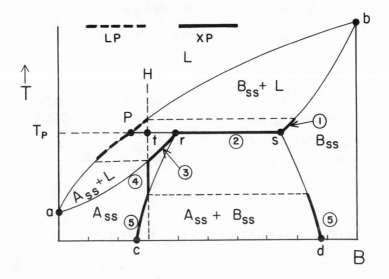

Figure 3.25 The hypothetical binary system A-B, which contains two solid-solution transition loops, a solvus, and a peritectic (point *P*).

Any melt whose composition is more A-rich than point P undergoes equilibrium or fractional crystallization similar to that described in Sections 3.3.1.4 and 3.3.1.5. Melts whose compositions fall between points P and s, however, experience reaction crystallization at T_P. For any melt whose composition is between points r and s, B$_{SS}$ reacts with the peritectic liquid at T_P to form A$_{SS}$ and is partially resorbed; the rock is completely solid below T_P.

We now consider in more detail a melt whose composition fall between points P and r, specifically melt H. The crystal path XP and liquid path LP for equilibrium crystallization of melt H are shown on Figure 3.25; LP is quite simple but XP is considerably more complex. With decreasing temperature equilibrium crystallization of melt H and subsequent subsolidus reaction of solid H takes place in five steps, which are numbered on XP. These steps are: (1) crystallization of a relatively B-rich solid solution (B_{SS}), (2) reaction of all the B_{SS} with the peritectic melt to form a relatively A-rich solid solution (A_{SS}) at the peritectic temperature T_P, (3) crystallization of A_{SS} until crystallization is completed, (4) cooling of the single crystalline phase A_{SS} until the solvus is reached, and (5) subsolidus exsolution of a B_{SS} guest from an A_{SS} host with continued cooling. With increasing temperature the same LP and XP hold for equilibrium melting. At temperature T_P, A_{SS} incongruently melts to form B_{SS} and a peritectic liquid.

We know that reaction crystallization (or incongruent melting) must take place at T_P if we read the P-t-s and P-t-r levers. Reading the P-t-s lever approximates conditions just above T_P; it yields about 10 percent B_{SS} of composition B76 coexisting with 90 percent of a peritectic melt (B24). Reading the P-t-r lever for conditions just below T_P results in approximately 33 percent A_{SS} (B40) coexisting with 67 percent of a peritectic liquid. These relationships can be seen mathematically:

Just above T_P Just below T_P

$$10(B76) + 0.90(B24) = 0.33(B40) + 0.67(B24)$$

$$7.6B + 21.6B = 13.2B + 16.1B$$

Considering rounding error, the two sides of the equation balance, as they must. Thus all B_{SS} (B76) must have reacted away at T_P to form A_{SS} (B40).

For fractional crystallization LP in theory continues along the liquidus all the way to point a. B_{SS} crystallizes while XP is along segment (1), then XP undergoes a rock hop to point r, where A_{SS} begins crystallizing. A_{SS} then continues to crystallize and becomes increasingly A-rich as XP moves along the solidus essentially to point a. While A_{SS} is crystallizing the melt's behavior is similar to that described in Section 3.3.1.5. Because fractional crystallization is operative, no reaction crystallization occurs at temperature T_P.

3.4 BINARY SYSTEMS AND P-T DIAGRAMS

In Section 3.2.3 *P-T-X* diagrams for binary systems were introduced, but we quickly went to *T-X* diagrams and to this point have spent virtually all our efforts on them. Equally important are *P-T* (Clapeyron; see Section 2.2.1) diagrams, so this section will deal with them. It is actually possible to construct a qualititative to semiquantitative version of a *P-T* diagram for a system, binary or otherwise, containing one invariant point. The compositions of all stable phases must be known and only a minimal amount of either experimental or thermodynamic data need be available. Lacking these data, field and petrographic relationships will commonly provide valuable but inconclusive evidence. This section will deal with the principles and techniques necessary to construct such a diagram.

3.4.1 Preliminaries

Before turning to the methods and axiomatic principles required to construct binary *P-T* diagrams, some background is necessary. First, some nomenclature heretofore not introduced must be covered. Then the concept of intersecting *Gibbs surfaces,* which provide the underlying basis for this entire procedure, must be explained. Last, an introduction to the *combinatorial formula* is required.

3.4.1.1 Nomenclature

Let us consider the subsolidus binary system W-Z with two intermediate phases X and Y. The equilibrium diagram for such a system would be:

Solution of the phase rule indicates that the maximum number of coexisting phases possible, at the invariant point, is C+2, or 4 for a binary system. Similarly, C+1 phases (three for a binary system) must participate in univariant reactions. It follows that C phases (two for a binary system) are stable in a divariant field.

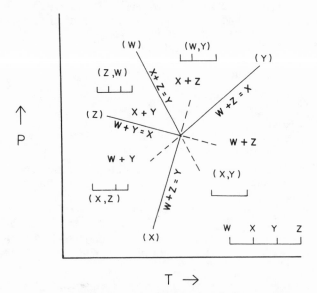

Figure 3.26 A *P-T* diagram for the hypothetical system W-Z. Univariant curves are labeled *(W)-(Z)* and single divariant fields are labeled *(X,Z)*, *(Z,W)*, etc. Metastable extensions are *dashed*. Appropriate reactions are shown on univariant curves and each single divariant field includes an equilibrium diagram and the unique assemblage.

In Section 3.2.2.2 we wrote all possible reactions for the system MgO-SiO_2, but only three phases can be involved in truly univariant reactions.

Utilizing the contact principle (Section 3.2.2.2), we will consider only those four reactions that involve three phases:

Univariant curve	Reaction
(Z)	W + Y = X
(W)	X + Z = Y
(Y)	W + Z = X
(X)	W + Z = Y

The above four reactions are the only reactions possible involving three phases. For the binary system W-Z, only these reactions can mark univariant curves. Each reaction is designated by the phase *not* participating in the reaction, i.e., the phase indicated in parentheses to the left of the reaction. This seems contradictory, but its utility will become apparent later. Thus we actually refer to the first reaction as reaction (Z), and so on for the remaining three. These reactions are applicable to the four univariant curves labeled (Z)-(X) on Figure 3.26. Each of these curves has a *metastable extension,* shown on the figure as a short, dashed line. The significance of these metastable extensions will be discussed shortly.

Divariant fields are similarly designated by the phases *not* present. For example, Figure 3.26 shows that the divariant field (Z,W) must fall between the univariant curves (Z) and (W). Similarly, all other (A,B) divariant fields fall between curves designated (A) and (B). According to the phase rule for a binary system if F = 2 then P = 2, so only two phases can be stable in any divariant field. Because phases Z and W cannot be stable in the divariant field (Z,W), X + Y must be the stable assemblage (referred to as the "unique assemblage"). Figure 3.26, therefore, has four diariant fields:

Divariant field	Unique assemblage
(Z,W)	Y + X
(W,Y)	X + Z
(X,Y)	Z + W
(X,Z)	W + Y

Each divariant field on Figure 3.26 has an equilibrium diagram, each of which is different. The explanation of these equilibrium diagrams must await the introduction to Schreinemakers' rules.

In the following example the assumption is made that all phases involved are crystalline solids, although the principles are the same if a melt is involved. For instance, as discussed in Section 3.2.6.8, phase Y might be a peritectic melt. If this is the case, however, all four univariant reactions written above are not physically possible.

3.4.1.2 Gibbs Surfaces

Equation 2.20 bears repeating here: $dG = VdP - SdT$. Differentiating with respect to P while holding T constant yields:

$$(\delta G/\delta P)_T = V > 0 \qquad\qquad (3.8)$$

This provides a thermodynamic definition of volume expressed in terms of G and P. Hence if we were to plot G versus P at constant T, the slope of the resultant line must always be positive, as V and S are always positive quantities. Differentiating with respect to T while holding P constant:

$$(\delta G/\delta T)_P = -S < 0 \qquad\qquad (3.9)$$

This equation is a thermodynamic definition of entropy expressed in terms of G and T. Furthermore, the slope of a line on a G versus T plot at constant P must always be negative.

We can put the above definitions to use in a three-dimensional plot of G-T-P (Fig. 3.27). Two planes are shown on this figure, one representing the hypothetical mineral assemblage A + B, and the other, the assemblage C + D. The slope of the line of intersection of each plane with the G-T plane is equal to $-S$ and therefore must be negative. Likewise, the slope of the line of intersection of each plane with the G-P plane is equal to V and thus must be positive. These planes are examples of Gibbs surfaces. The overall slope of a Gibbs surface is always constrained by the fact that its line of intersection with the G-T plane must have a negative slope, while the slope of its intersection with the G-P plane must be positive.

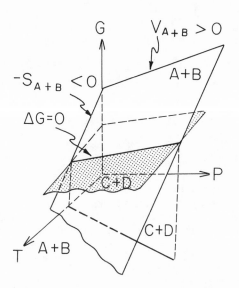

Figure 3.27 A G-P-T plot for the hypothetical reaction A + B = C + D. G-P-T plane for mineral assemblage C + D is *stippled;* plane for assemblage A + B is not. On the P-T projection, the assemblage A + B is stable on the front side of the equilibrium curve, while C + D is stable on the back side.

The line of intersection of these two Gibbs surfaces can be projected onto the plane P-T. This will produce a P-T (Clapeyron) diagram similar to Figure 3.4. The projection of the line of intersection is the phase boundary where $G_{A+B} = G_{C+D}$ and thus $\Delta G = 0$. We can determine which mineral assemblage is stable on each side of the phase boundary by simply determining which Gibbs surface is lower. Based on arguments developed in Section 3.1.5.3, the assemblage with the lower G will be relatively stable. The utility of these G-P-T plots will soon become apparent.

3.4.1.3 The Combinatorial Formula

In the combinatorial formula C, P, and F are defined as they are in the phase rule. If we are concerned about the number of univariant curves, separated by divariant fields, radial about an invariant point on a *P-T* diagram, we can determine the maximum number of possible equilibrium assemblages by use of the combinatorial formula. It is defined as:

$$A = (C + 2)! / (P! \, F!) \tag{3.10}$$

where A is the maximum number of possible equilibrium assemblages.

An example of a unary system (H_2O) with univariant curves separated by divariant fields radial about an invariant point is shown in Figure 3.1. Solving the combinatorial formula for this system:

Equilibria	C	P	F	A
Invariant	1	3	0	1
Univariant	1	2	1	3
Divariant	1	1	2	3

The values for A confirm Figure 3.1 -- only one invariant point exists, but there are three univariant curves and three divariant fields. Solution of Equation 3.10 for a binary system yields:

Equilibria	C	P	F	A
Invariant	2	4	0	1
Univariant	2	3	1	4
Divariant	2	2	2	6

Thus only one invariant point can exist with four univariant curves. The fact that six divariant assemblages can be present seems contradictory. If four univariant curves are radial about an invariant point and these curves are separated from one another by divariant fields, then why should more than four divariant fields exist? The answer to this question will shortly be forthcoming. To determine how these four curves are arranged around the invariant point, we turn to *Schreinemakers' rules*.

3.4.2 Schreinemakers' Rules

In a series of 29 articles that spanned 10 years (1915-1925), Schreinemakers developed a number of axiomatic principles based on geometric considerations that can be used to place constraints on the construction of *P-T* diagrams for multicomponent systems. The significance of this work was perhaps not fully appreciated by most petrologists until the classic paper by Zen (1966). A number of books referenced herein (e.g., Ehlers, 1972; Ernst, 1976; and Maaloe, 1985) discuss Schreinemakers' rules, but this book relies primarily on Zen's paper. First, the basic rules must be defined and explained by means of a hypothetical example and then some complications must be considered.

3.4.2.1 Basics

Schreinemakers' rules can be summarized as six axiomatic principles, but all the geometric arguments behind these rules are beyond our intent here. See Zen (1966) for more details. The six rules are:

1. *The fundamental axiom*
2. *The Morey-Schreinemakers rule*
3. *The overlap rule*
4. *The univariant scheme*
5. *The bundle theorem*
6. *The rule of metastable extensions*

The rule of metastable extensions applies to all binary systems, but is not applicable to all ternary or quaternary systems. As we are considering only binary systems, it is included herein. This rule and the bundle theorem are primarily used as checks. The fundamental axiom, although basic to the other rules, is of less practical use. The Morey-Schreinemakers and overlap rules, used in conjunction with the univariant scheme, are of more immediate practical value.

The fundamental axiom is a direct outgrowth of Gibbs *(G-P-T)* surfaces discussed above and shown on Figure 3.27. In fact, it is self-explanatory after that discussion. One mineral assemblage has a lower overall G and thus is more stable on one side of a univariant curve, whereas the opposite situation is true for the other assemblage. This holds for each side of stable as well as metastable extensions of univariant curves. The Morey-Schreinemakers rule is also easily explained. A divariant field must occur in a sector whose internal angle about the invariant point is 180° or less; i.e., a divariant field can never be larger than half the entire field around the invariant point. This is clear on Figure 3.26; all four divariant fields have internal angles that are less than 180°.

The overlap rule is less easily explained. In fact, the version of the overlap rule in this book is somewhat different from that in other sources. It has been developed over the years as a version more easily understood by introductory petrology students. First, it is necessary to explain the difference between single and composite divariant fields. A composite field, whose internal angle about the invariant point must also be 180° or less, is composed of two or more single fields. Refer to Figure 3.26. The (Z,Y) composite field (with the hypothetical unique stable assemblage W + X) consists of two single divariant fields (Z,W) and (W,Y), with respective unique assemblages X + Y and X + Z. Similarly, the (X,W) composite field consists of the two single fields (X,Z) and (Z,W). We can now state the rule: if a composite field contains (or "overlaps") a single field, then the hypothetical unique assemblage for the composite field must also be a possible assemblage for the single field. We refer to this type of assemblage as an "overlap" assemblage, to distinguish it from a unique assemblage, which has already been defined. Thus any single divariant field can contain two types of stable assemblages, unique and overlap.

We can now explain why A = 6 divariant fields for a binary system, as calculated from the combinatorial formula. All possible unique and overlap assemblages in Table 3.4 are: W + X, W + Y, W + Z, X + Y, X + Z, and Y + Z, for a total of six. In addition, we now have the tools necessary to explain the equilibrium diagram in each divariant field on Figure 3.26. Table 3.4 provides the explanation. Field (Z,W) has three divariant assemblages, one unique and two overlap, so all four minerals must be stable in this field. Fields (X,Z) and

(W,Y) each have one unique assemblage and one overlap assemblage, resulting in three stable minerals in each field. Only the unique assemblage W + Z is stable in field (X,Y).

A brief explanation of the overlap assemblages is required. Using the divariant field (Z,W) as an example, two overlap assemblages are present. The Y + Z overlap assemblage exists because of the overlapping composite field (W,X). The W + X overlap assemblage is the result of the composite field (Y,Z). Because no composite fields overlap field (X,Y), no overlap assemblages exist in this field. The two single fields (X,Z) and (W,Y) are each overlapped by one composite field, so they each have one overlap assemblage.

TABLE 3.4 Divariant Assemblages for a Binary System

Field	Diagram	Assemblages	
(Z,W)			
		X + Y	unique
		Y + Z	overlap
		W + X	overlap
(X,Z)			
		W + Y	unique
		Y + Z	overlap
(X,Y)			
		W + Z	unique
(W,Y)			
		X + Z	unique
		W + X	overlap

The univariant scheme is best explained by an example. Refer to univariant curve (W), representing reaction X + Z = Y, on Figure 3.26. We can write the following:

$$(W): X + Z = Y \quad \therefore \quad (X)\,(Z)\,|(W)|\,(Y)$$

This latter expression is the short-hand way of saying "(X) and (Z) are on one side of (W) and (Y) is on the other," where (W)-(Z) are univariant curves. Similarly:

$$
\begin{aligned}
(Z): W + Y = X \quad &\therefore \quad (W)\,(Y)\,|(Z)|\,(X) \\
(X): W + Z = Y \quad &\therefore \quad (W)\,(Z)\,|(X)|\,(Y) \\
(Y): W + Z = X \quad &\therefore \quad (W)\,(Z)\,|(Y)|\,(X)
\end{aligned}
$$

A careful examination of Figure 3.26 demonstrates that all four of these statements are correct. This is the univariant scheme, which can be used to arrange univariant curves in the proper sequence around the invariant point.

A bundle of one or more stable curves consists of all those stable curves that are not separated from one another by metastable extensions. A bundle can also be defined as all metastable extensions not separated from one another by stable curves. The angle subtended by a bundle must be no greater than 180°. The bundle theorem states that the number of bundles around an invariant point must be odd. Again refer to Figure 3.26. No matter if stable or metastable extensions are used to define the bundles, three bundles exist around the invariant point, which follows the rule.

Finally, the rule of metastable extensions states that the total number of stable phases plus the number of metastable extensions must be equal in all single divariant fields. This can be tested for Figure 3.26. Four stable phases exist in field (Z,W), but no metastable extensions. Two metastable extensions and two stable phases are in field (X,Y). The other two fields, (X,Z) and (W,Y), contain three stable phases and one metastable extension each. Thus a total of four stable phases and metastable extensions exist in each divariant field. Recall that this rule only applies in every case for binary systems.

3.4.2.2 Complications

Application of these rules seems quite straightforward, but some complications exist. Two will be considered, *enantiomorphic forms* and *degenerate systems*. We will consider enantiomorphic forms first. Upon reflection it should be apparent that all Schreinemakers' rules can be satisfied by other arrangements of univariant curves around the invariant point on Figure 3.26. For example, if we simply reverse the metastable and stable extensions on this figure, a different system will result. The original and reversed forms of this figure are termed enantiomorphic forms. Other enantiomorphic forms of this figure exist as well. Imagine revolving this array of univariant curves 180° around either a horizontal (E-W) or vertical (N-S) axis. Two more enantiomorphic forms would be the result.

How do we determine which form is correct? Lacking experimentally determined univariant curves, we could turn to thermochemical data. The four Clapeyron slopes could tell us which form is correct, assuming their respective error envelopes are sufficiently small to determine if they are positive or negative. Alternatively, we could examine relative entropies and molar volumes (or densities) of unique divariant assemblages. The W + Y assemblage in the (X,Z) field has a combined lower temperature and pressure than the other unique assemblages. Relative to the other unique assemblages, it should have a combined lower entropy and density (higher molar volume), assuming Figure 3.26 is the correct enantiomorphic form. The rocks in which these assemblages are present may also give us some clues about which assemblages are stable at different pressures and temperatures.

The second complication deals with degenerate systems. A system is considered degenerate if "chemical coincidences" exist in the system. The only chemical coincidence possible in a binary system is the presence of polymorphs. Four topologies are possible, as shown in the following equilibrium diagrams:

The symbol ∩ indicates the presence of one or more polymorphs for that composition. The system represented by Figure 3.26 is nondegenerate. Special rules apply to construction of arrays of univariant curves for degenerate systems. At first exposure degenerate systems appear to be related to non-degenerate systems as irregular verbs are to regular verbs in a language course. Upon further examination, however, they do follow certain rules, although not as straightforward as those followed by non-degenerate systems such as Figure 3.26. We will not consider degenerate systems further. Likewise, we will not deal with the application of Schreinemakers' rules to ternary systems. The details are more complicated, but the basic principles are identical. Refer to Zen (1966) for a comprehensive treatment of both these topics.

4

TERNARY AND
QUATERNARY SYSTEMS

A great number of fundamentals about melting and crystallization of igneous rocks have been learned from binary phase diagrams, but as mentioned many times, they do not approximate compositions of natural igneous rocks very well. Ternary systems are an improvement in this regard, but they also are limited. Although ternary systems can be represented by two-dimensional, triangular phase diagrams, some new techniques must be learned for their interpretation. These diagrams are actually two-dimensional projections of triangular prisms, where the third dimension is normally temperature. Quaternary systems are even better approximations of natural rocks. In fact, some quaternary systems seem to predict variations in major elements and rock-forming minerals fairly well. Unfortunately, a quaternary system is typically represented by a three-dimensional phase diagram, a tetrahedron, which greatly complicates its interpretation. A typical procedure when reading a quaternary phase diagram is to make two-dimensional projections from several different perspectives. This chapter is concerned with all these techniques. It will cover only the most basic ternary and quaternary systems. For more complete coverage, refer to Cox and others (1979) and Sood (1981). Ernst (1976) also provides a particularly good survey of ternary systems.

4.1 TERNARY SYSTEMS

4.1.1 A Simple Ternary Eutectic

For a ternary system $C = 3$, so some graphical method must be available to represent compositions of each of the three components. One way to do this might be a three-dimensional, orthogonal, X-Y-Z plot in which each mutually perpendicular axis represents mol fraction (or percent) of each component. However, other intensive parameters, such as temperature or pressure, are equally important. Moreover, a two-dimensional representation of the system is far more easily interpreted.

An alternative is to use a triangular prism, on which each of the three apices represents a component (AA, BB, and CC) and the height of the prism represents either temperature or pressure, most commonly temperature. A two-dimensional, orthographic projection of such a prism is shown in Figure 4.1a. The top of this triangular prism is contoured to represent the liquidus surface. Such a prism is less than satisfactory because it is a three-dimensional figure, which means that we would have to cast or mold a prism for each ternary phase diagram. Furthermore, two-dimensional projections such as Figure 4.1a are inadequate when we attempt to use such figures to describe magma behavior. We can avoid these problems by making a two-dimensional, triangular projection and depicting the liquidus surface by isothermal (or rarely isobaric) contour lines (Fig. 4.1b). Unfortunately, the solidus surface cannot also be represented on the same figure without creating confusion.

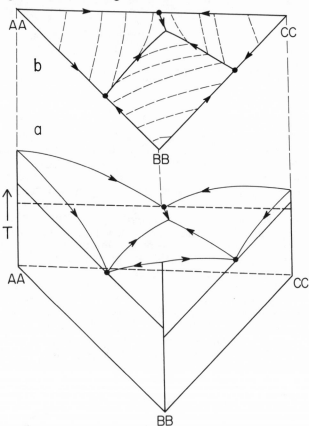

Figure 4.1 *a*. Three-dimensional perspective of the hypothetical ternary system AA-BB-CC, a triangular prism with a contoured upper surface (the liquidus surface); *b*. projection of this contoured triangular prism onto a flat surface. *Arrows* indicate decreasing temperature and *dashed curves* on *b* are temperature contours.

4.1.1.1 The System FO-SP-LC

The equilateral triangle in Figure 4.2 is similar to Figure 4.1b. The three components, forsterite-spinel-leucite (FO-SP-LC), that make up this ternary system can also be considered as three binary systems FO-SP, SP-LC, and LC-FO, all three of which are shown on Figure 4.2. This figure represents the ternary phase diagram for the system FO-SP-LC at 1 atmosphere pressure. If each of the three binary systems has only a binary eutectic, it is logical that the ternary system will have only a *ternary eutectic* (point E_T, Fig. 4.2). The system FO-SP-LC represents the simplest type of all ternary systems with immiscible solids. Again we can make the comparison to topography in that the triangular diagram is analogous to a topographic contour map and the liquidi on the binary diagrams are similar to topographic cross sections. Contour lines on the ternary diagram are on the liquidus surface only.

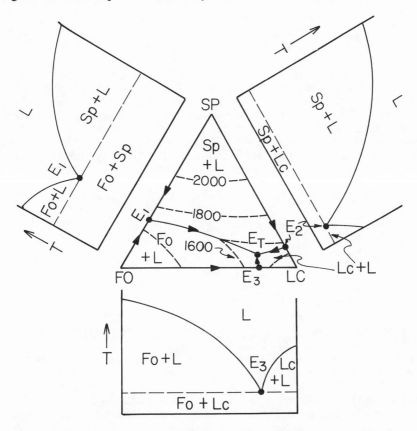

Figure 4.2 The ternary system FO-SP-LC at 1 atmosphere pressure (modified after Schairer, 1955; Jour. Am. Ceramics Soc., v. 38, p. 153-158). *Dashed lines* are temperature contours in °C. Also shown are the three appropriate binary systems. See text for abbreviations.

The three points labeled E_1 E_2, E_3 are binary eutectics and, as stated earlier, the point E_T is the ternary eutectic. The curves E_1-E_T, E_2-E_T, and E_3-E_T

are called *ternary cotectics* and are along the bottom of three *thermal valleys*. The isothermal contour lines on the liquidus surface "V" up temperature across thermal valleys just as topographic contour lines "V" upstream. The ternary eutectic E_T is at the lowest temperature on the entire diagram, so to continue this analogy with topography, E_T is similar to a sinkhole or playa lake in that all thermal valleys run down temperature toward it. Arrows on cotectics and side-lines of the triangular phase diagram are in the direction of decreasing temperature. Two cotectics bound a *primary-phase field*, of which there are three: Fo + L, Sp + L, and Lc + L. The mineral Fo will be the first to crystallize in the Fo + L field, and similarly for minerals Sp and Lc in their respective primary-phase fields. Melting points of pure minerals Fo, Sp, and Lc will be at points FO, SP, and LC, respectively. These melting points will be the highest temperature points in each of their respective primary-phase fields. Melting points for pure phases in this system are:

Fo: 1890°C
Sp: 2135°C
Lc: 1686°C

The solidus surface is an isothermal, flat plane at temperature E_T, which is the lowest temperature on the liquidus surface for the ternary system. Thus the ternary solidus intersects the ternary liquidus only at point E_T, analogous to the binary FO-SP solidus intersecting its liquidus only at point E_1. The ternary solidus does not appear on the ternary diagram. Because of this, solidi are dashed for the three binary systems. This does not imply that the three binary solidus (eutectic) temperatures are the same as the ternary solidus (eutectic) temperature. Because temperature must decrease from each binary eutectic down each cotectic to the ternary eutectic, binary solidus temperatures must be higher than the ternary eutectic temperature. Eutectic temperatures are:

E_T: 1473°C
E_1:   1725°C
E_2:   1553°C
E_3:   1493°C

4.1.1.2 Application of Phase Rule

Each ternary diagram for which temperature is contoured on the liquidus by necessity must be at constant pressure, so the phase rule can still be written as $P + F = C + 1$. Because $C = 3$, $P + F$ must equal 4. Application of the phase rule to a typical ternary system such as that on Figure 4.2 is summarized in Table 4.1. As for unary and binary systems (see Tables 3.1 and 3.3), on a ternary phase diagram an area is divariant, a line or curve is univariant, and a point is invariant. For the ternary diagram of Figure 4.2, these relationships can be summarized as follow:

Primary-phase field   divariant
Cotectic   univariant
Eutectic   invariant

In a primary phase field two phases coexist, the melt and one mineral, so only two of the four intensive variables $(X_{FO}, X_{SP}, X_{LC},$ and temperature) must be fixed to define the state of the system. If we know any two of these variables, we can determine the other two from the diagram. On a triangular diagram only two compositional variables are required to know the third (see Section 1.3.2.1), and temperature can then be determined from the contour lines. Likewise, if temperature and one compositional variable are known, the other two variables can be determined. Along any one cotectic only one variable needs to be known to determine the others, because the intersection of that cotectic with an isothermal or isocompositional contour line is a unique point. In fact, all the above arguments are based on the simple fact that the intersection of two lines or curves is a point, and thus a unique solution.

TABLE 4.1 Application of Phase Rule to a Ternary System

C	P	F	Equilibria	Variables	Geometry
3	2	2	divariant	any 2 of 4	area
3	3	1	univariant	any 1 of 4	curve
3	4	0	invariant	none	point

4.1.1.3 Recognition of Ideal Solutions

We will make no attempt to calculate a ternary phase diagram similar to that in Figure 4.2 from thermochemical data, although if ideal solutions are assumed it can be easily done. One could first construct the three binaries in a manner similar to that described in Section 3.2.4.6. This would define binary eutectics and intersections of contour lines around the sidelines of the ternary diagram. On Figures 4.3a-b the three cotectics have deliberately been constructed to fall along constant-ratio lines, each of which can be extended through an apex of the triangle. Only a straight line that passes through an apex of the triangle can be a constant-ratio line. For example, the Aa-Bb-L cotectic falls along an AA-BB constant-ratio line that extends to the CC apex.

   One requisite of an ideal solution is that it is not affected by changes in compositions of other components in the system. If components AA and BB behave as ideal components in the melt, then their coprecipitation as Aa and Bb should not be affected by the amount of CC in the melt. Stating this differently, the ratio of Aa to Bb along their mutual cotectic should not change, so the Aa-Bb-L cotectic, as well as the other two cotectics, fall along constant-ratio lines.

   A curved cotectic may not be a constant-ratio line, nor can a linear cotectic whose extension does not pass through the far apex of the triangle. For example, cotectic E_1-E_T on Figure 4.2 is slightly curved; cotectic E_2-E_T is slightly curved and its extension does not pass through point FO; and the extension of the short cotectic E_3-E_T does not pass through point SP. Consequently, the components FO, SP, and LC do not behave ideally in the melt, although their behavior does not deviate radically from ideality.

4.1.1.4 Relative Solubilities

An additional lesson can be learned from Figure 4.2. The Lc + L field is small relative to Fo + L, which in turn is smaller than Sp + L. The relative sizes of these fields are an indication of relative solubilities of components in the melt. The component LC is the most soluble; FO is next; and SP is least soluble.

Consequently, the size of a primary-phase field is inversely related to the solubility of the component that comprises its pure mineral. As an example, a melt with only about 15 percent SP (a melt composition in the Sp + L field just above E_T) will initially precipitate Sp on cooling, whereas a melt of at least 75 percent LC is required before Lc will crystallize initially. Relative to LC, SP must be quite insoluble in the melt. The solubility of FO is intermediate between that of LC and SP.

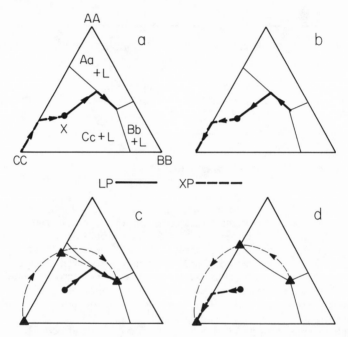

Figure 4.3 Hypothetical ternary system AA-BB-CC showing four different processes affecting composition X: a. equilibrium crystallization of melt X; b. equilibrium melting of rock X; c. fractional crystallization of melt X; and d. fractional melting of rock X. LP and XP represent liquid path and crystal path, respectively. Dashed curve for c indicates rock hop, and dashed curve for d signifies melt hop.

4.1.1.5 Equilibrium Crystallization and Melting

Liquid and crystal paths for equilibrium and fractional crystallization, as well as equilibrium and fractional melting, are shown on Figure 4.3 for the hypothetical system AA-BB-CC. At this point we will only qualitatively consider LPs and XPs; a quantitative treatment is given in Section 4.1.4.

Considering equilibrium crystallization first (Fig. 4.3a), a melt of composition X will initially precipitate Cc when its temperature cools to the liquidus surface. A corollary of the lever rule states that during crystallization of Cc, LP must move along a line directly away from CC until it reaches the cotectic, at which point the melt is saturated with both Cc and Aa. The two minerals Cc and Aa coprecipitate while LP moves down the cotectic toward the ternary eutectic. During crystallization of pure Cc, XP obviously stayed at CC, but XP moves along the AA-CC join toward AA during the coprecipitation of Aa and Cc. When LP reaches the ternary eutectic, all three phases coprecipitate and XP leaves the AA-CC join and moves toward X along a straight line that must pass through the ternary eutectic (again a corollary of the lever rule). When XP arrives at X, the final "rock" composition equals the initial melt composition and crystallization ceases. Arrows on LP and XP mark directions of decreasing temperature and increasing degree of crystallization.

Complete equilibrium melting (Fig. 4.3b) is the opposite of equilibrium crystallization. LP and XP are the same for both; only the directions of arrows have been reversed. A source rock of composition X will begin melting at the eutectic. During eutectic melting XP moves from X along a line toward the AA-CC join; the back projection of the line must go through the ternary eutectic. When XP reaches the join all Bb in the source rock has been melted, at which point LP departs the ternary eutectic and moves along the cotectic while XP moves along the AA-CC join toward CC. When XP arrives at CC, the melt composition, X, and CC must all fall along a straight line. Pure CC then begins melting and LP moves away from the cotectic toward X. When LP arrives at X the final composition of melt equals the initial source rock composition, so melting is complete.

4.1.1.6 Fractional Crystallization and Melting

The liquid path for fractional crystallization is identical to that for equilibrium crystallization, but XP is quite different (Fig. 4.3c). Remember that for fractional crystallization XP marks only the bulk composition of crystals presently crystallizing, not the bulk composition of all crystals removed since the beginning of crystallization. For this reason XP is represented by three discrete compositions separated by two "rock hops." The initial discrete composition is pure CC; the second is a point (for a curved cotectic, a short line) represented by the intersection of back tangents to LP along the cotectic with the AA-CC join; the third is the eutectic. Thus, initially pure Cc fractionates, then an Aa-Cc mixture fractionates, and finally a eutectic mixture Aa-Bb-Cc crystallizes. The liquid path is moving from X to the cotectic during fractionation of Cc, moving along the cotectic during cofractionation of Aa and Cc, and is stationary on the ternary eutectic during eutectic crystallization.

The XP for complete fractional melting (defined in Section 3.2.6.1) is identical to that for equilibrium melting, but LP is different (Fig. 4.3d). In this case XP is moving from X to the AA-CC join while a ternary eutectic melt is forming. When XP arrives on the AA-CC join all Bb has been melted and the remaining source rock is now binary. An interval of time passes while the source is being heated from the ternary eutectic temperature to the binary eutectic temperature. When the AA-CC binary eutectic temperature is reached, the melt composition hops from the ternary eutectic to the binary eutectic. Continued melting on the binary eutectic will cause XP to move along the AA-CC join toward CC. When XP arrives at CC the system becomes unary and another time interval passes until the system heats from the binary eutectic to the melting point of pure Cc. When this final temperature is reached, LP hops from the AA-CC binary eutectic to pure CC, until all Cc is melted.

4.1.1.7 All Possible Liquid and Crystal Paths

Figure 4.3 has concentrated on particular LPs and XPs, but all possible LPs and XPs for a similar system are easily constructed (Fig. 4.4). All LPs must fall along either cotectics or radiate out from the three apices of the triangle (Fig. 4.4a). Conversely, all XPs must be radial around the ternary eutectic or are along one of the three sidelines (Fig. 4.4b). A careful examination of Figure 4.3 reveals that indeed LP radiates from CC or is along the Aa-Cc cotectic, while XP radiates from the ternary eutectic or falls along the AA-CC join.

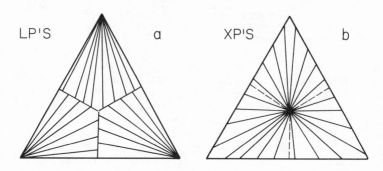

Figure 4.4 Examples of: *a.* all possible liquid paths for ternary eutectic system; and *b.* all possible crystal paths for ternary eutectic system.

4.1.1.8 Isothermal Sections

A great deal can be learned from the construction of isothermal sections through a ternary diagram. If the isothermal section is made above the liquidus (Fig. 4.5a), then of course a "one-phase field" of melt exists throughout the system. Stating this differently, no matter what the composition of the system, it will be 100 percent melt. If the section is made at a temperature that intersects the liquidus surface but not a cotectic (Fig. 4.5b), then the one-phase melt field is surrounded by three "two-phase fields": Aa + L, Bb + L, and Cc + L. Boundaries of the melt field are the appropriate contour lines for the isothermal section. In each of these two-phase fields radiating tie lines are shown connecting coexisting melt and crystal compositions. These radiating tie lines are constant-ratio lines that radiate out from the the three apices AA, BB, and CC, each apex representing the composition of Aa, Bb, and Cc, respectively. The opposite end of each tie line represents the melt composition that is in equilibrium with the appropriate mineral at the temperature given on the contour line that bounds the melt field.

   If the isothermal section is made so that it intersects the three cotectics as well as the liquidus surface, then in addition to a one-phase melt field and three two-phase fields, three three-phase fields also exist (Fig. 4.5c). In each three-phase field a melt of fixed composition (on a cotectic) and crystals of two minerals coexist. Note that only one melt composition can exist in a three-phase field, whereas any number of melts (on the contour line along the boundary of the melt field) can exist in a two-phase field. If the isothermal section represented by Figure 4.5c is made, for instance, at 1000°C, then the 1000°C contour line will bound the melt field. If the isothermal section is made below the solidus (Fig. 4.5d), then the entire diagram is a three-phase field (Aa-Bb-Cc); i.e., no melt remains and the system at every composition will consist entirely of some combination of minerals Aa, Bb, and Cc.

4.1.2 Other Ternary Systems with Immiscible Solids

To this point we have only considered one of the simplest of all ternary eutectic systems. Five more topologies for the liquidus are necessary to describe most ternary systems involving immiscible solids (Fig. 4.6). Contour lines are not shown on these diagrams. Arrows are used to indicate directions of decreasing temperature; these are normally used on complicated ternary phase diagrams rather than contour lines to avoid clutter. The dashed line is the subsolidus join Cc-Ab. Also to avoid clutter, only Figure 4.6a (equivalent to Figure 4.6c) is labeled. For every diagram on Figure 4.6, no matter the location of the Cc-Ab subsolidus join, the primary-phase field between Aa + L and Bb + L is Ab + L.

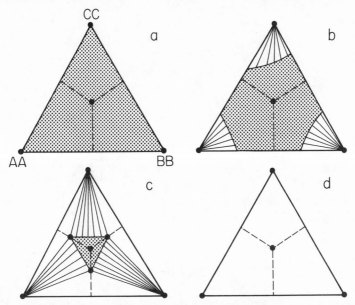

Figure 4.5 Four isothermal sections at progressively lower temperatures for a ternary eutectic system: *a*. above the liquidus (highest temperature); *b*. between the liquidus and solidus showing three two-phase fields (with tie lines); *c*. between the liquidus and solidus showing three two-phase and three-phase fields; and *d*. below the solidus (lowest temperature). *Stippled areas* represent one-phase melt fields.

4.1.2.1 A Thermal Divide

Figure 4.6b, with a thermal divide and two ternary eutectic points, is a simple extension of Figures 4.2 and 4.3. Four binary eutectics are located around the sidelines of Figure 4.6b. The subsolidus join Cc-Ab marks the crest of the thermal divide; temperature decreases away from this divide in both directions toward the ternary eutectics. Consequently, all melts to the left of this divide, upon crystallization, evolve toward the Aa-Ab-Cc ternary eutectic, whereas all melts to the right evolve toward the Bb-Ab-Cc ternary eutectic.

The Cc-Ab subsolidus join also marks the change in temperature on the solidus (compare with Fig. 4.7b). The solidus can be considered as two iso-thermal planes, each at a different temperature, joined along the Cc-Ab sub-solidus join. This is true for every diagram on Figure 4.6. The system depicted on Figure 4.6b can be treated as two ternary eutectic systems, AA-AB-CC and BB-AB-CC, each of which behaves in a manner identical to that described for Figure 4.3.

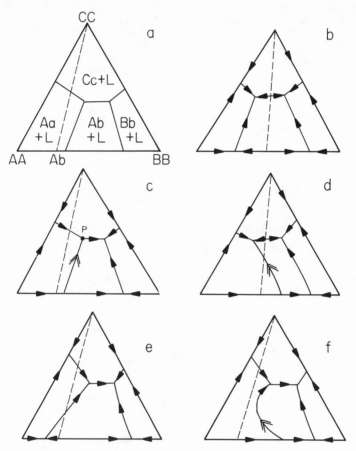

Figure 4.6 Other ternary systems with immiscible solids. Labels given in *a* are applicable for remaining diagrams. Reaction curves are shown with *double arrows,* while coprecipitational cotectics are labeled with *single, dark arrows.*

Figure 4.6b is the ternary analog of the binary system NE-Q (see Fig. 3.10). Consequently, the discussion in Section 3.2.5.2 regarding the relationship of thermal divides to igneous rock series can also be applied to ternary systems with thermal divides. In general, rocks resulting from parental magmas on the left side of the Cc-Ab subsolidus join will be in one rock series, while those on the right side will be in another.

4.1.2.2 Reaction and the Alkemade Theorem

Note the double arrows on the Aa-Ab cotectic of Figure 4.6c. These arrows signify a ternary *reaction curve*, the ternary analog to a binary reaction point or peritectic. This becomes apparent when examining the system FO-AN-Q at 1 atmosphere pressure in Figure 4.7. A cotectic with only one arrow, such as all cotectics on Figure 4.6b and the remainder on Figure 4.6c, is termed a *coprecip-itational cotectic,* because two minerals coprecipitate when the melt composition (LP) is moving along it. Along a reaction curve one mineral reacts with the melt to form another mineral. On Figure 4.6c Aa will react with the melt to form Ab as LP moves along the reaction curve. We will treat this process quantitatively in Section 4.1.4, but for the moment a question must be asked: if the arrows are absent and we wish to add them, can one determine which cotectics are copre-cipitational and which are reaction curves?

The answer to this question is affirmative, with the aid of the *Alkemade theorem.* Expressed as succinctly as possible, this theorem states that arrows along a cotectic depicting decreasing temperature must point directly away from either the join representing the cotectic or a linear extension of that join. If arrows point directly away from the join itself (i.e., a back-tangent from the cotectic intersects the join), then the cotectic is coprecipitational. If the arrows point away from a linear extension of the join (i.e., a back-tangent from the cotectic intersects the extension of the join but not the join itself), then the cotectic is a reaction curve.

Figure 4.6c will serve as an illustration of the application of the Alke-made Theorem. Any arrow drawn on the Ab-Bb cotectic must point directly away from the Ab-BB join, but cannot point directly away from an extension of the join, so the Ab-Bb cotectic must be coprecipitational. We indicate this by drawing one arrow on the cotectic. All back-tangents along the cotectic Ab-Bb intersect the Ab-BB join. Conversely, any arrow drawn on the Aa-Ab cotectic cannot point directly away from the AA-Ab join, but must point away from an extension (to the right) of the AA-Ab join. All back-tangents from the Aa-Ab cotectic intersect the extention of the AA-Ab join, but not the join itself. Hence the Aa-Ab cotectic must be a reaction curve and is labeled with double arrows. Using this test on the remaining cotectics on Figure 4.6c and all cotectics on Figure 4.6b, they all must be coprecipitational and are labeled with single arrows.

We can now apply the Alkemade theorem to the pseudoternary system FO-AN-Q (Fig. 4.7). It is pseudoternary because the composition of Sp (spinel, $MgAl_2O_4$) does not fall anywhere within the system, yet a Sp + L primary-phase field exists. Isothermal contour lines are omitted to avoid clutter. The system FO-Q was discussed in Section 3.2.8 in connection with liquid immiscibility. Liquid immiscibility within the pseudoternary system is also present and can be visualized as a "blister"-shaped two-liquid field (a solvus) along the EN-Q sideline near the Q apex labeled 2L. Another interesting feature of the ternary system is the phase boundary representing the phase transformation between the two silica polymorphs cristobalite and tridymite. Because this transformation takes place at a constant temperature (1470°C), the phase boundary must parallel the contour lines. The dashed line En-An is the subsolidus join separating the two solidi.   The solidus for Fo-En-An is at the peritectic temperature P_1 (not P_2), whereas the solidus for En-An-Q is at the eutectic temperature E_T. This becomes apparent when the binary analog FO-Q is closely inspected.

The binary system contains a binary eutectic E_B and a binary peritectic P_B. It should be no surprise that the ternary system has a corresponding ternary eutectic E_T and ternary peritectic P_1.   In fact, it has two ternary peritectics, P_1 and P_2.   The Alkemade theorem indicates that the cotectic Fo + Pren + L (the curve P_B-P_1; Pren is the abbreviation for protoenstatite) is a reaction cotectic because back-tangents all along its extent intersect the extension of the Fo-En join, rather than the join itself.   In contrast, the Pren + SiO_2 polymorph + L cotectic (curve E_B-E_T) must be coprecipitational because all back-tangents along it intersect the join En-Q.

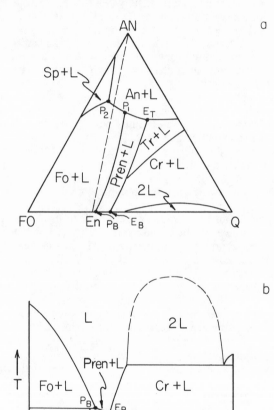

Figure 4.7 The ternary system FO-Q-AN (modified after Andersen, 1915). Also shown is the binary system FO-Q (see section 3.2.8 and Fig. 3.17). See text for explanation of abbreviations.

4.1.2.3 Ternary Invariant Points

We must also consider the various types of ternary invariant points and how to distinguish one from another.   Three types exist, two of which have already been illustrated. Cox and others (1979) is recommended reading for principles to distinguish different types of invariant points and cotectics. A ternary eutectic can be considered as a coprecipitational invariant point, because all three phases coprecipitate at a ternary eutectic. A useful device to distinguish it from the other two types is to remember it as a "three-in" point, referring to the fact that all three arrows on the surrounding cotectics point inward to a ternary eutectic. Note that arrows on all three cotectics surrounding each ternary eutectic on Figure 4.6b point down temperature toward the eutectic. Another important characteristic of a ternary eutectic is that it falls *inside* the three-phase triangle representing its subsolidus mineral assemblage. For instance, the Pren-Tr-An ternary eutectic E_T falls inside the En-An-Q three-phase subsolidus triangle of Figure 4.7. Reasons for this involve the contact principle (see Section 3.2.2.2) and will be explained in the next section.

The second type of ternary invariant point is shown on Figure 4.6c as the point P. This is termed a *monoresorptional* ternary reaction point (or ternary peritectic) because Aa is being resorbed by reacting with the melt to form Ab and Cc. It also can be considered a "two-in, one-out" peritectic in that two arrows on surrounding cotectics are pointed down temperature toward it and the third arrow is pointing away, down temperature toward the ternary eutectic. In contrast with a ternary eutectic, this peritectic does not fall inside its three-phase subsolidus field. The Aa-Ab-Cc peritectic clearly does not fall inside the Aa-Ab-Cc three-phase field. Point P_1 on Figure 4.7 is an example of a "two-in, one-out" peritectic.

The third type of invariant point, a *biresorptional* or "one-in, two-out" peritectic is uncommon and is not shown on Figure 4.6 or 4.7. In this case two minerals are being resorbed by reacting with the melt to form a third mineral. One down-temperature arrow on a surrounding cotectic points toward the biresorptional peritectic and two point away. It also falls outside its subsolidus three-phase field.

Armed with these techniques for distinguishing different types of cotectics and invariant points, we now can consider Figures 4.6d-f. In each case we are concerned only with the Aa-Ab cotectic and join, as well as the Aa-Ab-Cc invariant point and subsolidus three-phase field. For all three figures the remaining cotectics are all coprecipitational and the remaining invariant points are all ternary eutectics.

For Figure 4.6d, the Alkemade theorem indicates that cotectic Aa-Ab is a reaction curve. The Aa-Ab-Cc invariant point must be a ternary eutectic, because it falls inside the Aa-Ab-Cc subsolidus field and the "three-in" rule applies. For Figure 4.6e, the reverse situation prevails: according to the Alkemade theorem, the Aa-Ab cotectic is coprecipitational and point Aa-Ab-Cc is a "two-in, one-out" mono-resorptional, ternary peritectic. Figure 4.6f is an example of a system with an extremely curved cotectic (clearly not an ideal solution!). Point Aa-Ab-Cc is another "two-in, one-out" ternary peritectic, but curve Aa-Ab is more complex. The Alkemade theorem indicates that it is a reaction curve near the base of the triangle, for CC-deficient melts, but it changes to a coprecipitational cotectic near the ternary peritectic. It changes from resorptional to coprecipitational at the point on the cotectic where its back-tangent intersects point Ab. In summary, consideration of Figures 4.6b-f indicates that any combination of types of invariant points and univariant curves (cotectics) is possible.

4.1.2.4 The Contact Principle Revisited

In Sections 3.2.2.2 and 3.2.6.8 the contact principle was explained and applied to binary systems. This principle is equally applicable to ternary systems, although some apparently new techniques must be learned. In fact, these techniques are extensions of those used previously. We will now examine ternary invariant points in light of the contact principle.

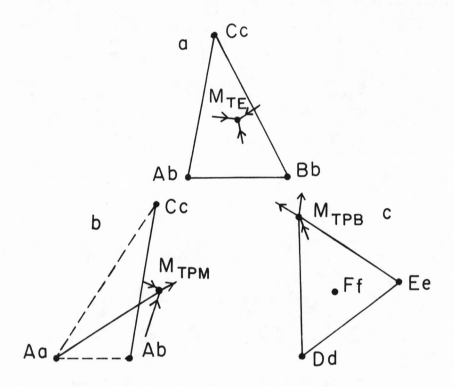

Figure 4.8 Three types of ternary invariant points: *a.* coprecipitational; *b.* mono-resorptional; and *c.* bi-resorptional. See text for explanation of abbreviations.

The three types of ternary invariant points are shown in Figure 4.8. Imagine each of these scalene triangles as lying inside a ternary phase diagram, so each apex of every triangle represents the composition of either a mineral or a melt. In Figure 4.8a the ternary eutectic melt of composition M_{TE} co-precipitates crystals of Ab, Bb, and Cc. This point is comparable to point E_T on Figure 4.7. This coprecipitation can be expressed by the chemical equation:

$$M_{TE} = Ab + Bb + Cc \qquad (4.1)$$

The point M_{TE} falls inside the triangle Ab-Bb-Cc, and thus the triangle and point can be thought to be in contact. In Figure 4.8b the point M_{TPM} represents a mono-resorptional ternary peritectic, analogous to point P_1 on Figure 4.7. A melt of composition M_{TPM} is reacting with crystals of Aa to form crystals of Ab and Cc. The chemical equation can be written:

$$M_{TPM} + Aa = Ab + Cc \qquad\qquad (4.2)$$

The M_{TPM}-Aa tie line crosses the Ab-Cc tie line, so the two tie lines can be considered to be in contact at a point. The point M_{TPB} on Figure 4.8c represents a biresorptional ternary peritectic. A melt of composition M_{TPB} is reacting with crystals of Dd and Ee to form crystals of Ff according to the reaction:

$$M_{TPB} + Dd + Ee = Ff \qquad\qquad (4.3)$$

This situation is similar to that in Figure 4.8a where the point Ff falls inside and thus is in contact with triangle M_{TPB}-Dd-Ee.

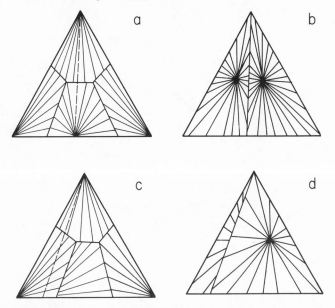

Figure 4.9. Examples of: *a.* all possible liquid paths for Figure 4.6b; *b.* all possible crystal paths for Figure 4.6b; *c.* all possible liquid paths for Figure 4.6c; and *d.* all possible crystal paths for Figure 4.6c.

Recall from the discussion in Section 3.2.2.2 that for a chemical reaction to occur in a binary system the line or point representing the reactants must be in contact with the line or point representing the products. This principle can be easily extended to ternary systems, as shown previously. One need only include a triangle and state that for a chemical reaction to occur in a ternary system the line, point, or triangle representing the products must be in contact the line, point, or triangle representing the reactants. If no contact occurs, a chemical reaction cannot take place. For example, in Figure 4.8b a melt of composition M_{TPM} cannot react with crystals of Cc to form crystals of Aa and Ab because the M_{TPM}-Cc and Aa-Ab tie lines do not cross. Likewise, a melt of composition M_{TPM} cannot coprecipitate crystals of Aa, Ab, and Cc because the point M_{TPM}

does not fall inside, and thus is not in contact with, the triangle Aa-Ab-Cc.

4.1.2.5 All Possible Liquid and Crystal Paths

Patterns made by all possible LPs and XPs for the systems depicted on Figure 4.6 are quite interesting. Only two examples are given (Figs. 4.9a-d; companion diagrams for Figs. 4.6b-c), but given these the others are easily envisioned. Trajectories for all possible LPs and XPs for a ternary system with a thermal divide and two ternary eutectics (equivalent to Fig. 4.6b) are shown on Figures 4.9a and 4.9b. All possible LPs (Fig. 4.9a) radiate out from all four phases and fall along all cotectics. Note that no LP crosses the thermal divide. All possible XPs (Fig. 4.9b) radiate out from the two ternary eutectics or fall the sidelines of the diagram.

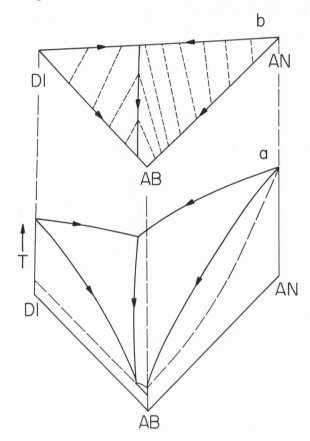

Figure 4.10 *a.* Representation of three-dimensional figure for system AB-AN-DI at 1 atmosphere pressure. *b.* Projection of *a* onto flat surface. *Dashed curves* on *b* are temperature contours. *Dashed line and curve* on *a* represent solidi for binary systems AB-DI and AB-AN, respectively. All *arrows* point down temperature.

The situation is somewhat more complex for Figures 4.9c and 4.9d (equivalent to Fig. 4.6c). All possible LPs on Figure 4.9c also radiate away from all four phases and fall along cotectics, but one complication exists. The point representing the composition of phase Ab falls outside the Ab + L primary-phase field, so LPs within the Ab + L field must radiate from a point outside the field. A similar complication exists for XPs on Figure 4.9d: XPs in the Aa-Ab-Cc three-phase field radiate out from the Aa-Ab-Cc peritectic, but this peritectic falls outside its subsolidus three-phase field. The remaining XPs radiate out from the ternary eutectic, inside its subsolidus three-phase field, or are along the sidelines of the diagram.

4.1.3 Ternary Systems with Solid Solutions

With the exception of quartz, all major igneous rock-forming minerals form at least limited solid solutions. We would be remiss if we failed to examine some basic ternary systems with solid solutions. Diagrams for the more basic of these systems commonly appear deceptively simple. As will become apparent, interpretation of these diagrams can be considerably more complicated than interpretation of ternary diagrams involving only immiscible solids.

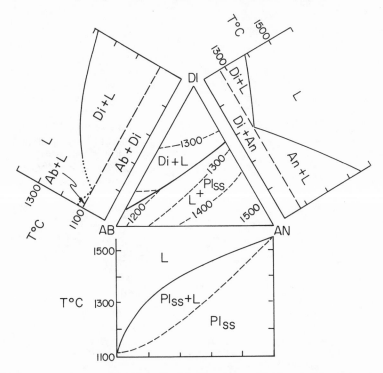

Figure 4.11 The ternary system AB-AN-DI at 1 atmosphere pressure (modified after Bowen, 1915). See also Osborn (1942), Schairer and Yoder (1960), and Kushiro (1973). Also shown are the appropriate three binary systems. Solidi are *dashed* on binary systems. *Dashed curves* on ternary system are temperature contours in °C.

4.1.3.1 The System AB-AN-DI

Perhaps the most basic, certainly one of the earliest, of all ternary systems with a solid solution was originally published in 1915 by Norman L. Bowen. Bowen, for many years Director of the Geophysical Laboratory in Washington, D.C., is generally considered as the father of modern igneous petrology. This is the system AB-AN-DI at one atmosphere pressure (Fig. 4.10). Because plagioclase and clinopyroxene (augite or less commonly pigeonite rather than diopside) are generally the most common rock-forming minerals in most basaltic rocks, this system is a crude approximation of a SiO_2-saturated basaltic melt crystallizing on the earth's surface.

Figure 4.10 is the solid-solution equivalent of Figure 4.1. The system AB-AN-DI as depicted on Figure 4.10 appears quite simple -- it contains only one cotectic and no eutectic. The cotectic lies in the bottom of a thermal valley that runs down temperature from near the middle of the DI-AN join to the AB-DI join near the AB apex.

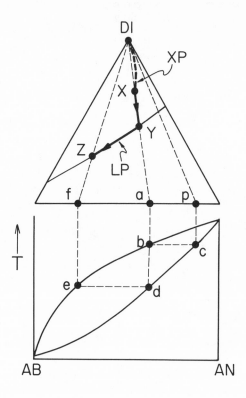

Figure 4.12 Equilibrium crystallization in the ternary system AB-AN-DI for magma *X* based on the simplifying assumption that all solutions are behaving ideally. Also shown is the binary system AB-AN. On ternary system crystal path *(XP)* is shown as *heavy dashed curve,* and liquid path *(LP)* is *heavy solid curve.*

The equilateral triangle in Figure 4.11 represents the standard projection of Figure 4.10b. In addition, the three binary systems, AB-DI, DI-AN, and AN-AB are also shown on Figure 4.11. Points AB, AN, and DI represent melting points for minerals Ab, An, and Di, respectively. Only two primary-phase fields exist in this ternary system: diopside plus liquid (Di + L) and plagioclase solid solution plus liquid (Pl$_{SS}$ + L). As expected, the phase rule indicates that the two primary-phase fields are each divariant and the cotectic is univariant. Two binary eutectics are shown on Figure 4.11. In fact, the system AB-DI is only pseudobinary, so the Ab-Di eutectic point does not truly exist (Morse, 1980). Assumption of an Ab-Di eutectic point, however, is adequate for our purposes. Because the system AB-DI is pseudobinary, the system AB-AN-DI is pseudoternary. We will interpret it, however, as if it were an ideal ternary system; this does not lead to any serious errors in our conclusions. As the solidus surface cannot be shown on the ternary diagram, the three solidi for the respective binary diagrams are dashed. The binary system AB-AN, of course, is the plagioclase solid solution transition loop discussed in Section 3.3.1.2.

4.1.3.2 Crystallization/Melting in the System AB-AN-DI

Equilibrium crystallization in the system AB-AN-DI is depicted in Figure 4.12. Given a magma of composition X in the Di primary-phase field, as pure Di crystallizes, the liquid path, LP, will move straight away from DI until it intersects the cotectic, at which time Pl_{SS} will begin crystallizing. While LP is moving from X to the cotectic, the crystal path, XP, obviously stays at DI.

If we assume that all components behave ideally in the melt, we can approximate the composition of the first plagioclase to crystallize when LP reaches the cotectic. Moreover, we can approximate the composition of the last plagioclase and the final liquid. This technique is in fact an oversimplification, because these components do not behave ideally in the melt (see Section 3.3.1.2). The assumption of ideality, however, does not create any large errors in our results and thus suffices for our purposes. For a more complete discussion resulting in more accurate determinations, see Ernst (1976).

The technique for these approximations is shown in Figure 4.12. The LP is extended along a constant-ratio line until it intersects the Ab-An join; this point of intersection is then extended vertically down until it intersects the plagioclase liquidus. A horizontal, isothermal tie line is then constructed through this liquidus composition and its corresponding solidus composition. These projections are represented by the path DI-X-Y-a-b-c-p. The solidus composition c, which is projected to point p in the ternary system, approximates the composition of the first plagioclase crystallizing when LP reaches the cotectic. If point Y represents the magma composition and p approximates the plagioclase composition when LP reaches the cotectic, then the triangle DI-Y-p represents a three-phase field, all phases of which are in equilibrium.

As temperature is lowered further, LP moves down temperature along the cotectic. Just how far it moves down toward the Ab-Di join can be approx-imated by the path DI-X-Y-a-b-d-e-f-Z. The final magma composition is approximated by Z, the final plagioclase composition by a. Thus the path representing the change in plagioclase composition on the ternary diagram is p-a; its analogous path on the binary diagram is c-d. The final three-phase field, just before crystallization ceases, is DI-a-Z. The liquid path is, therefore, X-Y-Z, while XP is the curved path DI-X. It must be curved because it initially starts moving toward p, but as the crystallizing plagioclase becomes more AB-rich, XP deviates from path DI-p and moves toward X. When crystallization ceases, XP must be at point X, the initial magma composition. In other words, because the system is closed, the final rock composition must be the same as the initial magma composition.

Any original magma whose composition initially falls in the Pl_{SS} field will have a different history (Fig. 4.13). Magma R will begin crystallizing Pl_{SS} and LP will move toward the cotectic along a curved path. This path will be curved because of the change in the crystallizing plagioclase composition from relatively calcic to relatively sodic with decreasing temperature. Initially, LP moves away from a relatively calcic plagioclase composition; later, it will move away from a more sodic composition. The result is a curved path for LP. When

LP reaches the cotectic, Pl$_{SS}$ and Di will coprecipitate while LP moves down temperature along the cotectic. The initial and final plagioclase compositions, in addition to the final magma composition, can be approximated in a manner similar to that for crystallization of magma X on Figure 4.12. While LP is moving across the Pl$_{SS}$ primary-phase field toward the cotectic, XP is moving along the An-Ab join toward AB. When LP starts moving down the cotectic while the magma coprecipitates plagioclase and diopside, XP will depart from the An-Ab join and move along a curved path until it arrives at the original magma composition R as crystallization is completed.

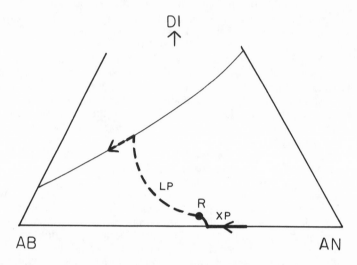

Figure 4.13 Crystal path *(XP)* and liquid path *(LP)* in the ternary system AB-AN-DI for an initial melt *R* whose composition is in the *Pl$_{SS}$* field, assuming equilibrium crystallization.

Not surprisingly, equilibrium melting behaves in exactly the opposite fashion. We can reverse the arrows on LP and XP on Figures 4.12 and 4.13 to describe equilibrium melting. For example, a rock of composition X will begin melting a liquid of composition Z (Fig. 4.12). As temperature is raised, the melt composition (LP) will move along the cotectic toward point Y. Meanwhile, the remaining rock ("restite") composition XP is moving along a curved path from X to DI. When all plagioclase in the source rock is melted, LP arrives at Y and XP arrives at DI. Only DI remains in the source, so continued melting causes LP to move straight toward DI. The final melt composition must necessarily be the same as the initial rock composition X.

Fractional crystallization is rather easily described in the system AB-AN-DI (Fig. 4.14), given the previous discussion of this process for simple binary solid solutions in Section 3.3.1.5. Recall from that discussion that both LP and XP respectively continue down the liquidus and solidus until, at least in theory, they simultaneously reach pure AB. Reasons for this behavior are explained in Section 3.3.1.5 and will not be repeated here. Similarly, during fractional crystallization LP will continue down the cotectic until it theoretically reaches the Ab-Di join at the binary cotectic composition (Fig. 4.14). Given a magma of composition X on Figure 4.14, XP will be at DI until LP reaches the cotectic, at which time XP undergoes a "rock hop" to point m on the line DI-P. This point will be at the intersection of line DI-P and the back-tangent of the cotectic curve at point Y. XP will then follow behind, but will not necessarily coincide with, LP; XP will catch up to LP when LP reaches the Ab-Di join and fractional crystallization ceases.

Fractional melting will not operate in reverse of fractional crystallization. A source rock of composition X will begin melting a liquid of composition Z (see discussion of Fig. 4.12), because incipient fractional melting will be indistinguishable from incipient equilibrium melting. The liquid path LP will then move up temperature along the cotectic until it reaches point Y, at which time all plagioclase is melted. An interval passes when no melting occurs until the temperature is raised to the melting point of Di, at which point pure Di begins to melt. Thus the liquid experiences a "melt hop" from point Y to DI.

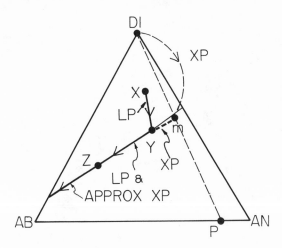

Figure 4.14 Liquid path *(LP)* and approximate crystal path *(XP)* for initial magma *X* in the system AB-AN-DI, assuming fractional crystallization.

4.1.3.3 The System AB-AN-DI Applied to Basaltic Rocks

This system only roughly approximates a very narrow range of naturally occurring basaltic compositions. It corresponds to compositions along the plagioclase-diopside join in the basalt tetrahedron (see Fig. 1.14, Section 1.3.3.2). As neither quartz nor olivine is involved, it is exactly SiO_2 saturated. However, because normative plagioclase and diopside are common to all three major types of basaltic rocks, we cannot state that rocks whose compositions approach this system are of any one particular type. DI-rich melts will initially crystallize clinopyroxene, which will likely, but certainly not necessarily, become phenocrysts. Likewise, AN-rich melts are likely to produce plagioclase phenocrysts. Neither of these observations is particularly surprising.

At least two observations of relationships in this system may have some merit when applied to natural rocks. The first deals with melting relations as interpreted from Figures 4.12 and 4.13. Although LP for partial melting of a basaltic source rock of composition X (Fig. 4.12) is Z-Y-X, it is unlikely that any partial melt compositions would ever depart the cotectic. This is because the percentage of partial melting would have to be prohibitively large for the partial melt composition to leave the cotectic and move toward DI by melting pure Di. As the composition of the partial melt approaches point Y, however, the "restite" approaches a rock with 100 percent pyroxene, a pyroxenite. In this manner partial melting of a pyroxene-rich basaltic source rock might leave a pyroxenite restite behind in the source area. Similarly, partial melting of a calcic plagioclase-rich source (point R; Fig. 4.13) might leave an anorthositic source.

The second observation has to do with the shapes of liquid paths (LPs) as seen on Figures 4.12 to 4.14. When LP reaches a cotectic and an additional phase begins precipitating, a sharp break appears in LP. If a solid solution is involved, LP will have a curved path. We commonly see sharp breaks and curved trends for supposed liquid paths on variation diagrams plotting natural rock analyses (see Section 5.2.2.3). These are most easily explained as entry of an additional phase or changes in solid solution compositions during fractional crystallization. Such explanations are reasonable in the light of crystallization in the system AB-AN-DI.

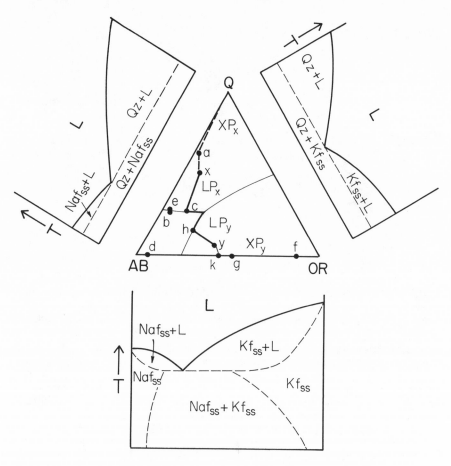

Figure 4.15 The system AB-OR-Q-H_2O at P_{H2O} = 5 kb (modified after Winkler, 1979). Also shown are the appropriate binary systems. Solidi on the three binary systems are dashed. Assuming equilibrium crystallization and melting, liquid path (LPₓ) for initial magma X is shown as *heavy solid curves,* and crystal path (XPₓ) is *heavy dashed line.* Analogous symbology is used for initial magma Y (LPᵧ and XPᵧ).

4.1.3.4 The System OR-AB-Q-H₂O

Because alkali feldspar and quartz are important rock-forming minerals in most granitoids, the system OR-AB-Q-H₂O at elevated fluid pressure is an especially important ternary system. Much of the basic experimental work in this system was done by Tuttle and Bowen (1958) and Luth and others (1964). It is the SiO_2-oversaturated portion of Bowen's "petrogeny's residua system;" the reasons for this term will become clear. Note that the system contains four components, although it is represented on a conventional ternary phase diagram. This is because H_2O is not a component in any of the minerals stable (feldspars and quartz) under conditions considered herein. Had muscovite been a stable phase, a ternary phase diagram would not have been adequate to explain all the complexities in the system without making several assumptions. In the strictest sense, then, this system is pseudoternary.

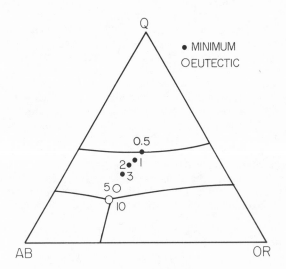

Figure 4.16 Effect of increasing P_{H2O} on the H_2O-saturated pseudo-ternary system AB-OR-Q (modified after Luth and others, 1964). Between 3 and 5 kb solidus intersects solvus and minimum point becomes eutectic point. Minimum and eutectic points migrate approximately toward the AB apex with increasing P_{H2O}.

The ternary system at 5 kilobars H_2O pressure $(P_{H2O} = P_T)$, with its attendant binary systems, is shown on Figure 4.15. Experiments have been conducted at a variety of H_2O pressures; 5 kb was chosen for this discussion because it is a resonable pressure for the crystallization of many granitoids and it can be easily tied to the quaternary haplogranite system (see Section 4.2.3). Superficially, the ternary system appears to be a simple ternary eutectic system, but an examination of the attendant binary diagram AB-OR indicates that this is not so. The binary system AB-OR exhibits partial solid solution and, as a result, a solvus (see Section 3.3.2.4). Consequently, although the ternary system appears to indicate that all three phases are immiscible, this is not the case for Na-rich (Naf$_{SS}$ or Ab$_{SS}$) and K-rich (Kf$_{SS}$ or Or$_{SS}$) alkali feldspars. Both feldspars are immiscible, however, with respect to quartz, as indicated by the simple binary eutectic diagrams for AB-Q and OR-Q (Fig. 4.15). Again, application of the phase rule demonstrates that the ternary eutectic is invariant, the three cotectics are univariant, and the three primary-phase fields are divariant.

Although Figure 4.15 is based on experiments at 5 kb P_{H2O}, it is informative to examine at least qualitatively the effects of changing H_2O pressure on the system (Fig. 4.16). Essentially, there are two main effects. The first has to do with eutectic points and minimum points. Recall the discussion in Sections 3.3.2.1 and 3.3.2.2 of the effect of H_2O pressure on a binary system exhibiting partial solid solution and a solvus. With increasing P_{H2O} the liquidus and solidus temperatures drop rapidly, while the solvus temperature increases slightly. The net effect is at some P_{H2O} the solidus and solvus intersect and an invariant eutectic point results; below this critical P_{H2O} a minimum point exists. This same phenomenon occurs in ternary systems. The critical P_{H2O} for the system OR-AB-Q-H_2O is between 3 and 5 kb (Fig. 4.16). Below this P_{H2O} only one alkali feldspar can crystallize, whereas above this P_{H2O} two, a comparatively AB-rich and an OR-rich, co-precipitate along their shared cotectic.

The second effect pertains to AB solubility in the melt and shift of the minimum or eutectic point with increasing P_{H2O}. We now know that the solubility of a component in the melt is inversely related to the size of the primary-phase field for that component's pure phase. We are not dealing with pure feldspar phases here, but the principle still applies. With increasing P_{H2O} the solubility of AB increases in the melt, resulting in a decrease in the size of the primary-phase field for the Ab-rich feldspar phase (this is especially apparent when a eutectic is present at or above 5 kb; Fig. 4.16). The effect is a shift of the minimum point or eutectic toward the AB apex of the diagram.

If components in a ternary eutectic system are behaving ideally in the melt, then constant-ratio lines through the eutectic should coincide with the cotectics. For example, the extension of an Kf_{SS}/Q constant-ratio line drawn through the eutectic and the AB apex (Fig. 4.15) should coincide with the Kf_{SS}-Q cotectic. To avoid clutter, no constant-ratio lines are drawn on Figure 4.15, but the results of doing so are easily envisioned. Allowing for slight curvature in the Kf_{SS}-Q cotectic, this cotectic and the Kf_{SS}/Q constant-ratio line are reasonably close to one another. In contrast, the other two cotectics and constant-ratio lines deviate from one another considerably. In the next section ideality is assumed in making certain projections, which, although necessary, is clearly an oversimplification.

4.1.3.5 Crystallization/Melting in the System OR-AB-Q-H_2O

As was our earlier procedure, we will first examine equilibrium crystallization, then equilibrium melting, and finally fractional crystallization and melting. Any magma whose initial composition is in the quartz primary-phase field (magma x; Fig. 4.15) behaves in a manner similar to, but not exactly as, a magma crystallizing in a ternary eutectic system with only immiscible solids. The melt precipitates quartz while LP moves straight away from the Q apex. During this time XP is at Q. When LP reaches the Q-Naf$_{SS}$ cotectic at point c, quartz and an AB-rich solid solution (Naf$_{SS}$) begin coprecipitating. LP then moves down the cotectic while quartz and Naf$_{SS}$ are coprecipitating until the ternary eutectic is reached, at which time all three phases, quartz, Naf$_{SS}$, and Kf$_{SS}$, crystallize together at the eutectic. While LP is moving along the cotectic, XP is moving down path Q-a; this path does not fall along the Q-AB sideline because a small amount of OR component is dissolved in the Naf$_{SS}$. Eutectic crystallization begins when LP arrives at the eutectic; XP moves from point a to x. When XP arrives at x, crystallization has obviously been completed.

As expected, equilibrium melting of a source rock of composition x operates in the reverse fashion; arrows can simply be reversed on LP and XP. The first melt has a composition at the eutectic. A eutectic melt continues to form until all Kf_{SS} in the source rock has melted. During this time XP is moving along path x-a. When Kf_{SS} is used up in the source, LP moves back along the cotectic toward point c and XP moves along path a-Q until all Naf_{SS} has melted. At this point only quartz remains in the source, so during the final stage of melting, LP moves from point c on the cotectic to x while XP is at Q.

For fractional crystallization LP is the same as LP for equilibrium crystallization, but the two XPs are different. As LP moves from the initial magma composition x to the cotectic, XP, as for fractional crystallization, stays at Q. When LP reaches the cotectic at point c, XP undergoes a rock hop to b, an approximated composition determined by the intersection of a back-tangent to the cotectic at point c and line d-Q. The composition b is approximated because line d-Q is only appoximate. Note that composition b does not necessarily fall on the cotectic; it does, of course, if the cotectic is a straight line. Finally, when LP arrives at the ternary eutectic, XP hops to the ternary eutectic as the melt coprecipitates Naf_{SS}, Kf_{SS}, and quartz.

Fractional melting results in an XP identical to XP for equilibrium melting; but the two LPs are different. For fractional melting of source rock x, LP stays at the ternary eutectic and XP moves from x to point a until all Kf_{SS} is consumed, at which time LP hops to point e (determined by intersection of d-Q and the cotectic). XP then moves from point a to Q while LP stays at point e. When XP arrives at Q, all Naf_{SS} has melted, so LP hops to Q and the remaining source rock consisting of pure Q melts.

Considerable time passes preceding each of the two melt hops, because there is a large temperature difference between the ternary eutectic and point e (the Naf_{SS}-Q eutectic), as well as between point e and Q, the melting point of pure quartz. Time must elapse as the source rock is first heated from the ternary eutectic temperature to the Naf_{SS}-Q eutectic temperature and then from the Naf_{SS}-Q eutectic to the melting point of pure quartz. In fact, this third melting event of pure quartz seldom, if ever, occurs in nature, because it calls for unrealistically large percentages of partial melting and high temperatures.

So far we have only been concerned with melting and crystallization involving a composition in the primary-phase field of quartz (point x, Fig. 4.15). We need not dwell at length on compositions in the two primary-phase fields of the two feldspars, but we must at least briefly consider them. A discussion of equilibrium crystallization and melting follows for composition y (Fig. 4.15). A useful exercise is to trace XP and LP for fractional crystallization and melting as well. For equilibrium crystallization, a melt of composition y initially crystallizes Kf_{SS} of approximate composition f. LP moves across the Kf_{SS} primary-phase field until it reaches point h. While LP is moving from y to point h, XP is moving from point f to g (approximately!). Then Naf_{SS} and Kf_{SS} coprecipitate as LP moves down the cotectic from point h to the ternary eutectic. During this time XP is moving from point g to k. When LP reaches the ternary eutectic, simultaneous crystallization of Naf_{SS}, Kf_{SS}, and quartz occurs and XP moves from point k to the original melt composition y.

As for previous examples, simply reverse the directions on arrows for LP and XP to describe equilibrium melting of a source rock y. The initial melt has the ternary eutectic composition. While LP is at the ternary eutectic, XP moves from y to point k. When all quartz has melted from the source rock, LP moves along the cotectic toward point h while XP is moving from point k to g. In the unlikely event that all Naf_{SS} is consumed in the source, then LP moves along a curved path from h to y and XP moves approximately from point g to f as only Kf$_{SS}$ melts. All the source rock is consumed when LP arrives at y and XP arrives at f.

4.1.3.6 The System OR-AB-Q-H$_2$O Applied to Granitoids

Any discussion of the application of this system to naturally occurring rocks is predicated on a number of simplifying assumptions. The most obvious assumption is that other components have little effect on the system, which is clearly not the case. For instance, AN component in the feldspar and melt can have profound effects.

Another assumption, that a separate fluid phase exists and thus $P_{H2O} = P_T$, may often be true at shallow levels in the crust, as evidenced by vesicles, amygdules, scoria, and pumice. Miarolitic cavities, evidence for a separate fluid phase in plutonic rocks, are less common. These differences are quite reasonable, because more fluid can be dissolved in the melt at high P_T than at low. Given a fixed amount of fluid, the more that can be dissolved in the melt, the less is available to create a separate fluid phase. Moreover, we must assume some pressure $(P_{H2O} = P_T = 5$ kb in Fig. 4.15), but granitic melts can form by partial melting and can crystallize over a wide range in depths and thus pressures.

Consequently, many assumptions that we must make to apply this system to natural rocks are frequently suspect. Some consolation, however, exists. Many of the breakdowns in these and other assumptions mean that we may not be able to predict quantitatively compositions, temperatures, or pressures, but we can make some qualitative comparisons. In this section we are content with this limited objective.

Let us use the example of quartz monzonite as a typical granitoid. It is an observational fact that most quartz monzonites contain two feldspars, AB$_{SS}$ (plagioclase) and an OR$_{SS}$-rich perthite. Both have limited AN component; usually the perthite has less AN than the Ab$_{SS}$. Because this system has no AN component, we cannot deal with it. Occasionally, however, these rocks only contain one feldspar, an alkali feldspar of intermediate AB-OR with minor AN composition. In quartz monzonites this feldspar is a perthite in which the two phases, Ab$_{SS}$ and Or$_{SS}$, are in approximately equal amounts. Tuttle and Bowen (1958) termed the more common two-feldspar granitoids *subsolvus granites* and those with only one feldspar as *hypersolvus granites*. In Section 4.1.3.4 we briefly examined the effect of changing pressure on this system. Above some $P_{H2O} = P_T$ between 3-5 kilobars, we discovered that the solvus and the solidus intersect, leading to a ternary eutectic and resultant crystallization of the two separate alkali feldspars found in subsolvus granitoids. Below this critical pressure a minimum point exists, which would cause one alkali feldspar to crystallize. With subsequent cooling and exsolution, a perthite of intermediate OR-AB composition would result, the single feldspar found in hypersolvus granitoids.

Does this argument imply that all hypersolvus granitoids crystallize above some critical P_{H2O} and all subsolvus below? The answer is negative, because of the effect of the AN component (Section 4.2.3.2). Only those granitic melts with relatively low fluid pressures *and* extremely low AN contents crystallize one feldspar; the remainder crystallize two, which explains why subsolvus granitoids are more common. Only at relatively low fluid pressures and very low AN contents can a ternary minimum point, rather than a ternary eutectic, exist.

   We mentioned earlier that this system is the SiO_2-oversaturated part of
N. L. Bowen's *petrogeny's residua system*. Figure 4.17 is one of the more
famous examples of the use of norms in petrology. It is a contoured plot of more
than 500 granitoid rocks for *or-ab-q* (Tuttle and Bowen, 1958). The contours
form a "bull's eye" around a region of about equal amounts of *or, ab* and *q*. The
outer contour line bounds an area that contains most of the minimum and
eutectic points shown on Figure 4.16. Note the contours are elongated in part
parallel to the trend of these minimum and eutectic points, and also parallel to a
line of constant ratio for *or/ab* through the *q* apex. The elongation parallel to the
trend of minimum and eutectic points may be a reflection of eutectic granitoid
melts crystallizing at different pressures. Either or both of two reasons could
explain this second pattern of elongation: (1) some analyses may reflect coprecipitation of Or_{SS} and Ab_{SS} along cotectics approximately parallel to this *or/ab*
constant-ratio line (Fig. 4.16); and (2) the error in calculating *q* is greater than
that for *or* or *ab* (Section 1.4.2.2), leading to more spread in this direction.

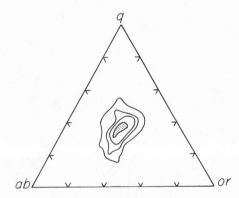

Figure 4.17. Normative
plot of *ab-or-q* showing
contours for more than 500
granitoids (after Tuttle and
Bowen, 1958; Geol. Soc.
America Mem. 74).

   Regardless of the reasons for the shapes of contour lines on Figure 4.17,
some valuable lessons can be learned. Bowen's term *petrogeny's residua* stems
from the fact that a melt composition anywhere on Figure 4.15 fractionally
crystallizes to produce a final residual melt on a minimum point or eutectic
point. Inasmuch as most of these points are within the compositional field of
most granitoids, at least some of these rocks can be considered as residua of
fractional crystallization for a wide compositional range of different melts.
Likewise, the first (lowest temperature) partial melt from a wide compositional
range of different source rocks is at a eutectic or minimum point, and thus has a
composition similar to that of a granitoid. Indeed, the limits on amount of partial
melting for any one melting event may be so low that partial melt compositions
never leave a eutectic.
   If partial melt compositions do leave the eutectic, however, an explanation for either very Na-rich or very K-rich quartz-bearing igneous rocks can be
found, provided fractional melting is operative. The second partial melt of a
source rock of composition x (Fig. 4.15) has a composition e, a Na-rich melt
somewhat analogous in composition to trondhjemites (plutonic rocks) or quartz
keratophyres (volcanic rocks). Similarly, the second fractional melting event of
any source rock near the Q-OR sideline forms a melt whose composition falls
near the Q-OR sideline and on the q-Or cotectic. Such a source rock might be a
K-rich shale. This leads to very K-rich granitic rocks. Perfect fractional melting

probably never occurs, but if batch sizes are small enough during equilibrium batch melting, the fractional melting process described may at least be approximated.

4.1.4 Quantitative Treatment

In Section 3.2.2 numerous examples were given to show that chemical reactions on binary phase diagrams can be represented by balanced chemical equations. The contact principle was employed to indicate which reactions are possible and which are not. In addition, the lever rule was demonstrated to allow quantitative calculations of melt compositions (in the case of solid solutions, also crystal compositions), as well as percentages of melt and crystals. Ternary systems can be treated in a similar fashion, although some seemingly new techniques must be learned. This section deals with these techniques, which are in fact only variations on an old theme.

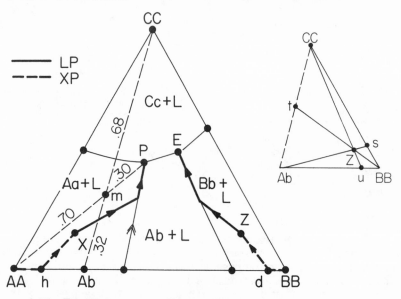

Figure 4.18. Equilibrium crystallization in the hypothetical ternary system AA-BB-CC for initial magmas X and Z. Calculations are given in the text.

4.1.4.1 Reactions at Eutectic and Peritectic Temperatures

The hypothetical system AA-BB-CC discussed in this section (Fig. 4.18) is somewhat similar to the system FO-AN-Q at one atmosphere (Section 4.1.2.2). We will consider the hypothetical system because it is simpler. If we wish to examine the state of the system at the ternary eutectic point E, the first step is to write the equation for point E and a magma of starting composition Z (Fig. 4.18). This enables us to calculate a eutectic composition, which then can be compared with the eutectic composition as determined from the phase diagram for verification. The magma initially crystallizes Bb and LP moves directly away from the line BB-Z until it reaches the Ab-Bb cotectic, at which time Ab + Bb coprecipitate as LP moves down the cotectic. Locating point Z within the scalene triangle Ab-BB-CC, the final rock composition for a melt of original composition Z is about 13Ab + 72Bb + 15Cc.

xxxxxxxxxxxxxxxxxxxxxxxxxxxxxxxxxxxxxxxxxxxxxxxxxxxxxxxxxxxxxxx

A brief digression is useful here. In Section 1.3.2.3 we learned a simple method for plotting a point in a scalene triangle, given its coordinates. Here we retrieve the Ab-BB-CC coordinates of point Z (inset of Fig. 4.18) by use of the lever rule. This can be easily done by reading the three levers BB-Z-t, CC-Z-u, and Ab-Z-s:

$$(Z\ s \times 100) / Ab\ s = \%Ab$$
$$(Z\ u \times 100) / CC\ u = \%Cc$$
$$(Z\ t \times 100) / BB\ t = \%Bb$$

where Z s, Ab s, etc., are lengths of line segments. Only two of these three levers need be read to determine the coordinates for point Z, but the third is useful as a check.

xxxxxxxxxxxxxxxxxxxxxxxxxxxxxxxxxxxxxxxxxxxxxxxxxxxxxxxxxxxxxxx

Construction of a tie line from the eutectic point E through the original magma composition Z to the Ab-BB sideline allows two levers to be read. The first lever is d-Z-E:

$$(d\ Z \times 100) / d\ E = \%L_E = 31 \quad \%XL = 100 - \%L_E = 69$$

where $d\ Z$ and $d\ E$ refer to the lengths of lines d-Z and D-E, respectively, and %XL represents percent crystals. The second lever is Ab-d-BB:

$$(Ab\ d \times 100) / Ab\ BB = \%Bb = 90 \quad \%Ab = 100 - \%Bb = 10$$

Only 69 percent of the melt has crystallized, therefore, percentages of Bb and Ab expressed in terms of the total system are:

$$\%Bb = 0.90 \times 69 = 62 \quad \%Ab = 69 - 62 = 7$$

The information from these two levers defines the state of the system just before LP reaches the eutectic E (or just above the eutectic temperature): about $31L_E$ + 7Ab + 62Bb, where L_E refers to the eutectic melt.

We can now write the following equation, which represents the state of the system at point E just above and just below temperature T_E:

Just above T_E Just below T_E

$$31L_E + 7Ab + 62Bb = 13Ab + 72Bb + 15Cc \qquad (4.4)$$

$$L_E = 0.20Ab + 0.32Bb + 0.48Cc$$

The composition on the right side of Equation 4.4 is the composition Z. More Ab and Bb are present in the final rock just below T_E than in the crystal mush above, confirming the fact that eutectic crystallization occured at T_E. One additional step is necessary -- we should express L_E in terms of components AA-BB-CC rather than minerals Ab-Bb-Cc. Reading the AA-Ab-BB lever indicates that the intermediate compound Ab is composed of 75 percent AA and 25 percent BB, thus:

$$L_E = 0.20(0.75AA + 0.25BB) + 0.32BB + 0.48CC$$
$$L_E = 0.15AA + 0.37BB + 0.48CC$$

The composition of L_E as read from the graph is 15AA - 36BB - 49CC, which given rounding errors and errors in reading lengths of levers, agrees remarkably well with the calculated L_E. Consequently, the graphical and numerical method are internally consistent.

We now turn to a similar calculation for a reaction point. A melt of composition X (Fig. 4.18) undergoing equilibrium crystallization initially crystallizes Aa as LP moves straight away from line AA-X until it encounters the Aa-Ab cotectic. At this point Aa begins reacting with the melt to form Ab, so as LP moves down the cotectic toward the ternary peritectic P, the amount of Ab crystals increases at the expense of Aa crystals. Just before the peritectic (i.e., just above the peritectic temperature) the system contains about $34L_P + 39Aa + 27Ab$ where L_P is the peritectic liquid. These percentages can be confirmed by constructing and reading the levers P-X-h and AA-h-Ab according to the method described previously. Just below T_P (the peritectic temperature) the rock has completely crystallized with the composition $25Aa + 60Ab + 15Cc$. These percentages can be measured by determining the location of point X within the scalene triangle AA-Ab-CC. We can now write the equation:

Just above T_P Just below T_P

$$34L_P + 39Aa + 27Ab = 25Aa + 60Ab + 15Cc \qquad (4.5)$$

Less Aa is present in the final rock just below T_P than in the crystal mush just above T_P, confirming the fact that Aa did indeed react with L_P to form some Ab at the peritectic. Substituting $(.75AA + .25BB)$ for Ab and solving Equation 4.5 for L_P:

$$L_P = 0.32AA + 0.24BB + 0.44CC$$

Composition of L_P as read directly from the graph (Fig. 4.18) is 30AA - 25BB - 45CC, which reconciles the graphical and numerical methods within limits of error of the calculations and measurements.

A second method to calculate L_P involves writing a chemical equation for the reaction crystallization occurring at the peritectic. Coefficients for the reacting phases are determined from crossing tie lines (joins). Recall that if tie lines cross, a chemical reaction is possible. In this case the reaction is between Aa and L_P to form Ab and Cc, so Aa-L_P and Ab-Cc tie lines are constructed on Figure 4.18. Then the AA-m-P and Ab-m-CC levers are read, which lead to the following reaction:

$$0.70L_P + 0.30Aa = 0.68Ab + 0.32Cc \qquad (4.6)$$

Parenthetically, if the forward reaction for Equation 4.6 represents reaction crystallization, then the back reaction must be incongruent melting (see Section

3.2.6.6). Remembering that Ab is composed of 75 percent AA and 25 percent BB, and substituting into Equation 4.6:

$$0.70L_P + 0.30AA = 0.68(0.75AA + 0.25BB) + 0.32CC$$

$$L_P = 0.30AA + 0.24BB + 0.46CC$$

This value for L_P also agrees well with the value read from the graph.

Calculations of the type discussed in this section are not necessary if a graphical solution to these ternary phase diagrams is employed. They confirm that any graphical solution can be written as a numerical solution.

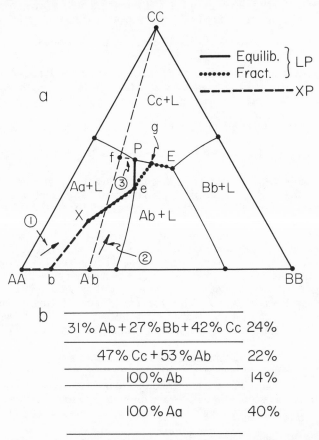

Figure 4.19 A comparison of equilibrium and fractional crystallization of an initial magma *X* in the hypothetical system AA-BB-CC: *a.* phase diagram *b.* sill of overall composition *X,* based on assumption of fractional crystallization caused by gravity settling. Calculations are given in text.

4.1.4.2 Additional Applications of the Lever Rule

If we can locate the composition of an original (parent) melt, a residual (daughter) melt, and the mineral(s) crystallized to form that daughter, they form a lever with the parent composition between the other two. This principle applies to ternary as well as other phase diagrams and even some variation diagrams (see Section 5.2.2.3). As in the case of binary diagrams discussed previously, we can use the lever rule to determine quantitatively compositions and percentages of all phases present at any given stage during crystallization (or melting). In this section we explore two examples, the equilibrium and fractional crystallization of magmas X and Y (Figs. 4.19 and 4.20). The hypothetical ternary system AA-BB-CC and magma X of Figure 4.19a are similar to those in Figure 4.18.

The liquid path (LP) and solid path (XP) for equilibrium crystallization of parent melt X are shown in Figure 4.19a. As LP moves from X to e, XP stays at AA as the melt crystallizes Aa. Just as LP reaches the Aa-Ab cotectic at point e, the lever AA-X-e can be read to determine percentages of melt and Aa:

$$(AA \ X \times 100) \ / \ AA \ e = \%L = 60 \quad \%Aa = 100 - \%L = 40$$

LP then moves along the reaction cotectic Aa-Ab from e to P as XP moves from AA to point b. Just as LP reaches the peritectic P the levers b-X-P and AA-b-Ab can be read:

$$(b \ X \times 100) \ / \ b \ P = \%L = 45 \quad \%(Aa + Ab) = 100 - \%L = 55$$

$$(AA \ b \times 100 \times 0.55) \ / \ AA \ Ab = \%Ab = 24 \quad \% \ Aa = 55 - 24 = 31$$

At point e 40 percent Aa had crystallized, but just as LP arrives at P only 31 percent Aa remains, so the cotectic Aa-Ab must be a reaction cotectic. Crystallization ceases at P, yielding a rock of composition 25Aa - 60 Ab - 15Cc (determined from reading the position of point X in the scalene triangle AA-Ab-CC).

For fractional crystallization, LP on Figure 4.19a follows the path X-e-g-E. The crystal path XP undergoes three rock hops: (1) AA to Ab when LP reaches point e, (2) Ab to point f when LP reaches g, and (3) f to E when LP reaches E. In other words, (1) LP follows X-e as the melt crystallizes Aa; (2) LP follows e-g as Ab crystallizes; (3) LP follows g-E as a mixture of Ab and Cc represented by the point f crystallize; and (4) LP and XP finally coincide at E during eutectic crystallization. Because this process is fractional crystallization, reaction of Aa with the melt does not occur, either along the Aa-Ab cotectic or at the peritectic P.

Let us imagine magma X fractionally crystallizing in a closed magma chamber that eventually becomes a sill. Let us further imagine that phases Aa and Ab are more dense than the melt, and that magma viscosity is low enough so gravity settling can occur. Initially we assume that no melt becomes trapped in the cumulus crystals; later we evaluate the effect of trapped liquid on fractional crystallization.

Layer 1 (the bottom layer) in this hypothetical sill must contain 40 percent Aa (Fig. 4.19b) from the AA-X-e lever:

$$\%Aa = (X \ e \times 100) \ / \ AA \ e = 100 - \%L = 40$$

Hence, this initial calculation is the same as that for equilibrium crystallization. If units in Figure 4.19a are in weight percent, then volume percent, and thus thickness in Figure 4.19b are assumed to be proportional to weight percent. Sixty percent of the original melt remains, so layer 2, comprised of pure Ab, is calculated to be 14 percent of the entire sill (Fig. 4.19b). This thickness of 14 percent is derived by:

$$\%Ab = (e\ g \times 100 \times 0.60)\ /\ Ab\ g = 14$$

from the Ab-e-g lever. Of the original magma, only 46 percent now remains. Layer 3 makes up 22 percent of the sill and is composed of 47Cc + 53Ab. The 22 percent is determined from:

$$\%(47Cc + 53Ab) = (g\ E \times 100 \times 0.46)\ /\ f\ E\ = 22$$

from the f-g-E lever. The percentages 47Cc + 53Ab are approximated from the Ab-f-CC lever.

These percentages of Cc and Ab are only approximated for the following reasons. If the Cc-Ab cotectic is a straight line, then f is a point and the Ab-f-CC lever can be read accurately. If the cotectic is a curve, f is a line and percentages of Ab and Cc from the Ab-f-CC lever can only be approximated. This is because f is defined as the intersection of a back tangent to line (or curve) g-E as LP moves along this path.

Now only 24 percent of the original melt remains, which crystallizes at the eutectic point. Hence, the top 24 percent in the sill (layer 4) has the eutectic composition 31Ab - 27Bb - 42Cc at point E as read in the scalene triangle Ab-BB-CC (Fig. 4.19a). Had the melt undergone perfect equilibrium crystallization, in theory the sill would be homogeneous from top to bottom, with a composition of 25Aa - 60Ab - 15Cc.

An inventory can be calculated to confirm our results:

Layer	AA	BB	CC
4	5.6	8.4	10.1
3	8.7	2.9	10.3
2	10.5	3.5	--
1	40.0	--	--
Total	64.8	14.8	20.4

These totals agree quite well with the composition for original melt X (Fig. 4.19a) of AA65 - BB15 - CC20. To calculate this inventory, one must remember that Ab is (0.75AA + 0.25BB). A useful exercise is to calculate the percentages in this inventory.

The effect of trapped liquid on fractional crystallization can now be evaluated. The chemical composition of layers 1, 2, 3, and 4 was determined to be at AA, Ab, f, and E, respectively. This assumes no trapped liquid between the cumulus crystals during crystal settling, which clearly cannot be true. We now examine the effect of 30 percent trapped liquid throughout fractional crystallization of melt X. The percent trapped liquid probably would not stay constant during the entire process (it would likely increase), but this simplifying assumption suffices.

The major effect of trapped liquid is that XP for layers 1-3 is represented by lines rather than points. In summary, layers 1-3 have compositions at lines (1)-(3), respectively, and the composition of layer 4 is at point E (Fig. 4.19a). Arrows on lines (1)-(2) represent change in composition from bottom to top in each layer. Line (1) is used as an example of the manner in which these compositions are determined. Assume point r (not shown) is at the back end of the arrow representing line (1), the end closest to apex AA. Point r, representing the composition at the lowermost part of layer 1, is at the fulcrum of the AA-r-X lever where:

$$(AA \text{ r} \times 100) / AA X = \% \text{ trapped liquid} = 30$$

Likewise, assume point s is at the front end of the arrow for line (1). Point s represents the composition at the uppermost part of layer 1 and is at the fulcrum of the AA-s-e lever; therefore:

$$(AA \text{ s} \times 100) / AA \text{ e} = \% \text{ trapped liquid} = 30$$

Liquid lines (2) and (3), respectively, for layers 2 and 3 can be similarly constructed. The composition of layer 4 is at the eutectic point E because the eutectic melt does not change composition during crystallization. Consequently, the main effect of trapped liquid is to cause each of the three layers 1-3 to grade upward in bulk chemical composition, rather than be homogeneous from bottom to top as is predicted by XP. However, as indicated by the short lengths of lines (1), (2), and especially (3), chemical changes within each layer are small.

The natural analog of such an intrusion (Fig. 4.19b) is called, for obvious reasons, a *layered intrusion*. Rocks within it are referred to as *cumulates*, because they are formed by crystals accumulating on the bottom of the chamber, which are known as *cumulus minerals*. Upon crystallization some of the trapped liquid is commonly referred to as the *mesostasis*, which is interstitial to the cumulus minerals. Because each of the four layers in the sill is approximately uniform in chemical and mineral composition from top to bottom, even allowing for compositional changes in trapped liquid, the layers are termed *unimodal* layers, and the boundaries between them are called *phase contacts*. Layered intrusions and cumulates are much more complicated than presented here and a good summary can be found in Cox and others (1979).

A second example, starting with parent magma Y (Fig. 4.20a), is equally informative. The most significant difference between parent magmas X and Y is the fact that X is on the left side of the subsolidus join Ab-Cc, whereas Y is on the right. This fact leads to some major differences in the manner in which these two melts crystallize.

First we consider equilibrium crystallization. Melt Y initially crystallizes Aa as LP moves from Y to point f (Fig. 4.20a). Just as LP reaches point f, the AA-y-f lever can be read to indicate about 15 percent crystals of Aa and 85 percent melt of composition f. At f LP starts moving down the cotectic toward P and XP moves from AA toward Ab while Aa is reacting with the melt to form Ab. When LP reaches point g and XP reaches Ab all Aa has been used up during reaction, so only Ab crystallizes as LP moves along the path g-k and XP stays at Ab. Just as LP reaches the Ab-Bb cotectic at point k, we can read the Ab-g-k lever to show that the system consists of about 44 percent melt of composition k and 56 percent crystals of Ab. LP then moves down the Ab-Bb cotectic and XP moves from Ab to m as coprecipitation of Ab and Bb occurs. When LP reaches E, the lever m-Y-E can be read to determine about 23 percent melt of composition E and 77 percent crystals of 5Bb + 95Ab (these latter percentages of Ab and Bb are calculated from the lever Ab-m-BB). Finally, eutectic crystallization occurs as LP stays at E while XP moves from m to Y. The final rock composition is determined from point Y in the scalene triangle Ab-BB-CC: 80Ab - 9Bb - 11Cc.

During fractional crystallization of melt Y Aa crystallizes first as LP moves from Y to f and XP stays at AA (Fig. 4.20a). As no reaction crystallization occurs, LP moves along the path f-h as only Ab crystallizes, so XP hops from AA to Ab when LP reaches point f. Layer 1, consisting of only Aa, as determined from the Aa-Y-f lever makes up the lower 15 percent of the second sill (Fig. 4.20b). Layer 2, pure Ab, is 52 percent of the sill, as read from the Ab-f-h lever, remembering to correct for the fact that 85 percent of the original magma is left after crystallization of layer 1. When LP reaches h at the Ab-Bb cotectic, XP hops to point s, then LP moves down h-E as the melt coprecipitates Ab and Bb. The width of layer 3 can be estimated from the s-h-E lever; allowing for only 33 percent remaining melt, layer 3 is about 10 percent of the sill. Lever Ab-s-BB is read to determine relative percentages of Ab and Bb in layer 3: 72Bb + 28Ab. Finally, when LP arrives at the eutectic point E, XP hops from s to E. The remaining 23 percent of the sill is comprised of the eutectic composition 31Ab + 27Bb + 42Cc, as explained. An inventory similar to that shown indicates that these calculations and measurements are correct. Please calculate the inventory and verify this fact.

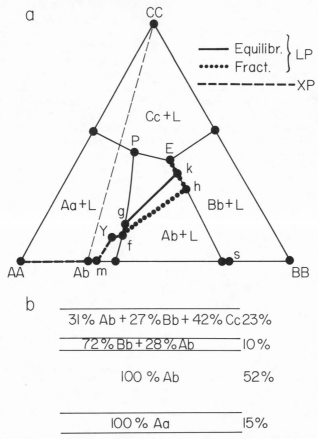

Figure 4.20. A comparison of equilibrium and fractional crystallization of an initial magma Y in the hypothetical system AA-BB-CC: a. phase diagram; b. sill of overall composition Y, based on assumption of fractional crystallization caused by gravity settling. Calculations are given in text.

Compare the two hypothetical sills in Figures 4.19b and 4.20b. Both have four layers, but thicknesses of corresponding layers in the first sill are quite different from those in the second. Layer 1 in both sills has the same mineralogy, as do layers 2 and 4. Layer 3 is different; it contains crystals of Cc in the first sill but Bb in the second. Other differences involve LPs for equilibrium (LP$_E$) and fractional (LP$_F$) crystallization (Figs. 4.19a and 4.20a). For initial magma X, LP$_E$ never goes beyond the peritectic, whereas LP$_F$ goes all the way to the eutectic. For parent magma Y, LP$_E$ and LP$_F$ both go to the eutectic.

Given some simplifying assumptions, the relative paths of LP$_E$ and LP$_F$ have interesting implications. As stated previously, the hypothetical system AA-BB-CC shown in Figures 4.18 to 4.20 can be considered as an extremely simplistic, distorted analog to the system FO-AN-Q (see Fig. 4.7) where Ab approximates En. Any magma in the primary-phase field of Fo and to the left of the En-An subsolidus join (comparable to melt X, Fig. 4.19a) that undergoes equilibrium crystallization produces a rock with minerals of Fo + Pren + An, a SiO_2-undersaturated rock. If this same magma fractionally crystallizes, the final eutectic residual magma crystallizes An + Pren + Tr and is SiO_2 oversaturated. Any magma to the right of the En-An subsolidus join and in the Fo primary-phase field (analogous to melt Y, Fig. 4.20a) also differentiates to a SiO_2-oversaturated end product. In contrast with a magma to the left of the join, if the magma to the right of the join and initially crystallizing Fo undergoes equilibrium crystallization, it also produces a SiO_2-oversaturated rock, despite the fact that its composition was originally in the Fo + L field. Granted these arguments are oversimplified when applied to natural rocks, but they illustrate the utility of thinking in terms of natural rocks when examining phase diagrams.

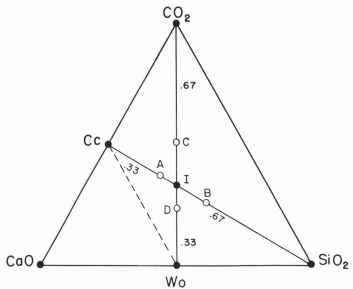

Figure 4.21. Equilibrium diagram for the system CaO-CO$_2$-SiO$_2$. The reaction calcite *(Cc)* + quartz *(SiO$_2$)* = wollastonite *(Wo)* + CO$_2$ is represented by crossing tie lines.

4.1.4.3 An Equilibrium Diagram for a Ternary System

Before continuing with quaternary systems, a brief digression into contact metamorphism is informative. In Chapter 2 we explored the decarbonation reaction whereby calcite and quartz react to form wollastonite and CO_2 (Equation 2.37

and Fig. 2.7). This reaction could occur in the metamorphic contact aureole of an igneous intrusion where the unmetamorphosed surrounding sedimentary country rocks contain quartz and calcite. Wollastonite is the diagnostic mineral that marks the aureole.

The components necessary to represent the phases in this reaction are CaO, SiO_2, and CO_2, so C = 3 for the phase rule. Pressure is not fixed $(P_T = P_{CO2}$ is a variable on Fig. 2.7), so P + F = C + 2, or P + F = 5. Along the univariant curve of Figure 2.7, F = 1, P = 4, and the phase rule is confirmed.

The equilibrium diagram for this system is shown in Figure 4.21. Calcite (Cc) plots half way between CaO and CO_2 because Cc can be written $CaO \cdot CO_2$, so the mol fraction of each component in Cc must be 0.5. The equivalent fraction is also 0.5, because moles and equivalents are the same in this case. Wollastonite (Wo) can be written $CaO \cdot SiO_2$, so its composition falls half way between CaO and SiO_2. The reaction Cc + Q = Wo + CO_2 is represented on the equilibrium diagram by crossing tie lines; the contact principle (Section 3.2.2.2) is thus applicable.

An interesting feature of this diagram is the relative lengths of line segments on either side of the point of intersection (point I) of the two crossing tie lines. The chemical reaction balanced in terms of moles is:

$$CaCO_3 + SiO_2 = CaSiO_3 + CO_2$$

Because 1 is the only coefficient for each phase required to balance this equation, we might expect point I to be at the midpoint for both tie lines. Figure 4.21 indicates that this clearly is not the case. Balancing the equation using equivalents (Section 1.2.2), we obtain:

$$2Cc + Q = 2Wo + CO_2 \quad or \quad 0.67Cc + 0.33Q = 0.67Wo + 0.33CO_2$$

The coefficients 0.67 and 0.33 represent the relative lengths of the appropriate line segments for tie lines on either side of point I. Thus if we use the standard equation balanced in moles, the coefficients are not proportional to these line segments, but if we balance the equation using equivalents, they are. In Section 1.3.2.3 we used equivalents to plot compositions of phases on the diagram *ne-fo-q*. For Figure 4.21 mol fractions and equivalent fractions are the same for CaO, SiO_2, and CO_2, so Cc and Wo plot at the same points in either case. The safe procedure on diagrams of this type is to plot compositions using equivalents.

The composition of an unmetamorphosed sedimentary rock containing only quartz and calcite must plot somewhere along the tie line Cc-SiO_2 on Figure 4.21. If it plots to the left of point I (point A, for example), the rock has an excess of Cc relative to the stoichiometrically balanced equation. Consequently, the metamorphic rock in the aureole contains the three-phase assemblage Wo-Cc-CO_2, represented by the appropriate scalene triangle on Figure 4.21. Assuming CO_2 eventually leaves the system, we could predict the relative abundances (in equivalents) of Wo and Cc in the metamorphic rock by determining the location of point A in the scalene triangle Wo-Cc-CO_2 (Section 4.1.4.1).

If the original composition of the sedimentary rock plots to the right of point I (point B), then the rock contains an excess of quartz and the same reasoning applies. The position of point B inside the right triangle Wo-SiO_2-CO_2 can be determined to predict the relative abundances of Wo and quartz in the metamorphic aureole.

This approach is useful for metamorphic systems, but it also can be applied to secondary (deuteric or hydrothermal) alteration of igneous rocks. We have only been considering the forward, or decarbonation, reaction. If we consider the back, or carbonation, reaction, this could simulate secondary alteration. Wollastonite, however, is very unlikely as an igneous mineral (other than a normative mineral), but it is a component in augite, a common igneous mineral. For the carbonation reaction, Wo reacts with CO_2 to form the alteration assemblage Cc and quartz. Composition of the system must fall along tie line Wo-CO_2. Excess CO_2 (point C) produces an alteration assemblage of Cc and quartz (crystalline phases only), but excess Wo (point D) produces the assemblage Cc-Wo-Q. Positions of points C and D can be determined in their respective triangles (Cc-CO_2-SiO_2 and Cc-SiO_2-Wo) to estimate relative proportions of crystalline phases.

4.2 QUATERNARY SYSTEMS

Our final treatment of experimental systems is a brief examination of quaternary (four-component) systems. Addition of another component to an experimental system allows this system to be even more useful in estimating melting of and crystallization to form naturally occurring rocks. A complication, however, exists. We cannot represent an undistorted phase diagram of a complete quaternary system in two dimensions. Four components require a tetrahedron, which is a four-sided, three-dimensional geometric form (occasionally referred to as a triangular pyramid, because each of the four sides is an equilateral triangle). Two-dimensional perspectives are commonly drawn, but they are difficult to interpret quantitatively and the third dimension is difficult to visualize, especially in complicated systems. A common approach is a two-dimensional projection from some point, generally on an edge or at an apex of the tetrahedron. In essence, this allows a quaternary system to be treated as a ternary or pseudoternary system. This section explores these and other aspects of quaternary systems.

4.2.1 System with a Quaternary Eutectic

As was the case for binary and ternary systems, the simplest possible quaternary system involves a quaternary eutectic, no peritectic, and all immiscible solids. No system approximating naturally occurring igneous rocks is this simple, but it provides a good point of departure.

4.2.1.1 Plotting and Projections

Before turning to a quaternary eutectic system, it is necessary to learn how compositional points are plotted on a two-dimensional perspective of a quaternary tetrahedron (Fig. 4.22). Either of two methods can be used and both are informative, but the second is simpler. The sides of a tetrahedron are equilateral triangles, but when drawn in two dimensions, they are scalene triangles. If we know the A-B-C-D coordinates of point P, we can construct its position.

The method shown in Figure 4.22a involves first plotting the position of point e in the scalene triangle A-C-D and then drawing the line e-B. An example is A13-B50-C13-D24. Point e is located by calculating the ratios:

A / (A + D) = 0.35 to locate point g
C / (C + A) = 0.50 to locate point i
D / (D + C) = 0.65 to locate point h

(see Section 1.3.2.3). Intersection of any two of the constant-ratio lines h-A, i-D, or g-C locates e; the third line can be constructed as a check. The next step is to plot the position of point f in the triangle A-B-D in a similar fashion and draw the line f-C. The intersection of lines e-B and f-C is at the point P. Similar points can be constructed for scalene triangles C-A-B and C-B-D. Lines drawn to the opposite apices of these latter two triangles should intersect lines f-C and e-B at point P, which provides a check of plotting accuracy.

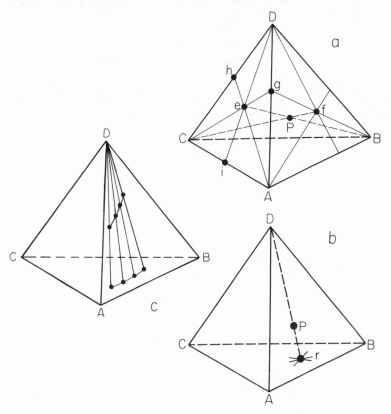

Figure 4.22 Graphical methods for plotting a point in a two-dimensional projection of a tetrahedron.

The second method first requires that C + A + B, which comprise the base of the tetrahedron, be recalculated to 100 percent. This yields C17-A17-B66, which is plotted as point r in the scalene triangle C-A-B (Fig. 4.22b). The line r-D is then drawn; point P must be on this line and 24 percent along its length up from point r. Obviously, point r can be plotted on any face of the tetrahedron, but using the base is convenient.

Figure 4.22 is equally useful to demonstrate projections from an apex of a tetrahedron onto an opposite face. In this book we do not deal with projections from a point along an edge or within a face. For an excellent summary of this more advanced technique, refer to Cox and others (1979). Projections from apices of the tetrahedron onto opposite faces suffice for our purposes. Such a projection is used to represent a compositional point within a quaternary tetrahedron as a point on an equilateral triangle representing one of the faces of the tetrahedron.

The principle can be learned from Figure 4.22, although in this two-dimensional perspective we are not dealing with equilateral triangles. Point P inside the tetrahedron is projected from apex C to the A-B-D face as point f (Fig. 4.22a). Mathematically, we need only to recalculate percentages of A + B + D for composition P to 100 percent; this locates the position of f in the triangle A-B-D. Likewise, point P is projected from apex B onto the A-C-D face as point e, so A + C + D can be recalculated to 100 percent to locate e in that triangle. In Figure 4.22b, P is projected from the D apex onto the C-A-B face as point r. As stated, recalculation to C17-A17-B66 locates r in C-A-B.

Figure 4.22c illustrates the projection of four points along a curve within the tetrahedron from the D apex onto the C-A-B face. The progressive increase in D with an increase in the B/A ratio is not apparent from either the curve within the tetrahedron or the projection from D, but becomes easily visualized when both curves are examined.

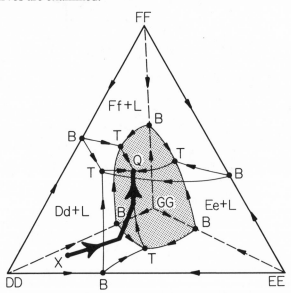

Figure 4.23 Two-dimensional perspective of the hypothetical quaternary eutectic system DD-EE-FF-GG (modified after Sood, 1981). *Heavy solid curve with arrows* is liquid path for initial magma *X*, assuming equilibrium crystallization. Point *Q* is quaternary eutectic; ternary eutectics are labeled *T*, binary eutectics, *B*. *Stippled area* represents the primary-phase volume *Gg + L*.

The quaternary system A-B-C-D has six binary systems, one for each edge of the tetrahedron. It has four ternary systems, one for each face, which can be considered as projections from the opposite apex:

projection from apex

A-B-C	D
A-B-D	C
C-A-D	B
C-B-D	A

Hence, if we project composition P from D onto the face A-B-C by recalculating A + B + C to 100 percent, we have represented composition P as point r in the ternary system A-B-C (Fig. 4.22b). This is a useful practice in dealing with quaternary systems, but it can lead to misleading results, particularly if the projection is from the wrong apex (Section 4.2.3.1).

4.2.1.2 Basics

Figure 4.23 is a two-dimensional perspective of the hypothetical quaternary system DD-EE-FF-GG. The difficulties in visualizing more complicated systems than this are obvious. It contains the quaternary invariant eutectic Q, four invariant ternary eutectics labeled T, and six invariant binary eutectics labeled B. A ternary eutectic is present in each of the four faces of the tetrahedron and a binary eutectic is present on each of the six edges. The four univariant curves T-Q connect each ternary eutectic T with the quaternary eutectic Q. Similarly, the 12 univariant curves B-T (ternary cotectics) connect each binary eutectic within each ternary system with the appropriate ternary eutectic. These univariant curves bound six divariant surfaces within the tetrahedron; each surface has four corners labeled Q-T-B-T. These divariant surfaces are referred to as *quaternary cotectics*. These six surfaces and the four faces of the tetrahedron bound four trivariant volumes, the primary-phase fields for Gg + L (stippled but not labeled to avoid clutter) Dd + L, Ee + L, and Ff + L.

On Figure 4.23 we have adopted the normal practice of showing arrows that indicate down-temperature directions. The quaternary eutectic Q is at the lowest temperature in the entire system, just as each ternary and binary eutectic is at the lowest temperature in its appropriate system. Within primary-phase volumes isothermal contour surfaces, rather than isothermal contour lines as for ternary systems, are present.

TABLE 4.2 Application of the Phase Rule to a Quaternary System

C	P	F	Equilibria	Variables	Geometry
4	2	3	trivariant	any 3 of 5	volume
4	3	2	divariant	any 2 of 5	surface
4	4	1	univariant	any 1 of 5	curve
4	5	0	invariant	none	point

As usual, in the previous paragraph we referred to points as invariant, curves as univariant, and surfaces as divariant. We must also be concerned about volumes, which are presumably trivariant. Let us now reconcile the phase rule with this quaternary system (Table 4.2). Because each quaternary system is at a fixed pressure, the phase rule is again written as $P + F = C + 1$, so $P + F = 5$. Within any primary-phase volume $P = 2$, so $F = 3$ and, as expected, the volume is trivariant. Any three of the variables DD, EE, FF, GG, and T must be fixed to determine the other two. Within the tetrahedron, two solid phases coprecipitate along each Q-T-B-T surface, so, including the melt, $P = 3$ and the surface is divariant. This requires that two of the five variables be fixed to determine the other three. Along the four curves T-Q, three solid phases and the melt coexist, so $P = 4$, $F = 1$, and the curves are univariant. In this case only one of the five variables must be fixed to know the other four. At point Q all four phases and the melt coexist, so $P = 5$ and the quaternary eutectic point is invariant. All five variables are now fixed.

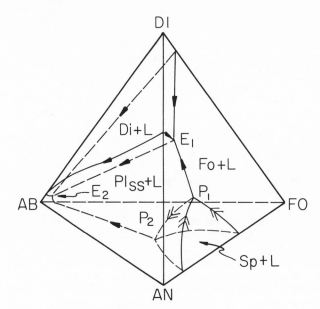

Figure 4.24 Two-dimensional per-spective of the quaternary system AB-AN-FO-DI at 1 atmosphere pres-sure (modified after Yoder and Tilley, 1962, and Schairer and Yoder, 1967.) Two ternary peri-tectics *(P₁ and P₂)* and eutectics *(E₁ and E₂)* are present. The ter-nary eutectic *E₂* is on the AB-FO-DI face of the tet-rahedron.

4.2.1.3 Crystallization and Melting

We will deal with melting and crystallization in the quaternary system DD-EE-FF-GG only in qualitative terms. A quantitative treatment is not impossible, but given the fact that we must deal with a two-dimensional perspective of a three-dimensional geometric figure, it is difficult. If quantitative results are desired, a numerical rather than graphical approach should be considered. In binary systems, during crystallization, an LP typically moves from a curve (the liquidus) to an invariant point; in ternary systems it moves from a primary-phase field to a curve (cotectic) to an invariant point. Hence, in a quaternary system an LP typically moves from a primary-phase volume to a cotectic surface to a univariant curve to an invariant point. System DD-EE-FF-GG is no exception to this generality.

As an example, given a parental magma in the Dd + L primary-phase volume, it initially crystallizes Dd and LP moves away from DD toward one of the two internal bounding surfaces Q-T-B-T, depending on its initial composition. The heavy, solid line in Figure 4.23 approximately marks the path of LP for parental magma X. When LP for melt X contacts the Dd-Ee cotectic surface, Dd and Ee coprecipitate as LP moves across the cotectic surface away from the Dd-Ee join. LP eventually intersects the Dd-Ee-Gg univariant curve and all three minerals begin precipitating; this process continues as LP moves down temperature toward the quaternary eutectic Q. When LP reaches Q all four minerals continue to precipitate until crystallization ceases.

For equilibrium melting, simply reverse this path. During fractional crystallization LP follows a path identical to that of equilibrium crystallization, but XP hops from pure Dd to a Dd + Ee mixture to a Dd + Ee + Gg mixture and finally to the quaternary eutectic. During fractional melting XP is the same as for fractional crystallization, but LP hops from the quaternary eutectic to the ternary eutectic DD-EE-GG to the binary eutectic DD-EE to pure DD.

4.2.2 A Haplobasalt System: AB-AN-DI-FO

Many *haplobasalt* quaternary systems could have been chosen for this discussion. The system AB-AN-DI-FO at low pressures was chosen because melting and crystallization trends within the system can be easily visualized. This system approximates SiO_2-undersaturated basaltic rocks containing plagioclase, olivine, and clinopyroxene ("olivine basalts"). All compositions within this system plot on the critical plane of SiO_2-undersaturation in the basalt tetrahedron (see Fig. 1.14). Because of this, rocks with similar compositions plot at the boundary between alkali olivine and olivine tholeiitic basalts (see Section 1.3.3.2). The system only approximates these basaltic rocks because no Fe is present, so there are no Fe-Mg solid solutions of olivine and clinopyroxene. In addition, the clinopyroxene in basaltic rocks is more likely to be augite or pigeonite than diopside. Despite these limitations, the system is quite instructive and thus is examined in this section.

4.2.2.1 A Two-dimensional Perspective and a Fold-out

A two-dimensional perspective of the AB-AN-DI-FO sytem at 1 atm pressure is given in Figure 4.24. The tetrahedron is divided into four trivariant primary-phase volumes: Fo + L, Pl_{SS} + L, Di + L, and Sp + L. The abbreviation Sp is for spinel, $MgAl_2O_4$. Because the composition of spinel cannot be plotted anywhere within this tetrahedron, this system is in fact pseudoquaternary. The ternary system AB-AN-DI was discussed in Sections 4.1.3.1 to 4.1.3.3.

Divariant cotectic surfaces bound the primary-phase volumes within the interior of the tetrahedron (Fig. 4.24). No quaternary eutectic or peritectic exists in this system, although there are two ternary eutectics (E_1 and E_2) and two ternary peritectics (P_1 and P_2). The ternary system FO-AN-DI contains P_1 and E_1, AB-AN-FO has P_2, and AB-FO-DI has E_2 very near its AB apex. The two ternary eutectics E_1 and E_2 are connected by a univariant curve along which the three solid phases Fo, Pl_{SS}, and Di coexist with the melt. This curve is cope-cipitational because all three phases crystallize as LP moves along it. This E_1-E_2 curve (dashed) along its entire length is extremely close to the univariant cotectic in the ternary system AB-AN-DI. In contrast, the curve P_1-P_2 is a univariant reaction curve. Moreover, the entire divariant surface bounding the Sp + L volume is a reaction surface. The remaining divariant interior surfaces are all coprecipitational quaternary cotectics. The ternary eutectic E_1 is at 1270°C, whereas E_2 is at 1135°C, so LPs moves down the curve from E_1 toward E_2.

xxxxxxxxxxxxxxxxxxxxxxxxxxxxxxxxxxxxxxxxxxxxxxxxxxxxxxxxxxxxxxx

Figure 4.25 is a fold-out of Figure 4.24 showing all four ternary systems. The base of the tetrahedron is the central triangle AB-AN-FO, so the top of the tetrahedron, represented by DI, is at every apex of the triangular fold-out. This fold-out is constructed in this way for a very particular reason. It provides an excellent learning device that aids immeasurably in visualizing quaternary systems. The fold-out can traced or reproduced, preferably onto some transparent material, and then folded along the interior joins and taped together to form the tetrahedron representing the quaternary system. Relationships that are unclear when viewing a two-dimensional perspective become easily visualized.

xxxxxxxxxxxxxxxxxxxxxxxxxxxxxxxxxxxxxxxxxxxxxxxxxxxxxxxxxxxxxxx

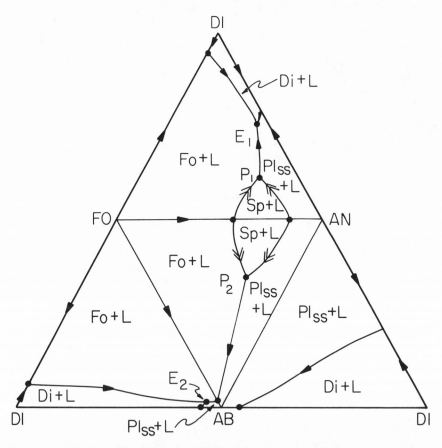

Figure 4.25 Fold-out of Figure 4.24 showing the four ternary systems for the quaternary system AB-AN-FO-DI at 1 atmosphere P. This figure can be folded along the three interior joins to form the tetrahedron.

4.2.2.2 Effect of Load Pressure

Before turning to applications of this system to basaltic rocks, we must consider the effect of load pressure. Changes in load pressure cause changes in relative solubilities of different components in the melt, which in turn affect sizes of primary-phase volumes. As a result, positions of univariant curves and divariant quaternary cotectics are a function of load pressure, which ultimately affects compositions of basaltic rocks.

The effect of load pressure can best be envisioned by examining a projection from the AB apex of Figure 4.24; i.e., the system AN-DI-FO. The 1 atmosphere system is shown as the top triangle in the fold-out (Fig. 4.25). Figure 4.26 shows the system at 7 kb. With increasing pressure, the solubilities of spinel and particularly diopside decrease, while the solubilities of forsterite and anorthite increase. This causes an increase in the sizes of the Sp + L and Di + L fields at the expense of the Fo + L and An + L fields. Consequently, the Fo-Pl$_{ss}$ cotectic present at 1 atmosphere disappears at high pressure. Presnall and others (1978) is recommended reading here.

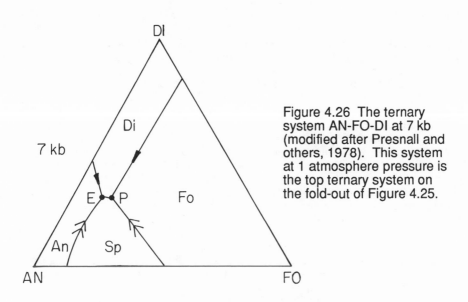

Figure 4.26 The ternary system AN-FO-DI at 7 kb (modified after Presnall and others, 1978). This system at 1 atmosphere pressure is the top ternary system on the fold-out of Figure 4.25.

One implication of these observations has to do with the order of crystallization. Given a relatively FO-rich melt in the Fo + L field, at 1 atmosphere pressure LP moves straight away from FO until it reaches the Fo-Pl$_{ss}$ cotectic, at which point Fo and Pl$_{ss}$ crystallize together as LP moves along the Fo-Pl$_{ss}$ cotectic (Fig. 4.25). When LP reaches the Fo-Di-Pl$_{ss}$ eutectic (actually, a *piercing point,* Section 4.2.3.2) diopside joins the other two and all three phases crystallize. In contrast, for a magma of similar composition at high pressures diopside crystallizes *before* plagioclase (see also Section 5.1.1.3), regardless of whether spinel or diopside crystallizes first (Fig. 4.26). At high pressure Fo and An never crystallize together. Another implication has to do with the expanding Sp + L and Di + L fields and the decreasing Fo + L field as pressure increases. Any parent magma in the Fo + L field but near the Sp-Fo or Di-Fo cotectic first crystallizes Fo at low pressure, but first crystallizes either Di or Sp at higher pressure.

Although we discussed these effects in terms of the ternary system AN-FO-DI rather than the quaternary system, the conclusions reached above also hold for the quaternary system. One reason is that the concentration of AB component is commonly fairly low in basaltic rocks, so many melt compositions plot not too far from the AN-FO-DI face within the quaternary tetrahedron. Perhaps even more important, it is clear from Figure 4.24 that no drastic changes occur proceeding inward into the tetrahedron from the AN-FO-DI face. This conclusion at least holds for low pressures. The main effect is a decrease in size of the Sp + L volume. In fact, with appreciable amounts of AB in the original melt, the Sp + L volume is never encountered at low pressures. Any time a projection of a quaternary is used, it is prudent to determine which of the four projections is the best. In this case, projection from AB is clearly the best.

4.2.2.3 Implications for Basaltic Rocks

In the previous section we learned that in some SiO_2-undersaturated basaltic rocks plagioclase should crystallize before clinopyroxene at relatively low pressures, and vice versa at higher pressures. What are the implications of this with regard to basalt petrogenesis? First, it is necessary to understand the possible importance of the phenocryst assemblage.

Early crystallizing minerals have been referred to by a variety of names: *cumulus crystals* in cumulate rocks, *near-liquidus phases, intratelluric crystals,* or *primocrysts* in general. Early crystallizing minerals are usually, but not necessarily, phenocrystic in volcanic and hypabyssal rocks. They are so commonly phenocrystic, however, that the mineralogy of the phenocryst assemblage, especially in volcanic and hypabyssal rocks, is often thought to be of prime importance. The reasoning is that minerals in the phenocryst assemblage are very likely involved in the process of fractional crystallization in that they are the fractionating phases. They supposedly have the best chance to fractionate because they are early formed and less inhibited by surrounding crystals. Presumably no process of fractional crystallization is so perfect that 100 percent of the early formed crystals is removed, so those phenocrysts observed in thin-section were the ones that escaped removal. Early crystallization and removal of their brethren would then cause fractional crystallization. This reasoning does not always hold because rocks that have formed primarily by equilibrium crystallization can also have phenocrysts, but it is a common working hypothesis.

If these arguments hold for phyric (phenocryst-bearing) basalts and diabases (dolerites), then the phenocryst assemblage may provide some clue as to the depth of fractional crystallization. Pioneering work in this area has been done by Green and Ringwood (1967) and O'Hara (1968). If the phenocryst assemblage contains olivine + plagioclase +/- clinopyroxene, the magma very likely fractionated at relatively shallow depths, perhaps in the upper part of the crust. If the assemblage is olivine + clinopyroxene, however, fractionation probably occurred at deeper levels, possibly in the lower crust or even upper mantle. Many mid-oceanic ridge basalts (referred to by the acronym *MORB)* apparently fractionated at relatively low pressures, because plagioclase can appear in the phenocryst assemblage before clinopyroxene (Grove and Bryan, 1983).

The expanding field of spinel with increasing pressure (Figs. 4.25 and 4.26) also means that spinel has a better chance of being preserved as a near-liquidus phase if fractionation takes place at relatively high pressures. Recall that the spinel volume is surrounded by a divariant reaction surface. Consequently, spinel is in a reaction relationship during later crystallization of a typical basaltic melt, but some spinel crystals may survive the reaction. Those that survive typically have reaction rims around them that are easily observed in thin section. If whole-rock chemical analyses, as well as microprobe analyses of the phenocrysts, are available, then the hypothesis of fractional crystallization can be tested by use of variation diagrams (Section 5.2.2.3) or numerical modeling (see Section 5.2.3).

O'Hara (1968) was among the first to introduce the concept of *polybaric fractionation*. The decreasing field of Fo with increasing pressure (Figs. 4.25 and 4.26) can be used to illustrate this phenomenon. It is based on the assumption that basalt magmas rise through the upper mantle in a diapiric fashion similar to the manner in which salt domes are emplaced upward through sedimentary rocks. As this diapir of basaltic magma rises, presumably some crystallization and thus crystal fractionation occurs. As the diapir moves upward, confining pressure is lowered. Hence fractionation is taking place through time under conditions of ever-decreasing confining pressure, thus the term *polybaric fractionation*.

The effect of polybaric fractionation on the crystallization of Fo (olivine in naturally occurring rocks) is caused by the expanding Fo + L field with decreasing pressure in the rising diapir. If Fo crystallization takes place at constant pressure, LP would move straight away from FO until it encounters a cotectic, where Fo and another mineral coprecipitate. Because of the expanding Fo + L field, however, the cotectic retreats in front of LP, so only Fo continues to crystallize. In essence this process can be considered a race between LP and the cotectic. If the cotectic stays ahead of LP during upward movement of the diapir and resultant decompression of the melt, then only olivine crystallizes. In this fashion olivine fractionation can be prolonged over a long time in the rising diapir. If LP moves rapidly enough to "catch up" with the retreating cotectic, however, then olivine and another mineral coprecipitate.

4.2.3 The Haplogranite System OR-AB-AN-Q-H$_2$O

Many haplobasalt systems are possible. This is not true for the haplogranite system; if we are limited to systems containing five components, the system OR-AB-AN-Q-H$_2$O (abbreviated OAAQW) is clearly the best choice. It accounts for the feldspars and quartz, which comprise at least 80 percent and commonly more than 90 percent of most granitoids. This is generally true for granites and quartz monzonites, particularly alaskites and leucogranites (granites commonly containing almost 100 percent quartz and feldspars). It is less true for granodiorites, and still less true for tonalites (quartz diorites). This is because the *color index* (volume percent oxides plus ferromagnesian silicates; i.e., minerals other than feldspars and quartz) generally increases in the order granite - quartz monzonite - granodiorite - tonalite. It is generally considered that the haplogranite system can be used as a predictive device for granitoids if the color index is less than 10 percent, and perhaps even less than 20 percent. Norms are used as the medium through which the haplogranite system is compared with granitoids, as is discussed later.

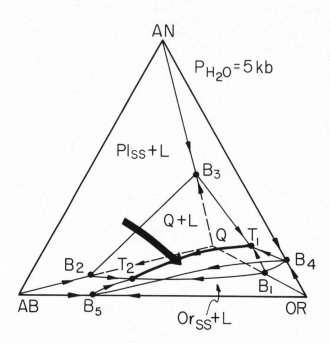

Figure 4.27 Two-dimensional perspective of the system OR-AB-AN-Q-H_2O at 5 kb P_{H2O} (modified after Winkler, 1979). Two ternary eutectics *(T$_1$ and T$_2$)* and five binary eutectics *(B$_1$-B$_5$)* are present. The primary phase field *Q + L* is present behind the surface *B$_1$-B$_2$-B$_3$*. H_2O is also stable throughout the system. *Heavy arrow* represents presumed liquid line of descent for some granitoids.

4.2.3.1 A Two-dimensional Perspective and a Fold-out

The haplogranite system OAAQW at approximately 5 kb water pressure *(P$_{H2O}$ = P$_T$)* is shown on Figure 4.27. It is represented as the quaternary tetrahedron OR-AB-AN-Q, but as H_2O is also included, the system is pseudoquaternary. It is divided into three trivariant primary-phase volumes: Or_{SS} + L, Pl_{SS} + L, and Q + L. These are bounded by three interior divariant surfaces that intersect along the interior univariant curve T_1-T_2. This curve connects the ternary eutectic T_1 in the system OR-AN-Q-H_2O with the eutectic T_2 in the system OR-AB-Q-H_2O (abbreviated OAQW). Along this curve T_1-T_2 all three solid phases are in equilibrium with the melt. The temperature at T_1 is higher than at T_2, so during crystallization LPs move along this curve from T_1 to T_2. Complete solid solution exists along the Ab-An join, but only partial solid solution is present along the Or-Ab join (see Section 3.3.2.4). Quartz and the feldspars are completely immiscible in one another, so simple binary eutectics are present at B_1-B_3. Another binary eutectic exists at B_4, although limited solid solution is present between AN and OR. The system contains no peritectics.

The system AB-AN-Q-H_2O is similar to the system AB-AN-DI (Section 4.1.3.1) in that it only has a cotectic and complete immiscibility between the feldspars and the quartz. The system OR-AB-AN-H_2O is superficially similar to AB-AN-DI in that they both have only a cotectic, but extensive solid solution between the feldspars causes the former to be considerably more complicated. The heavy, solid line (Fig. 4.27) approximates the best-fit curve for some norms of a granitoid suite (Section 4.2.3.3).

A fold-out of this haplogranite system OAAQW is given in Figure 4.28. As for the previous fold-out (Fig. 4.25), it can be reproduced or traced, and then folded along the interior joins to form the quaternary tetrahedron. As stated earlier, this is highly advisable. Because Q is at every apex of the triangle, the binary eutectics B_1-B_3 involving quartz fall along the edges of the triangle. The ternary eutectics T_1 and T_2 are labeled, as is the OR-AN binary eutectic B_4. Approximate fields of solid solution in the feldspars are shown as stippled areas. The heavy, solid curves with arrows are best-fit curves through some normative compositions for some granitoids (Section 4.2.3.3) and are presumed to represent a liquid line of descent. They are equivalent to the heavy curve on Figure 4.27.

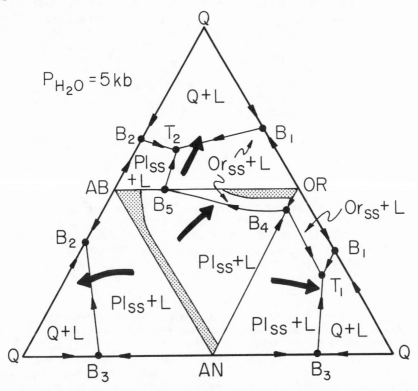

Figure 4.28 Fold-out of Figure 4.27 showing the four ternary systems for the haplogranite system OAAQW at 5 kb P_{H2O}. This figure can be folded along the three interior joins to form the tetrahedron. *Heavy solid curves* with arrows are presumed liquid lines of descent based on norms for some granitoids.

Plagioclase is least soluble in the melt at 5 kb P_{H2O}, quartz is next, and K-feldspar (Or_{SS}) is most soluble (Figs. 4.27 and 4.28). This is particularly obvious when the three-dimensional tetrahedron is observed. The Pl_{SS} + L volume is quite large and the Or_{SS} + L volume is relatively small. This means that for a wide variety of initial melt compositions plagioclase is the first mineral to crystallize and a considerable interval may pass before another mineral joins plagioclase and the two coprecipitate along a cotectic. Does this relationship hold for other water pressures and what effect does this have on petrogenesis of granitoids? We explore these two questions in the next two sections.

xxxxxxxxxxxxxxxxxxxxxxxxxxxxxxxxxxxxxxxxxxxxxxxxxxxxxxxxxxxxxxx

The observation is made earlier that the plagioclase volume is quite large, whereas the K-feldspar volume is comparatively small on Figure 4.27. Had we only examined the system AOQW (top triangle, Fig. 4.28), we would have reached a quite different conclusion. In this ternary system the Pl$_{SS}$ + L field appears quite small compared with the Or$_{SS}$ + L field. This is true when very small amounts of AN are present in the system, as in the case of most true granites. As more AN is introduced, however, the plagioclace volume greatly expands. This becomes quite apparent when observing the tetrahedral system made from the fold-out of Figure 4.28. AN-content of granitoids generally increases in the order granites - quartz monzonites - granodiorites - tonalites. If we wish this pseudoquaternary system to be applicable to all granitoids, we must be quite careful about projections (Section 4.2.3.3).

xxxxxxxxxxxxxxxxxxxxxxxxxxxxxxxxxxxxxxxxxxxxxxxxxxxxxxxxxxxxxxx

4.2.3.2 Effect of Fluid Pressure

In Section 4.1.3.4 we briefly discussed the effect of changing fluid pressure in the system OAQW. With increasing water pressure $(P_{H2O} = P_T)$ two main effects were noted: (1) a ternary eutectic rather than minimum point at relatively high pressures, and (2) a progressive increase in solubility of AB in the melt, with attendant decrease in size of the Ab$_{SS}$ + L primary-phase field. The system OAAQW now under discussion is simply that system OAQW with the additional component AN.

Of the two main effects cited for the OAQW system, only effect (2) remains for OAAQW. For all practical purposes, there is no minimum point in OAAQW, even at relatively low water pressures. Addition of AN component causes the solvus and the solidus to intersect, resulting in an apparent ternary eutectic (actually a *piercing point,* as explained later). Just as an increase in water pressure decreases the size of the Ab$_{SS}$ + L field in OAQW, a progressive decrease in size of the Pl$_{SS}$ primary-phase volume with increasing P_{H2O} is observed for OAAQW. Moreover, an increase in the AB/AN ratio of the original melt decreases the size of the Pl$_{SS}$ field.

These relationships can be visualized by projections from the AN apex onto the OR-AB-Q face of the tetrahedron (Fig. 4.29). The inset of Figure 4.29 shows the point of intersection P for the T_1-T_2 univariant curve (T_1 and T_2 are ternary eutectics, also shown on Figs. 4.27 and 4.28) with a plane of constant AB/AN ratio (plane Q-OR-X; AB:AN = 2.7). Point P, a piercing point, is then projected from AN onto the OR-AB-Q base of the tetrahedron, where it appears as a ternary eutectic P'. Thus in this case a piercing point is a *pseudoternary eutectic point.* The shaded quadrilateral encompassing all possible piercing points projected from AN onto the OR-AB-Q base is shown in Figure 4.29. This quadrilateral is occasionally referred to as the *minimum melt region,* because a melt whose composition coincides with a piercing point within this region forms at the lowest possible temperature for the appropriate pressure.

The distribution of piercing points for various water pressures and AB/AN ratios within the shaded quadrilateral is not random. The AB/AN ratio for piercing points increases from right to left across the quadrilateral, while P_{H2O} increases from top to bottom. Hence the upper right corner represents the piercing point for the lowest AB/AN ratio and P_{H2O}; the opposite situation exists for the lower left corner. The closer the piercing point is to the lower left corner of the quadrilateral, the smaller is the Ab$_{SS}$ + L field and the greater is the solubility of AB in the melt. As a consequence, increase of both AB/AN and P_{H2O} in the melt contribute to the decrease in size of the Ab$_{SS}$ + L field. This conclusion is important in the light of granitoid petrogenesis discussed next. A detailed discussion of these relationships is given in Winkler (1979).

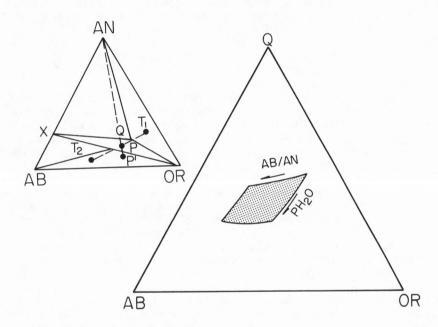

Figure 4.29 Effect of fluid pressure on the haplogranite system OAAQW. All possible AB-OR-Q piercing points fall in the stippled area. Point *P*, intersection of curve T_1-T_2 and plane *Q-OR-X* (AB/AN = 2.7), projected to point *P'* on inset is an example of a piercing point. See text for further explanation.

4.2.3.3 Use of Norms

In Section 1.2.5.1 the question was asked, "Why calculate a norm?" One of the main reasons cited was that they allow us to "bridge the gap" between field petrology and experimental petrology. In short, they proxy for components on phase diagrams. For example, on Figure 4.17 the normative minerals *or, ab,* and *q* proxy for the components OR, AB, and Q. An alternative is to convert LPs and XPs on phase diagrams to trends on elemental variation diagrams. The far more common practice is to do the opposite -- recast chemical analyses into norms and plot them on phase diagrams. This section explores the results of doing this for the normative minerals *or, ab, an,* and *q,* and the implications for granitoid petrogenesis.

We explore the use of norms by means of an example. The heavy, solid curves on Figure 4.28 are some best-fit curves for compositions from a series of granitoids, including tonalites, granodiorites, and quartz monzonites. The arrows point in the presumed direction of crystal fractionation (i.e., generally down temperature), tonalites being the least fractionated and quartz monzonites the most. Relative age relationships are not clear in this area, but limited evidence exists, based on cross-cutting relationships, that the quartz monzonites are the youngest and the tonalites are the oldest. These relative ages support a model of fractional crystallization.

Such trends are frequently referred to as *liquid lines of descent.* They are generally assumed to be equivalent in naturally-occurring rocks to LPs in experimental systems, although this is not always the case. In this section we assume that they do represent LPs. The four curves are actually projections of a single curve within the interior of the tetrahedron (heavy curve with arrow, Fig. 4.27) from each apex of the tetrahedron onto each opposite face. The difficulty in exactly locating the position of LP within a two-dimensional perspective of the tetrahedron is apparent.

If we assume that the quartz-poor end of LP (the least differentiated tonalites) represents a parental magma to the other rocks, then three of the four projections indicate that plagioclase should crystallize first. Only in the OAQW system does LP principally lie in the Or_{SS} field. Large, euhedral to subhedral plagioclases in the tonalites support the hypothesis that plagioclase was the first to crystallize. Furthermore, the trace of LP in the two-dimensional perspective (Fig. 4.27) indicates that plagioclase crystallized first. By apparently fractionating plagioclase (of approximate composition AN45), the melt changed composition as LP moved toward a quaternary cotectic surface, but which surface? Was quartz or K-feldspar the next mineral to crystallize? Which projection provides us with the truest picture of relationships inside the tetrahedron?

These questions cannot be answered unless we closely examine the three-dimensional OR-AB-AN-Q tetrahedron (Fig. 4.27). A two-dimensional perspective does not suffice. The most commonly used projections are from AN onto OR-AB-Q or from Q onto OR-AB-AN, but are they the most useful in this case? They clearly are not, as can be seen by observing the cotectic surfaces within the tetrahedron. Because of the slope on the PL_{SS}-Q cotectic surface, as well as the thin Or_{SS} + L volume, many compositions within the plagioclase volume plot in either the Q + L or Or_{SS} + L field on the OR-AB-Q projection. As stated earlier, the OR-AB-Q projection leaves a false impression about the size of the Pl_{SS} + L volume within the interior of the tetrahedron. The projection OR-AB-AN is not good for our purposes because Q is not represented as a phase. Likewise, AB-AN-Q is not useful because Or_{SS} is not present.

By the process of elimination we are left with the projection from AB onto the OR-AN-Q face. A close examination of cotectic surfaces within the tetrahedron reveals that this projection is quite sufficient for our purposes. Most compositions that plot in the Pl_{SS} + L volume (except those extremely enriched in AB and impoverished in AN, such as those for some trondhjemites) also plot in the Pl_{SS} + L field on the projection. Similarly, most compositions that fall into the Q + L and Or_{SS} + L volumes project into their appropriate fields on the OR-AN-Q face.

Thus the projection on Figure 4.28 that most likely represents the LP inside the tetrahedron is in the lower right corner -- OR-AN-Q. This projection indicates that LP should intersect the Pl_{SS}-Q cotectic very near the T1-T2 univariant curve. In fact, the point of intersection is so close to the univariant curve, one cannot determine which cotectic is intersected first. LP on Figure 4.27 appears to first intersect the Or_{SS}-Pl_{SS} cotectic, although its exact location is difficult to visualize. The important point is that whichever mineral, quartz or K-feldspar, that first coprecipitates with plagioclase, the other should follow shortly thereafter. Petrographic examination gives no indication as to whether quartz or K-feldspar crystallized first, although numerical modeling calculations (see Section 5.2.3.2) suggest that K-feldspar may actually have been first.

This example is intended to show how field and petrographic observations can be integrated with chemical analyses and a known experimental system to develop a model for crystallization of a rock suite. The use of norms is obviously critical to such an exercise. Utilization of such a system clearly has its limitations. For instance, because the OAAQW system contains no Fe and Mg components, we cannot determine the fate of ferromagnesian minerals. We also must assume $P_{H2O} = P_T$, which may not be the case unless miarolitic cavities are present, evidence for a separate fluid phase. Similar integrated studies (which also frequently include trace elements and isotopes), have been made to examine other processes, such as assimilation and magma genesis through partial melting. Unfortunately, we cannot explore all these areas, although some will be at least briefly examined in the next chapter.

5

IGNEOUS ROCKS

The previous four chapters of this book are preparation for this last chapter. Finally, we will be dealing with data from naturally occurring igneous rocks rather than abstract conversions of chemical data, complex thermodynamic formulae, or synthetic systems. The principal aim of this chapter is to explore practical ways in which experimental and analytical data on actual igneous rocks can be used to enlighten us about how these rocks may have formed. We must accept the fact, however, that in most cases we will never know with certainty.

5.1 MELTING AND CRYSTALLIZATION EXPERIMENTS

Synthetic, experimental systems are extremely valuable in helping us understand how partial melting of silicate rocks and crystallization of the resultant magmas operate. These systems have the advantage of being constrained by a precise law, the phase rule. They have one drawback, however. Experimental systems may proxy for naturally occurring rocks, but they can never be more than first approximations. In the previous two chapters we did not consider any system beyond a five-component (pseudoquaternary) system, yet the simplest igneous rock, even allowing for trace-element substitutions in major-element sites, has at least ten components.

One solution to this problem is to conduct experiments on naturally occurring rocks, rather than on mixtures of synthetic compounds, despite the fact that the phase rule is difficult to apply. Before the 1960s, experiments on natural rocks were not common, but development of the electron microprobe as an analytical tool meant that glasses and crystals less than 5 microns in diameter could be analyzed quantitatively. Thus experimental melting and crystallization studies on natural rocks became common, especially with regard to the genesis of basaltic magmas. The most intense research in this area was in the 1960s and 1970s. Classical early studies on basaltic magmas were by Yoder and Tilley (1962), as well as D. H. Green and Ringwood (1967). Other extremely significant papers in this area were by Mysen and Boettcher (1975), Walker and others (1979), and Stolper (1980). Because of the general interest in MORBs in recent years, a great deal of experimental work has been done on them (e.g., Bender and others, 1978; Fisk and others, 1980; Grove and Bryan, 1983; Elthon and Scarfe, 1984). A classic paper by T. H. Green and Ringwood (1968) on andesites utilized melting techniques, as did Piwinskii and Wyllie in a series of papers on more felsic rocks (for example, see Piwinskii and Wyllie, 1970). This section will briefly describe the results of some experiments and their implications for petrogenesis.

5.1.1 Dry Conditions

Dry conditions refer to experiments conducted in the absence of a fluid phase. For example, dry experiments are useful in understanding how partial melting of peridotite in the upper mantle can produce a relatively anhydrous and CO_2-free basaltic melt such as MORB (mid-ocean ridge basalts). All natural basaltic melts probably contain at least some H_2O, CO_2, and/or another volatile component, but some basaltic melts apparently are so nearly dry that these experiments are applicable. Melts for most other rock types, and even many basalts, contain sufficient quantities of volatiles such that these dry experiments are probably not adequate to describe their behavior. Results from 1-atmosphere experiments are similar to those obtained at pressures equivalent to shallow or even intermediate depths in the continental crust. Consequently, we can learn a great deal about *low-pressure* fractional crystallization of dry or nearly dry basaltic melts from these comparatively simple experiments.

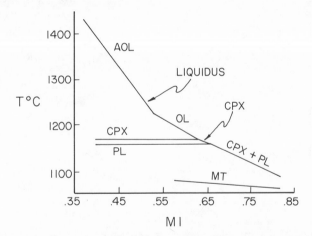

Figure 5.1 Summary of 1-atmosphere melting experiments on Kilauea basalts (modified after Thompson and Tilley, 1969).

Some experiments are conducted without buffers to control and maintain constant oxygen fugacity, but oxygen is normally buffered in the more exacting experiments. Because control of O_2 fugacity primarily affects the oxides, valuable information about the silicates also can be obtained from nonbuffered experiments. Experimental runs of short duration can be buffered by minerals present in the rock, but external buffers are preferred.

5.1.1.1 Nonbuffered One-Atmosphere Experiments

This and the following section will deal only with basaltic melts. Many of the classic papers that wholly or partly deal with 1-atmosphere experiments were written by C. E. Tilley and his co-workers (for example, Yoder and Tilley, 1962; Thompson and Tilley, 1969). A series of papers published in the Yearbook of the Carnegie Institute of Washington, D.C., are particularly important in this regard (e.g., Tilley and others, 1967). Thompson (1972) has summarized the experimental results based on samples from historical eruptions of the Kilauea volcano in Hawaii, while Thompson and Tilley (1969) provided chemical analyses of nine of these samples. This discussion will concentrate on results based on these samples from Kilauea and their implications for low-pressure fractionation of olivine tholeiite magmas.

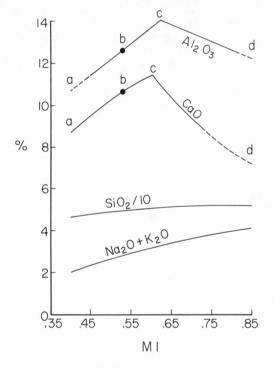

Figure 5.2 Oxide variation diagrams (in weight %) for Kilauea basalts (modified after Thompson and Tilley (1969). MI = mafic index = $FeO+Fe_2O_3$ / $FeO+Fe_2O_3+MgO$ (Section 5.2.2.2).

Results of 1-atmosphere melting experiments on Kilauea samples are summarized in Figure 5.1, a plot of the mafic index (MI) versus temperature at which a phase began crystallizing on cooling of the melt, or at which a phase had completely melted on heating of the solid sample. Oxygen fugacity was not rigorously controlled in these experiments, except to the extent that it was controlled by minerals in the rocks themselves. A line on Figure 5.1, such as line CPX, is often labeled as either "cpx in" or "cpx out," depending on whether the experimental petrologist is referring to appearance or disappearance of cpx during crystallization as temperature decreases. If cpx appears ("cpx in"), the line can be considered as equivalent to a coprecipitational cotectic; if cpx disappears ("cpx out"), then cpx is reacting with the melt and the line is resorptional.

All lines on Figure 5.1 are coprecipitational. On this graph mafic index = MI = FeO+Fe$_2$O$_3$ / FeO+Fe$_2$O$_3$+MgO (Section 5.2.2.2). A series of samples with MI values ranging from about 0.40 to 0.80 were used to compile this figure. An increase in MI normally indicates a greater degree of fractional crystallization.

The steeper line segment on the liquidus labeled AOL (Fig. 5.1) was based on samples that contained accumulative olivine, whereas the line segment OL contained no accumulative olivine. Consequently, the least fractionated aphyric sample has a mafic index of about 0.53, which marks the break in slope on the liquidus. Olivine is the first mineral on the liquidus for samples with MI < 0.62, followed by a very short interval (about 0.62-0.65) when clinopyroxene (augite) is the first mineral. For more fractionated rocks (MI > 0.65), both plagioclase and augite are on the liquidus. For olivine-rich compositions (e.g., for a sample with MI = 0.45), a long interval of olivine crystallization occurs before first augite (CPX), and shortly thereafter plagioclase (PL) begins to crystallize. In the more fractionated samples, magnetite (MT) does not begin crystallizing until temperatures are considerably below the liquidus.

Can these results be used to shed light on low-pressure fractional crystallization of an olivine tholeiite melt? The answer is affirmative and can be explained by the compositional trends on Figure 5.2 (data from Thompson and Tilley, 1969). These trends are based on many of the same samples used in the experimental studies used to develop Figure 5.1. The range in MI is the same for both figures. The alkalies and silica exhibit regular increases with fractionation (increasing MI). Trends for CaO and Al$_2$O$_3$ have sharp breaks in slope that correspond to the MI value where the CPX and PL lines intersect the liquidus (point c; Fig. 5.2).

The trends with positive slopes at lower values of MI on the CaO and Al$_2$O$_3$ graphs (curves a-b-c) are the result of *olivine control,* and can be referred to as *olivine control lines.* Similarly, the negative slopes at higher MI values (curves c-d) are caused by *augite plus plagioclase control.* The term *olivine control* in this case refers to the fact that either accumulation or removal of olivine will "control" this positive trend, as accumulation or removal of augite and plagioclase control the negative trend. Addition (accumulation) of olivine to a magma with a mafic index of around 0.53 (point b) depletes the resultant rocks in CaO and Al$_2$O$_3$, while lowering their MI values (curve b-a). Removal of olivine from that same magma momentarily enriches residual melts in CaO and Al$_2$O$_3$ (curve b-c), but augite and plagioclase soon begin fractionating, which depletes the residual liquids in CaO and Al$_2$O$_3$ as MI increases (curve c-d). Thus trend b-c-d is a liquid path (normally referred to as a *liquid line of descent),* while curve a-b represents the mixing of various proportions of magma b and olivine (a *mixing line).* Control lines, mixing lines, and liquid lines of descent are discussed further in Section 5.2.2.3.

These trends on Figure 5.2 are best-fit curves to individual data points, although there is very little scatter about the curves (Thompson and Tilley, 1969; Figs. 2-5). Note that on these best-fit lines the break in slope for CaO occurs at a lower mafic index than for Al$_2$O$_3$. The CaO/Al$_2$O$_3$ ratio is much greater in augite than plagioclase, so the break in slope for CaO marks the beginning of augite removal, whereas the break for Al$_2$O$_3$ marks the point at which both augite and plagioclase begin fractionating. Interestingly, there are no apparent slope breaks for SiO$_2$ or the alkalies.

5.1.1.2 A Buffered One-Atmosphere Experiment

One additional study utilizing 1-atmosphere experiments will be examined. This is a more recent study on the Lower Mesozoic tholeiitic basalts in the Hartford Basin, Connecticut, in which O_2 fugacity was rigorously controlled by use of buffers (Philpotts and Reichenbach, 1985). Three basalt flows are present in the Hartford Basin, from oldest to youngest, Talcott, Holyoke, and Hampden. We show results of 1-atmosphere melting experiments only for the Talcott flow (Fig. 5.3). Four oxygen buffers were used:

MW	magnetite-wustite
TFM	tridymite-fayalite-magnetite
LTM	liquid-tridymite-magnetite
NNO	nickel-nickel oxide

Of the buffers used in this experiment, NNO maintains the highest oxygen fugacity, whereas MW maintains the lowest (Fig. 5.3). Temperature has far more effect on the phase boundaries than does oxygen fugacity (fO_2); i.e., the boundary lines are more nearly horizontal than vertical. With increasingly oxidizing conditions the liquidus is slightly lowered; the middle boundary is essentially unaffected; and the lower boundary is slightly raised. This small effect of fO_2 on these phase boundaries is encouraging, because it means that experiments, such as those cited for which fO_2 was not so rigorously maintained, still have validity.

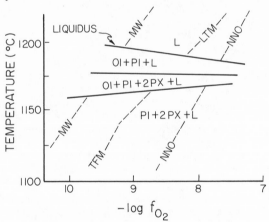

Figure 5.3 One-atmosphere, oxygen-buffered melting experiments on sample from Talcott basalt flow, Connecticut (modified after Philpotts and Reichenbach, 1985).

Olivine and plagioclase appear simultaneously on the liquidus, followed by a temperature interval of only about 15-25°C, at which point augite and pigeonite begin crystallizing (Fig. 5.3). Below the lower phase boundary only the two clinopyroxenes and plagioclase remain with the liquid. Hence olivine must have reacted away to form pigeonite and liquid, thus the lower boundary can be considered resorptional.

5.1.1.3 Low-pressure and High-pressure Fractional Crystallization

As explained earlier, some basaltic rocks are probably the only common rocks that may form under effectively dry conditions. In this section we deal with fractional crystallization of basaltic melts at various depths in the crust and upper mantle under dry conditions. Two terms commonly used by petrologists are *low-pressure fractionation* and *high-pressure fractionation.* These two processes can, under certain specific circumstances, lead to quite different chemical trends on variation diagrams. The terms became widely used particularly as a result of the pioneering work of D. H. Green and Ringwood (1967). Consequently, we will use their experimental results on the dry melting of an olivine tholeiite to examine fractional crystallization at different pressures.

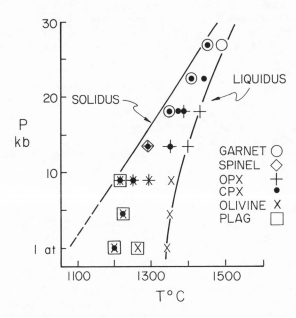

Figure 5.4 Results of melting experiments on olivine tholeiite (modified after Green and Ringwood, 1967).

Green and Ringwood's results of melting studies on an olivine tholeiite are summarized in Figure 5.4, a *P-T* plot showing the liquidus, solidus, mineral in points, and one mineral out point. An example of a mineral in point is olivine in at 1 atm pressure and about 1340°C. Mineral in points are marked by the first appearance of a mineral with decreasing temperature. The only mineral out point is opx out at 18 kb pressure and about 1370°C, where opx apparently reacts with the melt to form cpx.

An important fact to remember for the following discussion is that the base of the continental crust (depth of the "moho" under the continents) is typically at a depth of about 30-45 km, equivalent to approximately 10-15 kb pressure. In contrast, the moho under the oceanic crust is only at a depth of about 10-12 km (3-4 kb pressure).

The liquidus and solidus on Figure 5.4 both have positive slopes. This is because increasing confining pressure prevents melting but aids crystallization, whereas increasing temperature promotes melting and inhibits crystallization. We can also consider these slopes in terms of the Clapeyron equation (see

Section 2.2.1.2). Entropy increases with increasing temperature and the density of the molten peridotite melt is less than that of the solid peridotite, so dP/dT slopes on the liquidus and solidus are positive.

Melting studies on this olivine tholeiite can be summarized as follows:

1. At 5 kb or less pressure (equivalent to <15 km depth) olivine is on the liquidus, followed by a long temperature interval when only olivine crystallizes. Eventually either plagioclase or plagioclase + calcic clino-pyroxene (augite) begins crystallizing. These results are quite similar to those reported in the previous section for olivine tholeiite samples from Kilauea volcano. It is apparent that 1-atmosphere experiments have general applicability up to at least 15 km depth. Hence one-atmosphere experiments can be used to model fractionation at any depth in the oceanic crust, and even into the uppermost part of the oceanic mantle.

2. At 9 kb, (about 27 km depth, deep into the continental crust) olivine still crystallizes first, but is joined by orthopyroxene, rather than plagioclase or clinopyroxene. Only later does clinopyroxene (now subcalcic), and finally plagioclase very near the solidus, begin crystallizing.

3. At 13.5 kb, about 40 km (near the continental moho, either in the lowermost crust or uppermost mantle), olivine and plagioclase no longer crystallize. Orthopyroxene is on the liquidus, followed by subcalcic clinopyroxene, and finally spinel begins crystallizing near the solidus.

4. At 18 kb (about 54 km, normally well into the upper mantle), the situation is somewhat similar to 13.5 kb, until the solidus is approached, at which point garnet rather than spinel begins crystallizing. In addition, ortho-pyroxene reacts with the melt to form clinopyroxene.

5. At 22.5 kb (about 68 km) orthopyroxene no longer crystallizes and clino-pyroxene is on the liquidus, followed at lower temperatures by crystal-lization of garnet.

6. Finally, at 27 kb (about 81 km), garnet is on the liquidus, followed by clino-pyroxene at lower temperatures.

Because melting experiments were conducted at intervals of about 5-10 kb, the exact pressure at which a mineral coexists with the melt is not certain. Irrespective of whether these minerals occur near the liquidus or the solidus, plagioclase and olivine coexist with the melt at relatively low pressures, ortho-pyroxene and spinel at intermediate pressures, and garnet at high pressures. Clinopyroxene occurs throughout the pressure range. Plagioclase, spinel, and garnet ranges only minimally overlap, if at all. We will return to this point later in connection with "metamorphic facies" in upper-mantle peridotite.

Green and Ringwood (1967) concluded that four different patterns of fractional crystallization could occur, depending on depth. For fractional crystallization of an olivine tholeiite magma, these four patterns can be summarized by:

<15 km	early olivine control, followed by plagioclase and/or clinopyroxene control
15-35 km	early olivine control, followed by orthopyroxene and then clinopyroxene
35-70 km	early orthopyroxene control, followed by clinopyroxene

70-100 km   early clinopyroxene control followed by garnet, or at greater depths, early garnet control followed by clinopyroxene

It has become standard practice for petrologists to distinguish between low-P fractionation and high-P fractionation. Low-P fractionation is frequently taken to mean fractional crystallization at depths equivalent to the earth's surface or somewhere in the crust or uppermost oceanic mantle (i.e., about 35 km or less). Consequently, in this loose, twofold classification, high-P fractionation takes place at depths equivalent to the mantle below the continental crust (greater than about 35 km). It is apparent from the fourfold classification above that this twofold distinction is oversimplified. For example, distinct differences exist between high-pressure fractionation at less than 35 km in contrast to greater than 35 km. Fractionation at 70-100 km, involving primarily garnet and clinopyroxene, is occasionally referred to as *eclogite fractionation*. This is because eclogite is a rock, chemically similar to basalt, which contains dominantly garnet and omphacite, a Na-rich clinopyroxene.

Starting with an olivine tholeiite parent magma, fractional crystallization of a mineral extract assemblage at each of the four depths exerts a different control on the resultant liquid lines of descent. We should be able to plot chemical analyses of aphyric basaltic rocks on variation diagrams (Section 5.2.2.3) or use numerical methods (Section 5.2.3.2) and determine from the trends which minerals fractionated. This extract assemblage will give us a clue as to the depth where fractionation occurred.

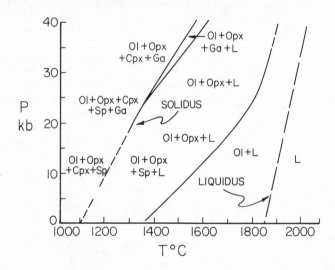

Figure 5.5 Results of dry melting experiments on a natural peridotite (modified after Ito and Kennedy, 1967).

5.1.1.4 Partial Melting

The mechanism for production of basaltic magmas is thought to be partial melting of peridotite in the upper mantle, perhaps under conditions that are effectively if not perfectly dry. Thus an experimental melting study at various pressures of a natural peridotite would be extremely useful in constraining the conditions under which basaltic partial melts form. In addition, the same study will enable us to learn more about the "metamorphic" mineral facies in the solid upper mantle.

Results of such a melting study on a natural peridotite are given in Figure 5.5. This is a considerably simplified version of the *P-T* diagram constructed by Ito and Kennedy (1967). Numerous stability fields are shown without phase boundaries around them. The exact locations of these boundaries were not determined in the original work, so for sake of simplicity they have been omitted on Figure 5.5. The critical boundaries, including the liquidus and the solidus, are shown. As for the dry melting of olivine tholeiite, both liquidus and solidus temperatures increase with increasing pressure. Interestingly, liquidus temperatures for the peridotite are much higher than those for the olivine tholeiite, but at lower pressures solidus temperatures are similar (compare Figs. 5.4 and 5.5). At low pressures olivine is the liquidus phase, followed by olivine + orthopyroxene, and then olivine + orthopyroxene + spinel near the solidus. At high pressures olivine is again the liquidus phase, again followed by olivine + orthopyroxene, but then followed by olivine + orthopyroxene + garnet. No plagioclase was reported by Ito and Kennedy at any pressure. Note that garnet is stable at higher pressure than spinel, in agreement with results on Figure 5.4.

We are not so concerned with crystallization patterns as temperature decreases as with melting trends as temperature increases. First let us consider melting at pressures less than about 20 kb (less than roughly 60 km depth). A necessary procedure is to examine carefully the mineral assemblage in the solid peridotite and then compare it with the assemblage coexisting with melt above the solidus. The peridotite below the solidus is a *spinel lherzolite,* a two-pyroxene, spinel-bearing peridotite (Fig. 5.5). After partial melting has occurred at the solidus, however, clinopyroxene has completely melted, so the remaining residuum contains only olivine, spinel, and orthopyroxene. Such a one-pyroxene peridotite is called a *spinel harzburgite.* The partial melt will be relatively enriched in components that make up clinopyroxene, SiO_2, CaO, Al_2O_3, TiO_2, and the alkalies; i.e., a rock of basaltic composition. It will also be enriched in any incompatible elements compared with the source rock (for definition of incompatible elements, refer to Section 5.2.1.1). For this reason, mantle composed dominantly of lherzolite is occasionally referred to as *enriched mantle,* whereas a harzburgite mantle is called *depleted mantle.* If sufficient additional partial melting occurs, spinel will also completely melt. It is doubtful that enough partial melting will ever occur such that all orthopyroxene melts and only an olivine residuum remains.

Let us now examine partial melting at a relatively high pressure, at 30-40 kb (about 90-120 km depth). The source rock is now a *garnet lherzolite* (Fig. 5.5). When solidus temperatures are reached, again all clinopyroxene melts, leaving a *garnet harzburgite* residuum. The fact that garnet rather than spinel is present in the source rock can have some profound effects on trace-element distributions in the basaltic partial melts (Sections 5.2.1.3, 5.2.1.4, and 5.2.2.6). At higher temperatures, all garnet has melted and only a harzburgite remains in the source. Ito and Kennedy (1967) report that the composition of this basaltic melt, however, is not a typical olivine tholeiite. Its composition is much higher in MgO and normative olivine than a typical olivine tholeiite. Such a basalt is referred to as a *picrite* or *picritic basalt.* Presumably typical olivine tholeiites can be formed from picritic basalts by fractional crystallization of dominantly olivine, in agreement with such workers as Yoder and Tilley (1962), O'Hara (1968), and Elthton and Scarfe (1984).

Presnall and others (1979) and Presnall and Hoover (1984) disagreed, concluding that olivine tholeiite primary magmas (MORB) could be directly generated by partial melting under mid-ocean ridges at 9-10 kb and about 1220°C. This would take place where an invariant point coincides with a "cusp" on the peridotite solidus (point I, at about 9 kb and 1220°C; Fig. 5.6). The ocean-ridge geotherm passes through point I, so partial melting to form MORB is reasonable according to this figure. Their work was based primarily on the experiments in the synthetic quaternary system CaO-MgO-Al$_2$O$_3$-SiO$_2$ (commonly known by the acronym CMAS). Stolper (1980) conducted some melting experiments on some primitive MORB encapsulated in a mixture of olivine and orthopyroxene at various pressures. His results, in contrast to those of Presnall and others, agreed better with earlier workers who proposed that MORB melts are not primary. Whether or not olivine tholeiites are primary or derivatives from a picritic primary melt is a moot point. The most important conclusion from all this work is that olivine-bearing basaltic rocks that make up a substantial part of the earth's crust are almost certainly formed by partial melting of mantle peridotite.

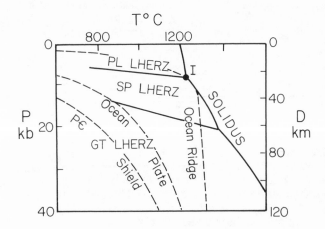

Figure 5.6 *P-T* diagram showing peridotite solidus, three geotherms *(dashed curves),* and three "metamorphic facies" in the mantle at different depths (modified after Morse, 1980).

Figures 5.5 and 5.6 clearly explain how basaltic melts can be formed by partial melting of either spinel or garnet lherzolite. A problem, however, remains. As the upper mantle is dominantly solid, a geotherm must intersect the lherzolite solidus for partial melting to occur. This reportedly can happen under a mid-oceanic ridge by dry melting. Excluding special circumstances, however, temperatures are probably too low in most parts of the mantle for completely dry melting to occur (Fig. 5.6). The typical subcontinental and suboceanic geotherms, shown on Figure 5.6, do not intersect the peridotite solidus. A small amount of H$_2$O, however, can lower the peridotite solidus to the shield or ocean-plate geotherm temperatures so that partial melting can occur.

5.1.1.5 Mineral Facies in the Upper Mantle

Experimental work on natural and synthetic peridotites led to the determination of subsolidus mineral assemblages at different pressures and temperatures. As a result, three *metamorphic facies* in the upper mantle are shown on Figure 5.6. With increasing confining pressure and thus depth they are plagioclase lherzolite, spinel lherzolite, and garnet lherzolite. The generalized reactions to form first spinel and then garnet with increasing pressures (Kushiro and Yoder, 1966; Morse, 1980) are:

$$PLAG + OLIVINE = aluminous\ OPX + aluminous\ CPX + SPINEL$$

$$aluminous\ OPX + aluminous\ CPX + SPINEL = GARNET$$

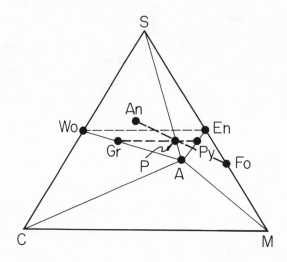

Figure 5.7. The subsolidus system $CaO-MgO-Al_2O_3-SiO_2$ (CMAS). See text for explanation of abbreviations.

We can examine whether these reactions are reasonable by applying the contact principle (see Section 3.2.2.2) to the subsolidus joins in the idealized quaternary system CMAS. It is customary to write chemical formulae for minerals in this system in abbreviated form, based on C = CaO, M = MgO, A = Al_2O_3, and S = SiO_2. Thus plagioclase (anorthite; $CaO·Al_2O_3·2SiO_2$) can be written CAS_2 and olivine (forsterite; $2MgO·SiO_2$) is M_2S. Because products of the first reaction are reactants of the second, we should be able to write the reaction:

$$PLAG + OLIVINE = GARNET$$

This garnet must contain both grossularite (Gr; C_3AS_3) and pyrope (Py; M_3AS_3) components. An implication of the contact principle is that for these two reactions to be possible, the grossularite-pyrope subsolidus join must intersect the anorthite-forsterite join at a point. As can be seen on Figure 5.7, this is indeed the case; the two joins intersect at point P. Hence, we can write the following balanced equation:

$$An + Fo = 0.33GR + 0.67PY$$
$$CAS_2 + M_2S = 0.33C_3AS_3 + 0.67M_3AS_3$$

Writing this equation in the more conventional manner:

$$CaAl_2Si_2O_8 + Mg_2SiO_4 = CaMg_2Al_2Si_3O_{12}$$

These equations should not be interpreted to mean that the garnet composition in every garnet lherzolite is exactly PY67-GR33. Remember that the system CMAS is an idealized, simplified system. For instance, no Fe minerals are in the CMAS system, yet almandine component ($Fe_3Al_2Si_3O_{12}$) is present in garnets from lherzolites. Garnets in peridotites, however, are generally PY-rich. In summary, the original two reactions explaining the existence of first spinel and then garnet lherzolite with increasing pressure are stoichiometrically possible.

Parenthetically, because the join Gr-Py falls in the plane Wo-En-A, point P also marks the intersection of this plane with the An-Fo join. Thus according to the contact principle we should be able to write a balanced reaction involving anorthite + forsterite = wollastonite + enstatite + corundum, although we have no reason to do so.

These arguments indicate that we would expect aluminous pyroxenes in the spinel lherzolite facies, but not in the plagioclase or garnet lherzolite facies. This is generally the case. The characterizing mineral for each of these mineral facies is the dominant aluminous phase, either plagioclase, spinel, or garnet. Moreover, orthopyroxene in lherzolites contains Al_2O_3 as *Mg-Tschermak's molecule* ($MgAl_2SiO_6$) and Al_2O_3 is present in lherzolite clinopyroxenes as *Ca-Tschermak's molecule* ($CaAl_2SiO_6$). For these reasons, the general term *aluminous lherzolite* is frequently used as a likely upper-mantle source rock for basaltic melts. Note that plagioclase lherzolite is stable at such low pressures it is doubtful that it ever exists under the continental crust.

5.1.2 Effect of Volatiles

In the examples of melting experiments cited volatiles were either completely absent or only trivially present. Except perhaps for formation of MORB under mid-ocean ridges or some tholeiites in continental rift valleys, effectively dry melting probably seldom occurs. As we have dealt only with basaltic rocks to this point, more felsic rocks deserve examination. Accordingly, after considering the overall role of water in magmatic systems, we will examine the effect of H_2O on fractional crystallization of a calcalkaline melt and partial melting of a deep crustal source. Then we return to basaltic melts and examine the combined effect of H_2O and CO_2 on their formation and fractionation.

5.1.2.1 Water

Water plays a vital role in most magmatic processes. Whitney (1988), for example, has reviewed the effect of water on the origin of granite and concluded that temperature and water content are the two most most important parameters. The role of water is so important that petrologists are concerned about developing *geohygrometers,* which allow the estimation of water contents in magma chambers (e.g., Baker and Eggler, 1983; Baker, 1987). In all P,T environments in which igneous activity occurs water will be a supercritical fluid rather than a vapor (Section 3.1.1.4). Consider an O^{2-} anion linking together two silica tetrahedra, or an alumina and a silica tetrahedron. When this O^{2-} anion reacts with water two OH^- ions are produced:

$$H_2O + O^{2-} = OH^- + OH^-$$

The effect of this reaction is to depolymerize the melt, which in turn depresses the liquidus and solidus to lower temperatures (Fig. 5.8a). This lowering of the liquidus and solidus is one of the most profound and significant effects of H_2O on silicate melts. Source rocks that cannot otherwise partially melt can commonly do so even if only a small amount of H_2O is present. The reaction above can be represented diagramatically (Burnham, 1975; Barker, 1983) for two silica tetrahedra by:

$$
\begin{array}{ccccccc}
\text{O} & & \text{O} & & & \text{O} & & \text{O} \\
| & & | & & & | & & | \\
\text{O--Si--O--Si--O} & + & H_2O & = & \text{O--Si--OH} & + & \text{OH--Si--O} \\
| & & | & & & | & & | \\
\text{O} & & \text{O} & & & \text{O} & & \text{O}
\end{array}
$$

This depression of liquidus and solidus temperatures occurs because a solid can melt to a depolymerized melt more easily (i.e., at lower temperatures) than to a polymerized melt. This phenomenon is related to *freezing point depression* (see Section 3.2.4.2), whereby introduction of another component (a network modifier) lowers the melting/freezing point of a solid compound by breaking the oxygen linkages.

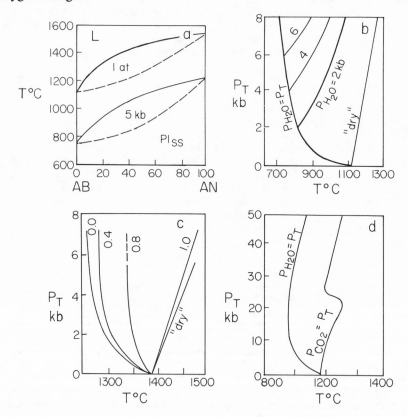

Figure 5.8 Effect of volatiles on: *a.* plagioclase (Ehlers, 1987); *b.* albite solidus (Burnham and Davis, 1974); *c.* diopside solidus (Rosenhauer and Eggler, 1975; numbers on curves are X_{CO2}); and *d.* peridotite solidus (Wyllie, 1979).

Another important effect of H_2O has to do with the slope on the liquidus and the solidus. In a dry system, such as the olivine tholeiite and peridotite discussed in the previous section, liquidus and solidus temperatures increase as confining pressure (P_T) increases (Figs. 5.4 and 5.5). In contrast, in a water-saturated $(P_{H2O} = P_T)$ system, liquidus and solidus temperatures are generally concave-up and decrease with increasing fluid pressure until extremely high H_2O pressures are reached (Fig. 5.8b-d). On Figure 5.8d an H_2O-saturated peridotite solidus has a negative slope until about 18 kb, at which point the slope is vertical. At higher H_2O pressures the slope becomes increasingly positive.

Again we can use the Clapeyron equation to explain these changes in slope. Imagine these reactions are going on in a chamber with a piston in one end so that V can change, rather than a confined chamber of constant V. Moreover, assume that the change in ΔS with increasing pressure is small compared with the change in ΔV. If the water-saturated phase boundary is traversed isobarically from low to high temperature, ΔS must be positive; because dP/dT is negative at less than about 18 kb, ΔV must be negative. Above 18 kb both ΔV and dP/dT are positive. Why is ΔV negative below about 18 kb, but positive above? For ΔV to be negative at relatively low pressures, V_{S+L+F} on the high-T side of the solidus must be less than V_{S+F} on the low-T side (S, solid; L, liquid; F, fluid). This is reasonable because considerably more fluid is dissolved in the melt than the solid. As pressure increases compressibility and total volume of the fluid drastically decrease and the solubility of the fluid in the melt increases. The net effect is for dP/dT to become positive because $V_{S+L+F} > V_{S+F}$.

A melt is saturated with H_2O and $P_{H2O} = P_T$ (Fig. 5.8b) when all the H_2O possible has dissolved in the melt; if additional H_2O is present, it will be in a separate fluid phase (i.e., bubbles in the melt). This situation probably seldom occurs until late stages of fractional crystallization or very early stages of partial melting. Because water has a greater affinity for the melt than the fractionating crystals, water will be increasingly enriched in the melt during fractional crystallization, or most concentrated in the earliest partial melts. At some point during fractional crystallization the H_2O content of the melt may become high enough so that the melt becomes supersaturated and a separate H_2O-rich fluid forms. This phenomenon is commonly referred to as *second boiling*. Parenthetically, this fluid normally carries with it a great deal of material, such as silica and the alkalies. These can form such features as mineralized quartz veins and pegmatite dikes that contain exotic minerals.

How does the solidus or liquidus appear in a system in which H_2O is present but the melt is not saturated? Figure 5.8b shows four solidi with positive slopes: $P_{H2O} = 0$ ("dry"), 2, 4 and 6 kb. Note that these four solidi intersect the H_2O-saturated solidus at 0, 2, 4, and 6 kb, respectively. Each solidus represents conditions in which P_{H2O} is constant but less than P_T; i.e., the melt is undersaturated with H_2O. The $P_{H2O} = 2$ kb solidus has been emphasized to illustrate this phenomenon. At total pressures up to 2 kb, $P_{H2O} = P_T$ and the solidus follows the H_2O-saturated solidus. Above $P_T = 2$ kb, however, P_{H2O} is maintained at a constant 2 kb, so the solidus becomes the curve with the positive slope labeled $P_{H2O} = 2$ kb. The main effect is that above $P_T = 2$ kb the H_2O-undersaturated solidus falls somewhere between the dry solidus and the H_2O-saturated solidus. Thus for a given P_T the H_2O-saturated solidus is at the lowest possible temperature; the H_2O-undersaturated solidus is at some intermediate temperature; and the dry solidus is at the highest possible temperature. This conclusion would also hold for H_2O-saturated, H_2O-undersaturated, and dry liquidi.

5.1.2.2 An Example: Melting of a Tonalite

As an example of the effect of P_{H2O}, we can now examine a melting experiment on a natural calcalkaline rock under conditions in which $P_{H2O} = P_T$. Results for H_2O-saturated melting of a tonalite are shown on Figure 5.9 (data from Gibbon and Wyllie, 1969). The liquidus mineral in this study is an Fe-Ti oxide (probably magnetite), but the exact position of the liquidus was not determined. A stability curve on Figure 5.9 marks the first appearance of that mineral with decreasing temperature and crystallization, or the disappearance with increasing temperature and melting. The biotite and hornblende curves have positive slopes, whereas the liquidus, solidus, plagioclase, quartz, and K-feldspar curves all have negative slopes. Biotite and hornblende are both hydrous minerals, but plagioclase, quartz, and K-feldspar are not. Increasing P_{H2O} expands the stability fields of hydrous minerals toward higher temperatures, while increasing P_{H2O} decreases the stability fields of anhydrous minerals.

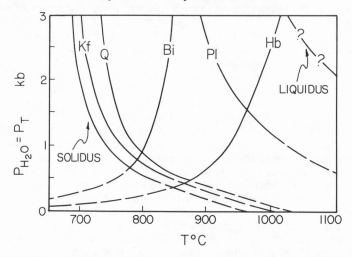

Figure 5.9 Results of a melting study on a natural tonalite: pressure-temperature diagram with excess H_2O, after Gibbon and Wyllie (1969). *Kf*, K-feldspar; *Q*, quartz; *Bi*, biotite; *Pl*, plagioclase; *Hb*, hornblende.

It is possible to use the experimental data of Gibbon and Wyllie and construct a phase-temperature diagram based on a model of ideal equilibrium crystallization when $P_{H2O} = P_T$. A diagram based on $P_{H2O} = P_T = 2$ kb was compiled by Ragland and Butler (1972) and is shown on Figure 5.10. Volume percentages are only approximate. Percent liquid was determined from percent glass in the runs. The final rock mineralogy (left edge of diagram) was estimated from modal analyses of chill-margin samples. Opaque minerals (Fe-Ti oxides, probably magnetite) begin crystallizing at slightly over 1000°C, hornblende at 985°C, plagioclase at 930°C, biotite at 835°C, quartz at 755°C, and K-feldspar at 717°C. The percentage of every mineral increased with falling temperature except hornblende, which began decreasing at about the temperature biotite began crystallizing. This suggests reaction of hornblende and melt to form biotite. The crystallization rate is clearly not proportional to the rate of temperature decrease. For instance, apparently the temperature dropped from about 800 to 755°C (the onset of quartz crystallization) without appreciable crystallization.

We will only qualitatively examine fractional crystallization for Figure 5.9. The order of crystallization depends on P_{H2O}. Magnetite will be the first mineral to crystallize at water pressures up to some value greater than 3 kb (off the graph), above which hornblende will crystallize first (Fig. 5.9). At P_{H2O} less than about 1.5 kb plagioclase will crystallize after magnetite, but above 1.5 kb hornblende will crystallize second. Biotite will crystallize at about the same temperature as quartz and K-feldspar at relatively low water pressures (less than 1 kb); at higher pressure (2 kb, for instance), biotite will crystallize at a considerably higher temperature. Ragland and Butler (1972) reported that a pluton from North Carolina whose initial magma was of this tonalitic composition fractionally crystallized by hornblende, plagioclase, and minor magnetite control. Early coprecipitation of both plagioclase and hornblende would seem to suggest a P_{H2O} of crystallization around 1.5 kb (approximate P_{H2O} where the Hb and Pl curves intersect), but unfortunately this is an oversimplification, as will be explored shortly.

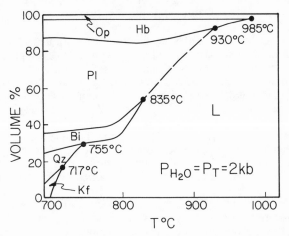

Figure 5.10 Phase-temperature diagram after Ragland and Butler (1972) based on data obtained by Gibbon and Wyllie (1969) to construct Figure 5.9. *Op,* opaques; *Hb,* hornblende; *Pl,* plagioclase; *Bi,* biotite; *Qz,* quartz; *Kf,* K-feldspar; *L,* liquid.

At these shallow depths (maximum of only about 9 km) it is unrealistic to think that extensive partial melting of a tonalitic source would take place, unless the circumstances were quite unusual. Massive invasion of basaltic melts from the upper mantle into intermediate or even shallow depths of the crust might provide sufficient heat for partial melting to occur. If it did occur, the components that constitute quartz and K-feldspars (silica, alumina, and the alkalies) would be enriched in the initial melts. The resultant melts would be more felsic than tonalite.

Figure 5.10 is useful in visualizing this process. If equilibrium melting occurred at $P_{H2O} = P_T = 2$ kb, then K-feldspar would first begin melting at the solidus temperature, followed by quartz at 717°C and biotite at 755°C. Because of the apparent reaction relationship between biotite and hornblende, biotite may actually melt incongruently (see Section 3.2.6.6) to hornblende and melt. If we assume that 40 percent is a reasonable upper limit for partial melting, then very little plagioclase would melt. Likewise, no hornblende or opaques would melt. If 40 percent partial melting occurred, the restite residuum would be a hornblende-plagioclase rock with minor biotite and opaques. Metamorphic rocks that contain dominantly hornblende and plagioclase are termed *amphibolites*. At 40 percent melting the partial melt would contain quartz, K-feldspar, biotite, and some plagioclase, i.e., a rhyolite or granite melt. A bimodal volcanic suite (Section 5.2.4.4), consisting of crustal-derived rhyolite and mantle-derived basalt, might be the result. Even though P_{H2O} = 2 kb is probably unrealistically low, extrapolation of the curves on Figure 5.9 to higher pressures does not invalidate the principles discussed. The order of melting and temperatures at which minerals melt would change, but the principles would still apply.

We must now consider an H_2O-undersaturated system. These scenarios for fractional crystallization and equilibrium melting are based on an H_2O-saturated system, yet most magmatic systems are probably H_2O-undersaturated. In the specific example shown on Figure 5.9, conclusions are probably still broadly valid. Near-liquidus phases should still be magnetite, hornblende, and plagioclase, although their exact order of crystallization may change. Likewise, any partial melt will probably still be rhyolitic or perhaps rhyodacitic in composition.

In general, however, differences between H_2O-undersaturated and H_2O-saturated equilibria must exist. Based on the concepts developed, the following H_2O-undersaturated curves on Figure 5.9 will occur at higher temperatures than their H_2O-saturated counterparts: solidus, Kf, Q, Pl, and liquidus. In contrast, the Bi and Hb H_2O-undersaturated curves will occur at lower temperatures than their H_2O-saturated equivalents. The net effect of an H_2O-undersaturated system will conceivably change the order of crystallization or melting. For example, during fractional crystallization, the hornblende-in curve may be lowered so much that only plagioclase and magnetite control the fractionation. This apparently was not the case for this pluton. Ragland and Butler (1972) found the variation diagrams could best be explained by fractionation of all three minerals.

Another factor must be considered, however. These H_2O-undersaturated curves will not remain stationary as crystallization proceeds. If an extraction assemblage of anhydrous minerals, such as plagioclase, quartz, or K-feldspar, is crystallizing from an H_2O-undersaturated melt, the water content and thus P_{H2O} increase in the residual melts. This causes the H_2O-undersaturated curves for hydrous minerals to migrate toward higher temperatures and similar curves for anhydrous minerals to migrate toward lower temperatures. If a mineral extraction assemblage with exactly the same H_2O content as the melt is removed, then P_{H2O} will not change and the curves will be stabilized. If the extraction assemblage contains more water, then curves will migrate in the opposite direction, but this is unlikely. We will attempt no more sophisticated evaluation of this phenomenon. It is apparent that any number of possibilities can occur, depending on the water content of the undersaturated melt and the extraction assemblage. Experimental studies on H_2O-saturated natural rocks are extremely valuable, even though most systems are probably H_2O-undersaturated and extreme caution is required when applying them to petrogenesis.

5.1.2.3 Carbon Dioxide and Other Volatiles

Carbon dioxide has a profoundly different effect than does water on silicate melts. In fact, this tendency is so strong that CO_2 in the fluid phase can negate the effect of H_2O. This clearly can be seen on Figure 5.8c. The dry solidus is shown as the line with positive slope at the extreme right side of the graph, and the H_2O-saturated solidus $(X_{CO2} = 0.0)$ is the usual concave-up curve with the negative slope on the extreme left side. Numbers on the solidi refer to X_{CO2} in the mixture of H_2O and CO_2 in the fluid phase. All solidi are fluid-saturated. The effect of increasing CO_2 in the fluid is striking. When the fluid is pure CO_2 $(X_{CO2} = 1.0)$, the solidus almost coincides with the dry solidus; the solidus temperature is only slightly lowered at elevated pressures. The effect of increasing X_{CO2} in the fluid is to raise the solidus temperature until it increasingly approaches that of the dry solidus. The effect on a liquidus is similar. In this way CO_2 tends to undo the effect of H_2O.

Another way in which CO_2 has a different effect from H_2O was discussed in Section 2.2.2.4. Most fugacity coefficients for CO_2 in the range applicable to igneous rocks have values greater than 1.0. Conversely, equivalent fugacity coefficients for H_2O are less than 1.0 (compare Tables A.1 and A.2 in Wood and Fraser, 1976). This results in fugacity-corrected univariant curves on CO_2-saturated P-T diagrams (see Fig. 2.7 and Section 2.2.2.4) at higher temperatures than uncorrected curves. On H_2O-saturated P-T diagrams (see Fig. 2.6) the reverse situation is true, with corrected curves at lower temperatures than uncorrected curves.

Why do CO_2 and H_2O have such profoundly different effects? CO_2 plays both an active and a passive role. It passively affects solidus and liquidus temperatures in that the more CO_2 present in a fluid, the less H_2O will be available. CO_2 plays an active role because an increase in X_{CO2} increases the degree of polymerization in the melt, whereas a higher X_{H2O} decreases the degree of polymerization. Barker (1983) explains this increase in polymerization with the following reaction:

```
        O              O                  O         O
        |              |                  |         |
     O--Si--O  +  O--Si--O  +  CO₂  =  O--Si--O--Si--O  +  (CO₃)²⁻
        |              |                  |         |
        O              O                  O         O
```

Consequently, two separate silica tetrahedra react with CO_2 and the resultant products are a carbonate complex ion and two tetrahedra linked together by sharing a common O^{2-} ion.

In contrast to H_2O, the solubility of CO_2 in the melt is quite low at low to intermediate fluid pressures. As a result, at these pressures CO_2 only minimally affects the position of a solidus or liquidus, other than its passive role in preventing H_2O from having an effect. Above confining pressures of about 25 kb (roughly 75 km-depth), however, the solubility of a pure CO_2 fluid markedly increases and it has a much more profound effect on a solidus or liquidus. This results in a "kink" in the solidus or liquidus at $P_{CO2} = P_T$ of about 25 kb (Fig. 5.8d). Even at pressures higher than this kink H_2O is more soluble and thus lowers the solidus or liquidus more than CO_2 (Fig. 5.8d).

Barker (1983) provides a good, succinct summary of the relative effect of other volatiles, in addition to a very helpful discussion of H_2O and CO_2. His conclusions on the effects of other volatiles are drawn from experimental work of Wyllie and Tuttle (1964), Burnham (1979a, 1979b), and Eggler and others (1979). At 2.75 kb confining pressure Wyllie and Tuttle added various other volatiles to an originally pure aqueous fluid in contact with a silicate melt and obtained the following results: (1) HCl and NH_3 raised the solidus temperature; (2) SO_3 had little effect; and (3) P_2O_5 and HF lowered the solidus temperature. Eggler and others reported that CO has an effect similar to CO_2. Burnham concluded, perhaps not surprisingly, that H_2S behaves as does H_2O, but SO_2 plays a role similar to that of CO_2. H_2S is the S-bearing volatile relatively enriched in more reducing magmas, whereas SO_2 is more abundant in relatively oxidizing magmas.

Whether or not any particular volatile has an effect on liquidi and solidi depends mainly on its solubility in the melt. If a volatile has high solubility, it tends to break O^{2-} linkages between tetrahedra, reduce polymerization, and lower liquidus and solidus temperatures. In a sense it behaves as a network modifier. If it has low solubility, it plays a more passive role in that it simply has relatively little effect, although some volatiles, such as CO_2, can actually increase the degree of polymerization. The more polymerized an original dry melt is, the more dramatic will be the effect of adding a volatile such as H_2O. Because dry silicic melts are more polymerized than are dry basaltic melts, H_2O and kindred volatiles will have a more profound effect on relatively silicic melts.

5.1.2.4 Water, Carbon Dioxide, and Basalt Types

Using the Yoder and Tilley (1962) normative criteria (see Section 1.3.3.2), three main types of basaltic melts exist: (1) quartz tholeiites, *q*-normative and *hy*-normative; (2) olivine tholeiites, *ol*-normative and *hy*-normative; and (3) alkali-olivine basaltic rocks, *ne*-normative. Factors that reportedly affect the type of basalt produced are:

1. Depth of dry melting. Alkali-olivine basaltic melts generally seem to form at deeper levels, and thus higher pressures, than do tholeiites.
2. Extent of partial melting. Alkali-olivine basaltic magmas are thought to be formed by a smaller degree of partial melting than tholeiites.
3. Role of CO_2. Alkali-olivine basaltic melts are considered to have been generated in a relatively CO_2-rich environment compared to tholeiites.
4. Role of H_2O. Quartz-tholeiitic magmas apparently form under more H_2O-rich conditions than do the other types. Olivine tholeiites seem to be the result of relatively dry melting.

Observations in the field and under the petrographic microscope, as well as analytical geochemical studies and experimental evidence, have been used to draw these conclusions. Some of the experimental evidence is summarized in Figure 5.11, which is based on work in synthetic systems rather than natural rocks. A particularly valuable summary of this work can be found in Morse (1980). Figure 5.11a shows the effect of increasing confining pressure (dry conditions) on a critical peritectic in the system NE-FO-Q (base of the basalt tetrahedron, projected from diopside; see Figs. 1.12 and 1.14). Point P marks the position of a typical peridotite source rock. Initial partial melts for a particular pressure will have a composition at the appropriate peritectic. Composition of a peritectic melt becomes increasingly FO-rich and particularly NE-rich as pressure increases; i.e., the higher the confining pressure, the more SiO_2-undersaturated is the peritectic melt. Hence conclusion 1 is strongly supported.

Conclusion 2 is supported by an argument based on equilibrium melting. Using the 20 kb peritectic (point R; Fig. 5.11a) as an example, we can draw a lever from this peritectic through the peridotite composition P to the FO-En sideline (point S). Reading this lever (S P / S R) indicates that about 8 percent melting at the peritectic will take place before all NE component is in the melt and the liquid path LP begins moving up the cotectic. This peritectic melt will have an approximate composition of an alkali-olivine basalt. After LP moves off the peritectic, it moves up-temperature along the cotectic; if sufficient partial melting takes place it will cross the critical plane of silica undersaturation (CPSU; Fig. 5.11a) at point B and become tholeiitic. Another lever (T P / T B) indicates that only approximately 17 percent or more partial melting is necessary for a partial melt to be an olivine tholeiite. This diagram indicates that tholeiitic melts form as a result of higher degrees of partial melting than do alkali-olivine basalts. This is persuasive evidence to support conclusion 2.

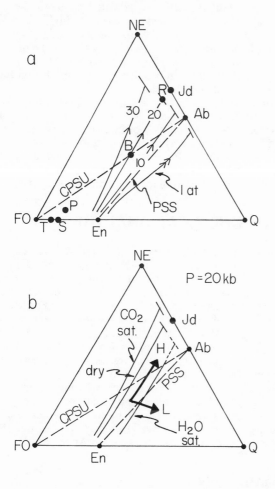

Figure 5.11 *a.* Effect of dry melting at different pressures on genesis of basalt magmas (modified after Morse, 1980). Numbers on cotectics in multiples of 10 are kilobars. *CPSU,* projection of critical plane of silica undersaturation; *PSS,* projection of plane of silica saturation. Points are defined in text. *b.* Effect of H_2O and CO_2 saturation on genesis of basalt magmas (cotectics and peritectics from Eggler, 1974, and Kushiro, 1973). *H,* high-P fractionation; *L,* low-P fractionation.

Figure 5.11b can be examined to explore the roles of CO_2 and H_2O in basalt genesis. Cotectics and peritectics under dry, CO_2-saturated, and H_2O-

saturated conditions $(P_T = 20$ kb) are shown on this figure. The effect is striking. The most SiO_2-undersaturated peritectic melt is under excess-CO_2 conditions, whereas the most SiO_2-saturated peritectic melt is H_2O-saturated. The peritectic for dry conditions is between the other two. Although the excess-H_2O peritectic is not yet in the quartz-tholeiite field, it will be at lower confining pressures. Conclusions 3 and 4 are supported by this evidence.

The critical plane of silica undersaturation (Fig. 5.11b) can be considered a thermal divide at low pressure. Moreover, we have seen that a common extraction assemblage at low pressure is plagioclase, olivine, and clinopyroxene. Consider an olivine tholeiite whose composition falls approximately in the center of the FO-En-AB field on Figure 5.11b. A reasonable low-pressure extraction assemblage at less than 15-km depth (Section 5.1.1.3) is made up of approximately equal amounts of olivine, plagioclase, and clinopyroxene, represented respectively by FO, Ab, and DI (which is not shown because the projection is from DI). Extraction of this mineral assemblage will drive compositions of residual liquids into the quartz-tholeiite field, as indicated by the arrow L on Figure 5.11b. This thermal divide is no longer present at higher pressures (Section 3.2.7.3), and orthopyroxene (En on the diagram) plus clinopyroxene become important minerals in the extraction assemblage at 35-70 km depth. Thus relatively high-pressure fractionation involving dominantly ortho- and clinopyroxene will drive residual magma compositions away from En on Figure 5.11b (arrow H) into the alkali-olivine basalt field.

In summary, experimental evidence can be cited to support all the above mechanisms for genesis of different basalt types. Moreover, any one basalt type can exist as a result of several factors, which may or may not be correlated. For example, at least four effects can be involved in generation of alkali-olivine basalts: (1) high-pressure fractionation of an olivine tholeiite, (2) partial melting at relatively great depths, (3) small percentages of partial melting (correlated with 2?), and (4) CO_2-rich conditions of partial melting (correlated with 2?).

5.2 TREATMENT OF CHEMICAL ANALYSES

Throughout this book numerous examples have been cited of ways in which chemical analyses are used as evidence to aid in the interpretation of igneous rocks. For example, variation diagrams were introduced in Chapter 1. Heretofore these and related topics have not been treated in any systematic manner, and the basis for some of these procedures has not been discussed. Furthermore, many of the pitfalls in dealing with chemical data have yet to be mentioned. Petrologists are forever looking for significant patterns or trends in chemical data sets. In doing so, they may either (1) conclude that some artifact of the manner in which the data are treated has petrologic meaning, or (2) do the opposite, overlook an important pattern in the data by failing to make the proper calculation or plot the most revealing graph. This section will deal with a wide variety of numerical and graphical techniques and point out some pitfalls associated with a number of them. Extensive compilations of chemical analyses are not included in this book, but are readily available (e.g., Le Maitre, 1982). A few representative rock and mineral analyses are given in Appendix E.

5.2.1 Trace Elements

In recent years the use of trace elements has become one of the principal tools of the igneous petrologist. Perhaps this era began when the potential for use of rare-earth element patterns in petrology was first realized. Trace elements seldom provide a unique solution to a petrologic problem, but, used in conjunction with other types of data, they allow constraints to be placed on petrogenetic models. In earlier days, led by the pioneering work of V. M. Goldschmidt, geochemists were content to describe and explain the distribution of trace elements. They were less concerned about whether trace-element patterns followed rigorous mathematical functions and thus could be used to model igneous processes. The principal basis for these numerical methods is the partition coefficient. This section will deal with some of the most elementary of these functions and models. More details and a number of more complex models are discussed in Arth (1976) and Maaloe and Johnston (1986). Cox and others (1979) and McBirney (1984) are also recommended reading in this area.

5.2.1.1 Partition Coefficients and Compatibility

Partition coefficients have been discussed several times previously. In this section we are only concerned about crystal-melt partition coefficients and their role in generating numerical models for igneous processes. At this point we must also introduce the concept of compatibility as it relates to crystal-melt equilibria. The term *incompatibility* was used in Section 1.3.1.2 to describe normative minerals that could not coexist in a standard norm, but in this section the term is used in a different context.

One basic law that governs partition coefficients is the *Berthelot-Nernst equation:*

$$K_D = C_i{}^A / C_i{}^B = f(P,T) \tag{5.1}$$

where K_D is the partition coefficient (or "distribution coefficient"); $C_i{}^A$ and $C_i{}^B$ are concentrations of component i in phases A and B, respectively. The phases can be two minerals, a mineral and a liquid, two immiscible liquids, a gas and a liquid, etc. The chemical species can be an atom, a simple ion, a complex ion, or a molecule. Concentration units can be in molar or weight units; in this section weight units, normally ppm, are used. Note that K_D is a function of P and T (see Sections 2.1.5.5 and 2.2.1.7).

If the chemical species is not in an ideal solution, K_D is also dependent on the major-element composition of the phase. The lower the concentration of the component, the greater will be the chances that it will behave either ideally (Raoult's Law) or at least predictably (Henry's Law; see Section 2.2.2.6). For these reasons, trace elements are normally preferred over major elements for these types of calculations.

Using crystal fractionation as an example, K_D is:

$$K_D = C_i{}^M / C_i{}^L \tag{5.2}$$

where M refers to a specific mineral and L to the liquid (magma). Concentrations are in weight units. A K_D value refers to a specific mineral, but during fractional crystallization two or more minerals can simultaneously precipitate, as along a cotectic or at a eutectic. Consequently, we must calculate a *bulk distribution coefficient (D),* which is simply a weighted average of the K_Ds for all the precipitating phases:

$$D = \Sigma w_i K_{Di} = w_1 K_{D1} + w_2 K_{D2} + \ldots \ldots \tag{5.3}$$

where w_i is the weight fraction of each fractionating phase. The sum of all w_is must be 1.0. If some magma is also removed by being trapped along with the fractionating crystals, it will have a K_D of 1.0. For instance, given the K_Ds for Rb in a fractionating assemblage from a mafic magma of 50 percent plagioclase, 30 percent hornblende, and 20 percent trapped liquid to be 0.1, 0.3, and 1.0, respectively:

$$D = 0.5 \times 0.1 + 0.3 \times 0.3 + 0.2 \times 1.0 = 0.34$$

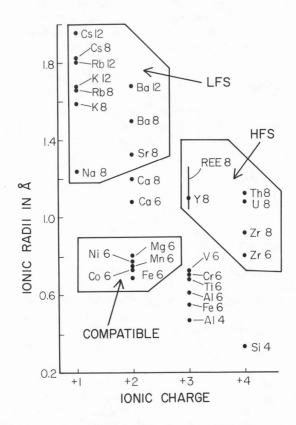

Figure 5.12 Plot of ionic charge versus valence for some common cations in igneous rocks (data from Whittaker and Muntus, 1970). The number following the name of each element is the co-ordination number. For every cation the ionic radius increases with co-ordination number. *LFS*, low field strength; *HFS* high field strength. For ultramafic through intermediate rocks the LFS and HFS elements are generally incompatible; many of these elements are compatible for felsic rocks. Likewise, the compatible group primarily applies to ultramafic through intermediate rocks.

An element with a D value >1.0 is considered to be a *compatible* element; it is preferentially partitioned into the fractionating mineral assemblage. For a compatible element, the ratio C_L/C_O decreases with increasing fractional crystallization. C_L is the concentration of an element in the residual liquid (the *daughter magma)* and C_O is its concentration in the original liquid (the *parent magma)*. Any element with a *D* value <1.0 is considered to be *incompatible;* if *D* = 0, it is completely excluded from the fractionating assemblage and is "perfectly incompatible." The term *hygromagmatophile* is essentially synonymous with incompatible. Complete exclusion from the bulk assemblage is rare because of trapped liquid. In the case of an incompatible element, the ratio C_L/C_O increases with the degree of fractionation.

TABLE 5.1 Crystal-Melt Partition Coefficients (K_D)

	Rb	Sr	Ba	Ce	Eu	Eu*	Yb	Cr	Ni
AMPH									
M	0.3	0.5	0.4	0.3	1	1	1	5	3
F	0.01	0.02	0.04	1	4	7	7	–	–
BIOT									
M	3	0.08	1.1	0.03	0.03	0.03	0.03	7	3
F	3	0.2	8	0.3	0.3	0.3	0.3	20	–
CPX									
M	0.01	0.1	0.01	0.3	0.9	0.9	1	9	2
F	0.05	0.5	0.1	0.9	2	3	2	–	–
GARN									
M	0.001	0.001	0.002	0.05	0.9	0.9	30	1	–
F	0.01	0.02	0.02	0.6	0.7	5	40	–	–
KFLD									
F	0.4	6	6	0.04	1.1	0.02	0.01	–	–
OLIV									
M	0.004	0.005	0.003	0.01	0.01	0.01	0.01	1	12
OPX									
M	0.01	0.01	0.005	0.02	0.05	0.05	0.3	6	4
PLAG									
M	0.1	2	0.2	0.14	0.3	0.09	0.07	0.0	0.01
F	0.06	5	0.4	0.3	2	0.1	0.05	0.2	–
SPIN									
M	0.01	0.01	0.01	0.08	0.03	0.06	0.02	5	10

Abbreviations:
    M, mafic rocks; F, felsic rocks; AMPH, amphibole;
    BIOT, biotite; CPX, clinopyroxene; GARN, garnet;
    KFLD, K-feldspar; OLIV, olivine; OPX, orthopyroxene;
    PLAG, plagioclase; SPIN, spinel

Refer to section 5.2.2.7 for explanation of Eu*.

     During partial melting incompatible elements are preferentially enriched in the early-formed melts, while compatible elements tend to remain behind in the unmelted source rock. Partial melting of peridotites to form basaltic melts can be used as an example to explain compatibility. Geochemists often consider elements as existing in one of two groups: *high field strength* (HFS) and *low field strength* (LFS). In general, incompatible elements during partial melting can be either HFS or LFS, but compatible elements are normally not considered as either. The primary control is ionic size and charge, as shown in Figure 5.12. Compatible cations are normally divalent, octahedrally coordinated, and have a rather restricted range in ionic size. Trivalent ions of similar size, especially Cr^{3+}, can be compatible as well. The compatible elements best fit into lattice sites of minerals that are stable in the upper mantle, such as olivine and pyroxenes. Both LFS and HFS elements are too large to fit readily into these lattice sites, so they tend to be incompatible. As a consequence, during partial melting of peridotite, elements such as Ni $(D > 1)$ are not enriched in early-formed melts, but stay behind in the unmelted source rock. In contrast, LFS or HFS elements $(D < 1)$ are incompatible and thus enriched in early-formed melts.

An element can be compatible under certain conditions but incompatible under others. For instance, if plagioclase were stable in the upper mantle, then Sr would be compatible, because Sr readily substitutes for Ca in the plagioclase lattice. At deeper levels in the upper mantle, where spinel rather than plagioclase is stable, Sr would be incompatible. Likewise, if phlogopite mica is stable, LFS elements such as K, Rb, Cs, and Ba, generally strongly incompatible, become compatible. Which elements are incompatible depends on which minerals are present in the source rock.

Field strength can be roughly considered as electrostatic charge per unit surface area of the cation. Hence, relatively small, highly charged cations tend to be HFS, whereas larger cations with lesser valences are most likely LFS (Fig. 5.12). A measure of these tendencies is *ionic potential,* defined as valence divided by ionic radius. Cations in the upper left part of Figure 5.12 have relatively low ionic potentials, whereas those on the right have higher. Consequently, HFS cations have large ionic potentials in relation to LFS cations. We will return to the significance of field strength in Section 5.3.1.1.

Table 5.1 gives some K_Ds for selected trace elements in a number of the most common rock-forming minerals. The two main sources for these data are Henderson (1982) and Cox and others (1979). Ranges are quoted in Henderson (1982 Table 5.2); they commonly exceed one order and occasionally even two orders of magnitude. For this reason, the accuracy of any single K_D value in Table 5.1 is suspect and should be considered only as an approximation. Because of this large range, most of the K_Ds in this table are reported to only one significant figure.

The trace elements in Table 5.1 were chosen to represent three groups of elements, the groupings based on their significance in igneous rocks. The first three (Rb, Sr, and Ba) are considered incompatible in many, but certainly not all, rock-forming minerals; Cr and Ni are compatible with respect to several minerals. The other group (Ce, Eu, and Yb) are representatives of the rare-earth elements (commonly abbreviated REE). The light rare earth elements (LREE) are represented by Ce and the heavy rare earths (HREE) by Yb. The REE Eu was included because of its special behavior in igneous systems (Section 5.2.2.7). These REE are commonly incompatible, with the exception of Yb in garnet (Table 5.1). Some other exceptions to these generalities are:

Rb and Ba in biotite	compatible
Sr and Ba in K-feldspar	compatible
Sr in plagioclase	compatible
Cr and Ni in plagioclase	incompatible

Cr and Ni are most certainly incompatible in K-feldspar as well, but no data are available. One surprise in Table 5.1 is the mildly incompatible nature of Rb in K-feldspar. Rb readily substitutes for K in mineral lattices, thus Rb might be expected to be compatible in K-feldpar.

No fractional crystallization or melting process will take place so perfectly that 100 percent separation of liquid and crystals occurs. At the time liquid is trapped, concentration of an element in the trapped liquid must be the same as that in the untrapped, so $K_D = 1.0$. Trapped liquid plays the same role as a fractionating crystalline phase in that it removes material from the remaining, "untrapped" melt. If appropriate, a term for trapped liquid can be included in any trace-element modeling calculation.

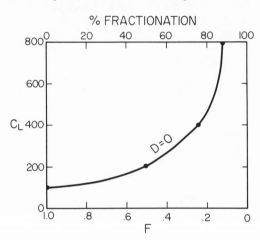

Figure 5.13 Graph showing increase in the melt composition (C_L) of a perfectly incompatible element $(D = 0)$ with increasing percentage of fractional crystallization. F, weight fraction remaining liquid.

5.2.1.2 Rayleigh Fractional Crystallization

Rayleigh fractional crystallization (RFC) is named for Lord Rayleigh, the English lord who in the late 1800s mathematically described similar processes, such as fractional distillation. Insofar as igneous petrology is concerned, RFC is perfect fractional crystallization in a closed chamber. Actual crystal fractionation processes in nature can be considerably more complicated, in that they may take place in open systems where magma mixing, contamination, and perhaps other processes are simultaneously operative. RFC, however, is a good point of departure. It can be understood intuitively through an example. Imagine a hypothetical closed magma chamber with 100 ppm Sr undergoing perfect fractional crystallization. Further imagine that Sr is completely incompatible in the crystallizing phases and no trapped liquid is present. As the crystallizing phases have no Sr, all Sr in the system must stay in the residual melt. As the melt reduces in volume through crystallization, Sr content must increase.

Evaluating this process quantitativelty is easily done. If the original melt contained 100 ppm, after half the melt has crystallized, the Sr content must be 200 ppm (Fig. 5.13). When half of this remaining half crystallizes, the Sr content in the residual melt must now be 400 ppm. When half of this remaining quarter crystallizes, Sr content is up to 800 ppm, and the process goes on. Figure 5.13 shows that Sr content in the residual melt apparently exponentially increases as percent fractionation increases. On this figure F is the weight fraction of remaining melt, so:

$$\% \text{ fractionation} = (1 - F) \times 100$$

The equation that describes this process is:

$$C_L = C_O / F \tag{5.4}$$

As above, C_L is defined as the concentration in the residual melt, C_O is the concentration in the original melt (100 ppm Sr in this example) and F is the melt fraction. For instance:

$$F = C_O / C_L = 100 / 800 = 0.125$$

Because C_O is a constant, as F approaches 0 (100 percent fractionation), C_L approaches infinity.

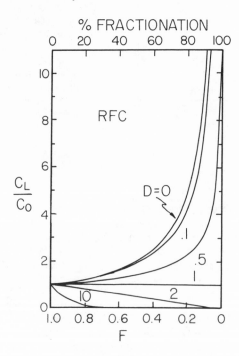

Figure 5.14 Plot of percent fractionation and weight fraction remaining liquid (F) versus C_L/C_O for various bulk distribution coefficients (Ds), assuming Rayleigh fractional crystallization (RFC). Modified after Cox and others (1979).

This example clearly explains RFC where the trace element is perfectly incompatible, but how do we mathematically describe this process when the trace element has some K_D other than zero? The general equation for RFC is:

$$C_L = C_O \exp [(D - 1) \ln F] = C_O F^{D-1} \tag{5.5}$$

where D is the bulk distribution coefficient. Solution of equation 5.5 for a perfectly incompatible element $(D = 0)$ yields $C_L/C_O = 1/F$ (equation 5.4).

Fractionation curves based on equation 5.5 for different values of D are shown in Figure 5.14. The ratio C_L/C_O for any element with $D = 1.0$ must remain at 1.0 throughout the fractionation.

An example calculation of RFC is informative. Assume fractionation of an olivine tholeiite at depths less than 15 km (Section 5.1.1.3) of an assemblage of 58 percent plagioclase, 25 percent olivine, and 17 percent trapped liquid from a basaltic melt. This approximate ratio of 7:3 = plagioclase:olivine is based on crystallization along the Pl_{SS}-Fo cotectic in the system AB-AN-DI-FO at low pressures (Figs. 4.23 and 4.24). Based on K_Ds for mafic melts in Table 5.1, the bulk distribution coefficient D for Sr is calculated as:

$$D = (0.58 \times 2) + (0.25 \times 0.005) + (0.17 \times 1.0) = 1.33$$

Hence, Sr should be mildly depleted in the residual liquids with increasing fractionation. For instance, based on equation 5.5, Sr content of a residual melt after 60 percent fractionation from a parent melt originally with 200 ppm Sr is:

$$C_L = C_O \, F^{D-1} = 200 \times 0.4^{(1.33-1)} = 148 \text{ ppm}$$

A BASIC computer program to calulate the composition of the residual melt for various increments of F is given in Section 1.6.2.5.

Figure 5.14 can be used to determine which trace elements will be more useful in modeling crystal fractionation. Clearly, any element with $K_D = 1.0$ is useless. At high percentages of fractionation (low values of F) incompatible elements are more useful because they change quite rapidly with small decreases in F. At low fractionation percentages incompatible element contents increase, but at a much slower rate. The opposite situation exists for highly compatible elements. They are generally more useful at low fractionation percentages; at high percentages their concentrations have become quite low and undergo minimal change.

Partition coefficients vary as a result of changes in temperature and pressure (see Section 2.2.1.7). Fractional crystallization clearly takes place under conditions of decreasing temperature, yet in equation 5.5, D is a constant. How can we justify using a term as a constant that may vary with temperature, particularly over wide ranges in percent fractionation? For highly incompatible elements we can justify this simplification because incompatible elements probably do not fit into lattice sites of the mineral at any temperature, thus their K_Ds will remain low.

Any melt crystallizing at a eutectic will do so at constant temperature, so D may not change during this interval of crystallization. This is late in a melt's crystallization history, and as explained, incompatible elements are useful at this stage, regardless of temperature. The problem is with the use of compatible elements at low fractionation percentages, because their K_Ds are temperature dependent. Thus how can we quantitatively monitor RFC of a melt with a highly compatible element such as Ni in olivine? Perhaps the answer is to do so with great caution.

5.2.1.3 Rayleigh Fractional Melting

Rayleigh fractional melting (RFM) is the process by which each infinitely small increment of melt that forms is removed from the source rock. Defined in this manner, RFM can never occur because a certain critical mass of melt is required before the partial melt can be removed from its source. Five percent partial melting is a reasonable critical mass, although if fluid pressures are sufficiently high, as low as 1 percent partial melting may suffice. The lower this percentage, the closer the actual partial melting process will approach ideal fractional melting.

It should be no surprise that the equation describing RFM is not simply the inverse of equation 5.5 for RFC. Recall that LPs on phase diagrams are different for fractional melting and crystallization (e.g., Section 4.1.1.6). The equation is:

$$C_L = [C_S (1 - F)^{(1/D - 1)}] / D \qquad\qquad (5.6)$$

where C_L, and D are defined as before, F is weight fraction of produced melt, and C_S is the composition of the original source rock. D is considered a constant, although as stated this is an oversimplification, especially for compatible trace elements. For a perfectly incompatible element $(D = 0)$, C_L/C_S is infinity. When $D = 1.0$, $C_L/C_S = 1.0$ and no fractionation of the element occurs.

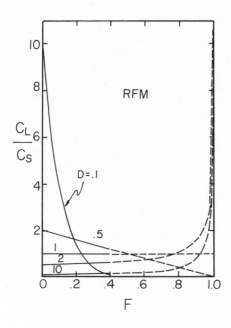

Figure 5.15 Plot of weight fraction of produced melt (F) versus C_L/C_S for various bulk distribution coefficients (Ds), assuming Rayleigh fractional melting (RFM). Modified after Cox and others (1979). Dashed curves indicate degree of partial melting normally not considered to be practical.

Figure 5.15 shows curves for various D values based on RFM and equation 5.6. As more than 40 percent partial melting is unlikely, curves are dashed beyond that limit. Highly incompatible elements are very strongly enriched in the early partial melts, but their concentrations drop fairly quickly in later melts to levels below that of the original source rock. Total percentages of partial melting probably never exceed 30-40 percent, so incompatible elements are extremely useful in monitoring RFM at reasonable percentages of partial melting. Additional evidence for the usefulness of incompatible elements is (1) many of these partial melts will be eutecticlike melts at constant temperature, especially for low percentages of partial melting, so the assumption that D is a constant may be reasonable in some instances; and (2) in any case, K_{DS} for incompatible elements should not be strongly temperature dependent.   In contrast, concentrations of highly compatible elements are very unsatisfactory for monitoring fractional melting.   Aside from their K_{DS} being relatively temperature dependent, during RFM their concentrations stay relatively constant until unreasonably high percentages of partial melting (Fig. 5.15).

As a test of these above conclusions, we will contrast the behavior of an incompatible and compatible element under similar conditions of RFM. Assume 5 percent perfect fractional melting of a garnet lherzolite source containing 40 percent olivine, 30 percent clinopyroxene, 20 percent orthopyroxene, and 10 percent garnet. We will choose 10 ppm Sr as the incompatible element and 2000 ppm Ni as the compatible element, which are reasonable concentrations for upper mantle peridotites. What are the Sr and Ni concentrations in the basaltic melt after 5 and 10 percent partial melting? For perfect RFM there can be no trapped liquid, so we will ignore this term. We do not know K_D for Ni in garnet, but considering the fact that it is incompatible and other terms in the equation for D_{NI} are so large, we can assume $K_D = 0$ for Ni in garnet. The bulk distribution coefficients are:

$$D_{SR} = (0.4 \times 0.005) + (0.3 \times 0.1) + (0.2 \times 0.01) + (0.1 \times 0.001)$$
$$D_{SR} = 0.034$$

$$D_{NI} = (0.4 \times 12) + (0.03 \times 2) + (0.2 \times 4) = 6.2$$

From equation 5.6, calculations for 5 percent melting result in:

$$C_{L\text{-}SR} = 10 \times [(1 - 0.05)^{(1/0.034 - 1)}] / 0.034 = 68 \text{ ppm}$$

$$C_{L\text{-}NI} = 2000 \times [(1 - 0.05)^{(1/6.2 - 1)}] / 6.2 = 337 \text{ ppm}$$

After 10 percent partial melting Sr = 14.7 ppm and Ni = 352 ppm, so a difference of 5 percent partial melting resulted in a drop in Sr concentration by a factor of 4.6, but only a 4 percent rise in Ni concentration. The utility of Sr, an incompatible element, over Ni, a compatible element, is clearly demonstrated. A word of warning, however, is necessary. A simplifying assumption in these calculations is that all these minerals are melting simultaneously. If the melt is a eutectic melt, which is probable for 5 and perhaps even 10 percent partial melting, this is likely a reasonable assumption. For larger percentages, however, clinopyroxene and garnet are consumed before orthopyroxene and olivine, so this assumption is not reasonable.

If we define C_R as the composition of a particular element in the residuum ("restite") left behind during RFM, then C_R can be defined by the relationship:

$$C_R = D \cdot C_L \tag{5.7}$$

which is simply a restatement of equation 5.2 in terms of D rather than K_D. In the example above, solution for C_R yields:

	5% melting	10% melting
ppm Sr	2.3	0.5
ppm Ni	2089	2182

Not surprisingly, Sr is rapidly depleted in the source, while Ni is only mildly enriched.

5.2.1.4 Equilibrium Batch Melting

Equilibrium batch melting (EBM) as a viable process in nature is more probable than RFM because EBM does not depend on all the melt being removed from the source as quickly as it is formed. During EBM the partial melt stays in chemical equilibrium with the source rock until a critical percentage is reached and the melt "batch" can separate. If the critical percentage remains at, for instance, 5 percent of the source for 30 percent partial melting, then six batches will separate. The equation that describes EBM is:

$$C_L = C_S / [D (1 - F) + F] \qquad (5.8)$$

where D is the bulk distribution coefficient at the time the melt batch is removed from the source rock, F is the batch size expressed as the melt fraction, C_L is the batch composition, and C_S is the composition of the original source rock.

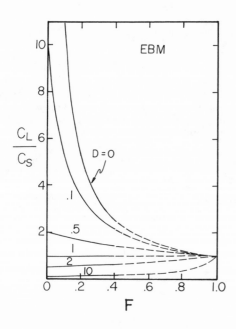

Figure 5.16 Plot of batch size (expresssed as melt fraction F) versus C_L/C_S for various bulk distribution coefficients *(Ds)*, assuming equilibrium batch melting *(EBM)*. Modified after Cox and others (1979). *Dashed curves* indicate degree of partial melting normally not considered to be practical.

For a perfectly incompatible element $(D = 0)$, $C_L/C_S = 1/F$, which is analogous to the case for RFC. For example, if 25 percent is removed as a batch $(F = 0.25)$, and all the incompatible element originally in the source is now in the melt $(D = 0)$, then:

$$C_L / C_S = 1 / 0.25 = 4$$

When $F = 1.0$, all the source has melted, thus $C_L/C_S = 1.0$ for all elements, both compatible and incompatible. As for both RFC and fractional melting, when $D = 1.0$, $C_L/C_S = 1.0$ and no compositional change in the element occurs.

Figure 5.16 shows curves for different values of D; the curves are dashed beyond the reasonable limits of partial melting. As was the case for RFM, for reasonable percentages of partial melting incompatible elements are apparently much more useful for monitoring EBM than compatible elements. Any element with a D value of less than about 0.1 is especially useful.

Equation 5.7 for the residuum composition C_R is applicable to EBM as well as fractional melting. In EBM, however, it takes on new importance. As suggested above, it is very likely that EBM takes place as several discrete events in which the batches are separated from the source (six events of 5 percent each was suggested). Thus it is necessary to know C_R for one event, which becomes C_S for the next event. This process has been referred to as *incremental batch melting*. It is based on three assumptions: (1) chemical equilibrium is maintained throughout the partial melting event; (2) when the batch is removed from the solid residuum, it is completely removed; and (3) D remains constant throughout the process. We will use the data for Sr in the previous example and incrementally melt three batches of 5 percent each $(F = 0.05)$ to demonstrate how this process operates. The Sr content of the source area is 10 ppm and $D = 0.034$; therefore, using equations 5.7 and 5.8:

batch 1: $C_L = 10 / [0.034 \times (1 - 0.05) + 0.05]$
            $C_L = 10 / 0.0823 = 122$ ppm
            $C_R = 0.034 \times 122 = 4.15$ ppm

batch 2: $C_L = 4.15 / 0.0823 = 50.4$ ppm
            $C_R = 0.034 \times 50.4 = 1.71$ ppm

batch 3: $C_L = 1.71 / 0.0823 = 20.8$ ppm
            $C_R = 0.034 \times 20.8 = 0.71$ ppm

Because equation 5.8 for EBM is a linear equation, error analysis (Section 1.4.1) can be used to calculate the uncertainty on the above calculations. For example, assuming errors (standard deviations) of 1 ppm for C_S and .01 for D, then the error for C_L and C_R in batch 1 can be determined by:

$$e_{DEN} = [.01^2 + (.01 \times .05)^2]^{1/2} = .0100$$

$$e_{CL} = 122 [(1/10)^2 + (.0100/.0823)^2]^{1/2} = 19 \text{ ppm}$$
$$e_{CR} = 4.15 [(.01/.034)^2 + (19/122)^2]^{1/2} = 1.38 \text{ ppm}$$

where e_{DEN} is the error on the denominator of equation 5.8; e_{CL} and e_{CR} are errors on C_L and C_R, respectively. A similar calculation cannot be made for RFC or RFM because equations 5.5 and 5.6 are not linear.

RFC, RFM, and EBM are all single processes. In nature several processes can be going on simultaneously. For instance, it is unlikely that any process of partial melting works so efficiently that all of any melt batch is removed from the source rock. Some melt probably stays behind and mixes with the next batch. One model for such a combined process has been termed *dynamic melting* (Langmuir and others, 1977). Likewise, fractional crystallization probably seldom takes place in a closed chamber. O'Hara (1977) has developed a model for a combined process of fractional crystallization and magma mixing, where some daughter magma stays in the magma chamber to mix with a new batch of parent magma. Another combined model is that of DePaolo (1981), which is based on open-system crystal fractionation combined with contamination of country rocks.

These combined models are complex and are not discussed here. A simplified model, however, for a combined incremental EBM-magma mixing process is easily determined using the data from the above example for Sr and Ni. This model calculation is somewhat similar to, but considerably simpler than, dynamic melting. Assume that 25 percent of each 5 percent batch (1.25 percent of the total) stays in the source peridotite to mix with the next batch. The composition of the mixed next batch, including new melt that formed by the standard EBM process and old melt from the previous melting event, can be determined by a simple weighted average calculation. A simplifying assumption is critical: the 1.25 percent melt remaining after a batch is removed is not involved in the next EBM event in any way except to mix with the next 5 percent batch. This probably is not true, but will suffice for our purposes.

Compositions for C_L and C_R are shown above. The composition C_{LM} is the melt composition after mixing:

batch 1: $C_L = 122$ ppm
            $C_R = 4.15$ ppm

batch 2: $C_L = 50.4$ ppm
            $C_R = 1.71$ ppm
            $C_{LM} = (0.20 \times 122) + (0.80 \times 50.4) = 64.7$ ppm

batch 3: $C_L = 20.8$ ppm
            $C_R = 0.71$ ppm
            $C_{LM} = (0.20 \times 64.7) + (0.80 \times 20.8) = 29.6$ ppm

The factors 0.20 and 0.80 in the weighted average solutions for C_{LM} arise from the fact that during each mixing event 1.25 percent old melt mixes with 5 percent new melt, therefore:

$$1.25 / (1.25 + 5) = .20 \qquad 1 - 0.20 = 0.80$$

A necessary consequence of this simplified model is that only 3.75 percent melt was separated from the source as the first batch, but 5 percent was separated for each successive batch. The effect of this mixing for an incompatible element such as Sr is to decrease the rate at which Sr content is lowered in each successive melt. Without this mixing, Sr is lowered from 122 to 20.8 ppm; with 25 percent mixing it is only lowered from 122 to 29.6 ppm, an appreciable difference.

We can now easily determine if an incompatible element, such as Sr, is more useful than a compatible element such as Ni for modeling simple incremental EBM, as well as incremental EBM plus mixing. Data for Ni from the example cited in Section 5.2.1.3 $(C_S = 2000$ ppm; $D = 6.2)$ are treated in the same manner as those for Sr and yield the following results:

batch	% melting	C_L	C_{LM}	C_R
1	5	337	–	2089
2	10	352	349	2182
3	15	367	363	2275

A useful exercise is to confirm these results. Relative to Sr compositions, those for Ni change very little from batch to batch, confirming our assertion that incompatible elements, such as Sr, are more useful than compatible elements, such as Ni, in modeling EBM. One significant conclusion that can be drawn from the discussion of all three processes of RFC, RFM, and EBM is that incompatible elements are generally more useful than compatible elements in modeling these processes. A possible exception is RFC for small percentages of fractionation, but K_Ds for compatible elements may well be variable during this process because of decreasing temperatures. In more complex modeling calculations, however, compatible elements are quite useful (e.g., Maaloe and Johnston, 1986).

xxxxxxxxxxxxxxxxxxxxxxxxxxxxxxxxxxxxxxxxxxxxxxxxxxxxxxxxxxxxxxxxx

A common practice is to attempt a fit of some combined model, such as EBM and magma mixing, to the chemical trends observed in the actual rocks being studied. This, of course, is the main purpose of modeling calculations, but it can be misleading. Let us consider EBM-magma mixing as an example. Henderson (1982) has shown that wide ranges exist for all K_D values, and one can choose any reasonable number of batches, batch size, and percent melt to leave in the source to mix with the next batch. Consequently, the permutations are virtually endless! Eventually some combination of variables that fits the observed trend very likely will be discovered. The fact that the model fits the observed data does not mean it is correct because it is probably not unique. It is only one line of evidence and should be considered along with other evidence, such as that based on isotopes, petrography, major elements, and field observations.

xxxxxxxxxxxxxxxxxxxxxxxxxxxxxxxxxxxxxxxxxxxxxxxxxxxxxxxxxxxxxxxx

As the batch size *(F)* approaches zero, C_L/C_S approaches $1/D$ (equation 5.8). This relationship holds for fractional melting as well (equation 5.6). Hence, for both EBM and RFM, C_L/C_S approaches infinity as the batch size approaches zero for a perfectly incompatible element. A more general comparison, however, can be made between EBM and RFM. It is best explained by means of an example. We compare EBM with 1 percent batch sizes *(F = .01)* and no mixing with RFM through 35 percent partial melting *(D = 0.1)*. Batch sizes of only 1 percent are probably near the lower limit of a critical mass that can be mechanically removed from the source rock. In fact, they very likely require a fluid-rich environment to be removed. Results are:

% melting	C_L/C_S - EBM	C_L/C_S - RFM
5	6.50	6.30
15	2.74	2.31
25	1.16	0.75
35	0.49	0.21

The ratio C_L/C_S stays consistently higher for EBM, which is reasonable because batch sizes for EBM are not infinitely small, as they are for RFM. The overall similarity in the two patterns, however, is apparent. In general as the batch size for EBM approaches zero the two processes become similar. After considering their basic definitions (Section 3.2.6.1), this is reasonable. A consequence of this conclusion is that RFM, despite the fact that ideally it cannot occur, can be used to model partial melting *if some a priori reason exists to assume that batch sizes are quite small*.

5.2.1.5 Equilibrium Crystallization

In Chapters 3 and 4 we always first considered equilibrium crystallization in any discussion of an experimental system. This was because other processes could not be truly understood unless one knew how to trace LPs and XPs on phase diagrams for equilibrium crystallization. During this consideration of trace elements equilibrium crystallization is the last topic considered, perhaps because it is the least applicable. This is true because most magmas undergo at least some fractional crystallization, so RFC is generally more realistic. Furthermore, concern about the effect of equilibrium crystallization on trace elements is not necessary to understand the other three processes. For sake of completeness, however, we will briefly consider equilibrium crystallization.

Because we have seen numerous examples on phase diagrams where LP and XP are the same for equilibrium melting and crystallization, it should be no surprise that equation 5.8 for EBM can be slightly modified for equilibrium crystallization. A slight rewrite of equation 5.8 yields:

$$C_L = C_O / [D (1 - F) + F] \tag{5.9}$$

where C_O is the composition of the original melt and C_L is the composition of the melt after $(1 - F)$ of the system has crystallized. Note that F does not equal batch size, as it does for EBM. This equation can be used to generate a set of curves identical to those on Figure 5.16 for EBM. Interpreting this figure for equilibrium crystallization, F is the remaining melt fraction and C_O in equation 5.9 is C_S in equation 5.8 and on Figure 5.16. When this figure is used to represent equilibrium crystallization, percent crystallization proceeds from right to left across the diagram. Note that compositions of highly compatible elements change most rapidly during early stages of crystallization, whereas concentrations of highly incompatible elements change rapidly during the latter stages.

5.2.2 Variation Diagrams Revisited

Variation diagrams were briefly introduced in Section 1.3.2 and have been frequently mentioned throughout this book. They are one of the favorite means by which chemical analyses of igneous rocks are expressed. Different types of diagrams have extensively proliferated. Some plot mathematical combinations of chemical compositions that are so far removed from the original data that their possible significance becomes doubtful, or at least difficult to interpret. Variation diagrams generally serve two purposes: (1) they are constructed in such a manner so possible "trends" (i.e., correlations, see Section 1.1.3.3) can be observed between the variables; and/or (2) they are used to discriminate between groups, such as different lithologic groups. The purpose of this section is to explore uses, as well as pitfalls if warranted, of several variation diagrams.

5.2.2.1 The Closure Problem

The closure problem, or *principle of constant sums,* can best be introduced by a worst-case example. Figure 5.17 shows two bivariate scattergrams, both based on the following data:

X	%X of (X+Y)	Y	%Y of (X+Y)
2	22	7	78
5	36	9	64
5	50	5	50
8	53	7	47
1	20	4	80
4	80	1	20
9	75	3	25

Figure 5.17a, based on the original X,Y data, shows no correlation between X and Y (a classic "shotgun plot"). However, when X and Y are recalculated so that they add to 100 (a constant sum), Figure 5.17b shows that they are perfectly negatively correlated *(r = -1;* Section 1.1.3.3) and all plot on a line with the formula $Y = -X + 100$. We have transformed a shotgun plot into a plot with a perfect negative correlation by simply recalculating the data to a constant sum, in this case 100 percent. This example, albeit extreme, is the essence of the closure problem. Major-element analyses, as well as modal analyses, add to 100 percent. Because there are more than two variables, perfect negative correlations are unlikely, but the negative correlations are stronger than they would be if there were no constant-sum problem. The smaller the number of variables, normally the stronger is the effect. Chayes (1964) has discussed this problem at length.

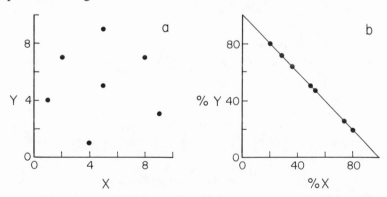

Figure 5.17 *a.* Hypothetical X-Y scattergram showing shotgun plot. *b.* Percent X - percent Y scattergram plotted after recalculating X and Y to 100 percent.

Excepting cases such as binary systems, no petrologist would recalculate two variables to 100 percent, as is done in the example on Figure 5.17. Three variables must be recalculated to 100 percent, however, to plot data on a triangular diagram (see Section 1.3.2.1). If a trend is desired among three seemingly noncorrelated variables, one method that may achieve such a trend is to recalculate the three variables to 100 percent and plot them on a triangular diagram. Is recalculation to a constant sum to achieve a "trend" in the data set legitimate? No simple answer to this question exists; perhaps it is a matter of personal conscience.

In most instances, however, the only motive is to consider three variables simultaneously and the simplest way to do this graphically is to use a triangular diagram. An example of this is shown on Figure 5.18, a classic *AMF diagram* (A = Na_2O + K_2O; M = MgO; F = FeO^*). The AMF diagram, originally popularized by Kuno (1965), is one of the most widely used triangular variation diagrams in igneous petrology. Frequently one can observe closure effects on AMF diagrams, and Figure 5.18a is no exception. Most commonly these effects are noticed for the most primitive (low alkali content) and most evolved (high alkalies) rocks, which is the case for Figure 5.18a. Note the strong trend, exhibiting very little scatter of points, near the A-F sideline of the graph. A moderately strong trend is shown by the points on the M-F sideline. The points in the upper middle of the triangle, however, are quite scattered.

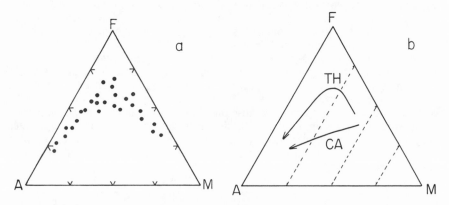

Figure 5.18 Two examples of an AFM diagram (A, Na_2O, F, FeO^*, M, MgO; all data in weight percent). a. Diagram showing possible closure effects. b. Diagram showing typical tholeiitic (TH) and calcalkaline (CA) igneous rock series.

These different degrees of scatter are very likely caused by closure. MgO is extremely low in the samples whose compositions plot along the A-F sideline, so the alkalies and FeO^*, when all three variables are recalculated, add to almost 100 percent. A bivariate plot of recalculated A and F will exhibit a strong negative correlation, but the original data may not show such a strong correlation. Likewise, the samples whose compositions plot near the M-F sideline are low in alkalies, so recalculated MgO and FeO^* add to almost 100 percent. The relatively high-FeO^* samples which plot near the upper middle of the diagram contain appreciable amounts of all three compositional variables, so closure is less critical for those samples. One test to determine if closure is affecting the plot is to plot all three bivariate scattergrams.

Closure can affect bivariate scattergrams as well, although normally not as severely as triangular plots. Major elements plotted on such bivariate plots as Harker diagrams (Sections 5.2.2.3) commonly exhibit closure problems, because about 12 oxides add to 100 percent. Silica, for example, is frequently negatively correlated with every oxide but K_2O and perhaps Na_2O in a group of samples from an igneous rock series. Many of these negative correlations are undoubt-edly real, but closure probably causes them to be stronger than they would be if the major elements could be expressed in concentration units that did not add to a constant sum. Even trace elements, especially those that readily substitute for major elements, can be affected by closure. In general, the more elements that add to 100 percent, the more likely it is that closure effects will be minimal. However, relative variances (see Section 1.1.3.2) of the elements are also important. If, for instance, all but two elements exhibited almost no variation, no matter how large is the total number of elements, those two would add to an approximately constant sum and necessarily be negatively correlated.

What do we do about the closure problem? Many variation diagrams are affected by closure, but we cannot abandon them. Although perhaps flawed, they remain one of the best ways to depict and evaluate chemical analyses of igneous rocks. No adequate method has been developed to recast percentage variables into other variables that are not affected by closure. Perhaps fortu-nately, petrologists normally do not attach confidence levels to correlation coefficients (see Section 1.1.3.3; see any standard statistics book for a discussion of confidence levels). When percentage data are used, confidence levels for negative correlations are unrealistically high. If we only use correlation coef-ficients in a relative sense, we are on much safer grounds. We should be aware of the problem and look upon strong trends associated with negative correlations with a jaundiced eye. However, if we can discover the petrogenetic process that causes that trend (e.g., Section 5.2.2.3), a closure problem may exist but will not matter. Stating this differently, if we can show that the trend is not solely an artifact of the graphical or data-reporting method, but rather has a reasonable petrologic explanation, then the trend very likely has validity.

5.2.2.2 Fractionation Indices

A *fractionation index* is a chemical variable, or a numerical combination of two or more chemical variables, that is used as a measure of the degree of fraction-ation. In this case fractionation generally refers to fractional crystallization, or less commonly some type of partial melting (RFM or EBM). Because fraction-ation indices are most often used for fractional crystallization, in this section we will use the term in that context. The terms *fractionation* and *differentiation* are frequently used synonymously in igneous petrology, so *fractionation indices* are also referred to as *differentiation indices*.

Fractionation indices were first used in the early part of this century (Harker, 1909; Larsen, 1938). Harker simply used SiO_2 as a fractionation index by plotting it on the X-axis of a bivariate scattergram, with increasing SiO_2 denoting an increasing degree of fractionation (e.g., Fig. 5.19). Samples with relatively low SiO_2 values are considered to be more "primitive" or "less fractionated," whereas those with higher values are more "evolved" or "more fractionated." This correlation of SiO_2 content with degree of fractionation primarily applies to calcalkaline rock series and Harker diagrams, which, along with AMF diagrams, are still widely in use (e.g., Frost and Mahood, 1987). In other series SiO_2 commonly does not vary greatly (tholeiitic) or even decreases (alkaline).

The *Larsen index* (LI) is defined as:

$$LI = 1/3SiO_2 + K_2O - (FeO + MgO + CaO) \qquad (5.10)$$

In general, LI increases with increasing fractionation. A quite similar index is the *Nockolds index* (NI; Nockolds and Allen, 1953):

$$NI = 1/3Si + K - (Mg + Ca) \qquad (5.11)$$

where the units are weight percent cation (see Section 1.2.1.2). Both LI and NI have lost favor in recent years and are seldom used now. The reason for this returns to our previous discussion about trends being an artifact of the plotting method. If we plot SiO_2 or K_2O versus LI, a tendency will exist for a positive correlation, whereas a similar plot for FeO, MgO, or CaO will tend to yield a negative correlation. Any correlations are, at least partly, an artifact of the manner in which LI is calculated -- the positive sign on $1/3SiO_2$ plus K_2O and negative sign on the other three oxides. The same reasoning applies for NI. Another difficulty with indices, such as LI and NI, is that relative error is larger on these more complex expressions than on simple indices such as SiO_2 (see Section 1.4.2.3).

For these reasons, Harker's use of only SiO_2 for a differentiation index is preferred. Although SiO_2 varies considerably and thus is a good index for calcalkaline series, it is generally less useful for tholeiitic series. With increasing fractionation in a tholeiitic series, FeO can increase at the expense of MgO (the "Fe-enrichment trend" of tholeiitic rock series). As a result, MgO alone or some combination of the Fe/Mg ratio is generally used for tholeiitic series. In these indices the term for total Fe may be $FeO + Fe_2O_3$, FeO^*, or $Fe_2O_3^*$ (see Section 1.2.1.3). For brevity we will assume FeO^*. Two variations on the Fe/Mg ratio are widely used, the *mafic index* (MI) and the *Mg number* (or M-ratio):

$$MI = FeO^* / (FeO^* + MgO) \qquad (5.12)$$

$$Mg \ number = M\text{-}ratio = mol \ [Mg / (Mg + Fe^{2+})] \qquad (5.13)$$

Units for the Mg number are in molar or cation percent (or cation proportion; see Section 1.2.3.4). As MI increases the degree of fractionation increases, and vice versa for the Mg number.

All three indices (MgO, MI, and Mg number) are widely used for tholeiitic rock series, which dominantly consist of basaltic rocks. The indices MI or Mg number are preferred by some petrologists because they are primarily sensitive to the Fe/Mg ratio in the ferromagnesian solid solutions, and not a function of the ratio of ferromagnesian minerals to feldspars. A complicating factor is that the Fe/Mg ratio is also affected by oxide crystallization. Some variation of Fe/(Fe+Mg) or Mg/(Fe+Mg) is preferred to Fe/Mg because the first two ratios vary between 0 and 1, whereas Fe/Mg varies between 0 and infinity. Other petrologists prefer simply MgO; a principal reason is MgO when plotted against another element yields a straight mixing line or control line (Sections 5.2.2.3), but MI and the Mg number do not because they are ratios (Section

5.2.2.5).   Results for 1-atmosphere melting experiments on samples from Kilauea volcano (see Fig. 5.1) were plotted against MI. These results indicate that MI is strongly negatively correlated with liquidus temperatures.

Another widely used index, particularly for igneous rocks associated with convergent plate boundaries, is the *Solidification index* (SI; Kuno, 1965). It is:

$$SI = 100 \times MgO / (MgO + FeO^* + Na_2O + K_2O) \tag{5.14}$$

As SI increases the degree of fractionation decreases.   SI is the numerical equivalent of an AFM diagram (Fig. 5.18). All terms in the denominator of SI form the apices of this diagram. Thus the contour lines for increasing MgO (Fig. 5.18b) are numerically equal to SI. A typical curved "Fe-enrichment" tholeiitic trend (similar to the trend on Fig. 5.18a; also known as a *Fenner trend)* is contrasted with a comparatively straight calcalkaline trend (commonly referred to as a *Bowen trend)* on Figure 5.18b. The most primitive samples of both suites are similar to one another, as are the most evolved. The two fractionation paths, however, are quite different. The calcalkaline trend and the Fe-enrichment part of the tholeiitic trend cut across constant-MgO lines, so SI systematically changes along these trends. Along the alkali-enrichment part of the tholeiitic trend the curve approximately parallels the MgO line and SI would be less satisfactory.

Several different ratios have been used to characterize fractionation in the feldspar solid solutions. Variations are:

$$Na_2O / (Na_2O + CaO) \quad or \quad ab / (ab + an)$$

Another variation adds *or* to both numerator and denominator of the second equation. In general, Na_2O and K_2O increase in concentration at the expense of CaO in fractionating feldspars. One of these ratios is typically referred to as the *Felsic index* (FI):

$$FI = (K_2O + Na_2O) / (K_2O + Na_2O + CaO) \tag{5.15}$$

These indices are assumed to be largely unaffected by variations in compositions or amounts of the ferromagnesian minerals. This is a simplification, because of the effect of such oxides as K_2O in biotite, or CaO and Na_2O in augite or hornblende.

Another index based on norms is the *Differentiation index* (DI) of Thornton and Tuttle (1960). This index is defined as:

$$DI = q + or + ab + ne + lc + kp \tag{5.16}$$

All these normative minerals cannot occur together (see Section 1.3.1.2). For instance, *q* cannot be present with *ne, lc,* or *kp*. In SiO_2-oversaturated rocks, DI reduces to *q + or + ab*. In most alkaline rocks, *q* is not present in DI but one or more of the feldspathoids are (Section 1.3.3.3). The index DI is a numerical expression of Bowen's "petrogeny's residua" system; as DI increases the degree of fractionation increases. Table 5.2 shows the calculations for several of these indices for the basalt analysis in Table 1.1. Three indices for several samples from an igneous rock series are given in Table 5.3.

All these indices have one common failing.   They allow qualitative statements to be made about the degree of fractionation, but they do not allow us to determine quantitatively the percent fractionation. We can say that one sample is more evolved than another, but we cannot express these degrees of evolution in terms of percent fractionation. Doing so obviously requires that we know the original magma composition, which we can assume to be the composition of the most primitive sample, provided that it does not contain too many phenocrysts. We want an estimation of the composition of the most primitive *liquid*, not crystal mush.

In this case we are interested in elements that behave almost perfectly incompatibly. If RFC applies and $D = 0$, equation 5.4 can be rewritten:

$$P_F = 100 \times [1 - (C_O / C_L)] \tag{5.17}$$

where P_F is percent fractionation. C_O is a constant, the composition of the most primitive sample (presumably of the parent magma), and C_L is the composition of the evolved (daughter) magma. As an example, if the most primitive aphyric sample in a rock suite contains 100 ppm Zr, and a more evolved aphyric sample contains 300 ppm, the evolved sample represents 67 percent fractionation (assuming $D = 0$ for Zr and RFC is operative). Because of trapped liquid, however, D can seldom be assumed to be zero.

TABLE 5.2 Fractionation Indices for Basalt Analysis in Table 1.1

---

1. LI = 1/3 x 48.7 + 0.47 - (8.29 + 6.63 + 10.7) = -8.9

2. MI = 10.1 / (10.1 + 6.63) = 0.60

3. SI = (100 x 6.63) / (6.63 + 8.29 + 2.05 + 2.83 + 0.47) = 32.7

4. FI = (0.47 + 2.83) / (0.47 + 2.83 + 10.7) = 0.24

5. DI = 2.85 + 26.0 = 28.8

Norm calculated in Section 1.2.5.5 for DI.

Which fractionation index is the best to use? If a priori evidence exists for highly incompatible behavior of a trace element, equation 5.17 is preferred, although it depends on $D = 0$. If mixing or control lines are of interest, one-element indices, such as MgO or SiO_2, are preferable. If variations in solid-solution compositions are critical, then some ratio index such as MI or FI should be used. Indices should be avoided that create a trend which is an artifact of the calculation of the index, such as LI and NI. In general, the more complex the index, the less it is useful. If an index is used simply as a indication of the degree of fractionation and is not plotted against other chemical variables, any one of the above indices might suffice.

5.2.2.3 Application of the Lever Rule

The lever rule is just as useful for certain types of variation diagrams as it is for phase diagrams. Although not stated, it was implicit in the treatment of variation diagrams (see Fig. 5.2) used with experimental melting studies of Kilauean basalts. It normally cannot be used, however, on variation diagrams involving ratios. Hence, only bivariate scattergrams involving single oxides or cations are used in this section. An excellent discussion of this topic can be found in Chapter 6 of Cox and others (1979). The principal conclusion regarding the closure problem was that if some reasonable petrologic explanation can be found for a trend on a variation diagram, the trend probably has validity. This section will explore ways to arrive at reasonable geologic explanations.

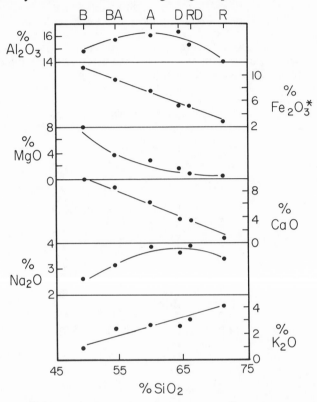

Figure 5.19 Stacked Harker diagrams for a calcalkaline, volcanic igneous rock series (all data in weight %). *B*, basalt; *BA*, basaltic andesite; *A*, andesite; *D*, dacite; *RD*, rhyodacite; *R*, rhyolite.

Major-element analyses and three differentiation indices for a typical volcanic rock series are given in Table 5.3. The rocks range in composition from basalt to rhyolite. Analyses have been plotted on "stacked" Harker diagrams in Figure 5.19. The mafic rocks are generally older than the felsic rocks and percentages of phenocrysts are less than 5 percent. Consequently, we will test the hypothesis that (1) the rocks are related to one another by crystal fractionation; (2) the trends on Figure 5.19 represent liquid lines of descent; and (3) the basalt is parental to the remaining samples. Note the negative correlations between SiO_2 and $Fe_2O_3^*$, MgO, and CaO, as well as the positive correlation for SiO_2-K_2O. The $Fe_2O_3^*$ and CaO trends with SiO_2 are apparently linear, but the MgO-SiO_2 trend is a curve. Both Al_2O_3 and Na_2O reach a maximum in the middle of the series. Because LOI was determined on the samples before the oxide analyses, total Fe is reported as $Fe_2O_3^*$. The LOI determination oxidizes most or all of the ferrous iron. All these trends are typical of calcalkaline rock series.

If we wish to test our hypothesis that the trends on Figure 5.19 are caused by fractional crystallization, we can do so by utilizing the lever rule. Before proceeding, however, some fundamentals concerning the use of the lever rule on variation diagrams must be introduced. Actually, these are straightforward extensions of the use of the lever rule on phase diagrams. These relationships are summarized on Figure 5.20, five plots of two hypothetical elements (or oxides) X and Y. The letters on the figure denote the following: D, most evolved sample; P, most primitive sample (also composition of parent melt); line P-D, liquid line of descent (LLD); E, composition of bulk mineral extract; A, B, C, compositions of minerals in extract assemblage.

TABLE 5.3 Analyses of a Typical Calcalkaline Volcanic Rock Series

Oxide	B	BA	A	D	RD	R
SiO_2	50.2	54.3	60.1	64.9	66.2	71.5
TiO_2	1.1	0.8	0.7	0.6	0.5	0.3
Al_2O_3	14.9	15.7	16.1	16.4	15.3	14.1
$Fe_2O_3^*$	10.4	9.2	6.9	5.1	5.1	2.8
MgO	7.4	3.7	2.8	1.7	0.9	0.5
CaO	10.0	8.2	5.9	3.6	3.5	1.1
Na_2O	2.6	3.2	3.8	3.6	3.9	3.4
K_2O	1.0	2.1	2.5	2.5	3.1	4.1
LOI	1.9	2.0	1.8	1.6	1.2	1.4
Total	99.5	99.2	100.6	100.0	99.7	99.2
MI	0.58	0.71	0.71	0.75	0.85	0.85
SI	34.6	20.3	17.5	13.1	6.9	4.6
FI	0.26	0.39	0.52	0.63	0.67	0.87

Abbreviations: B, basalt; BA, basaltic andesite; A, andesite; D, dacite; RD, rhyodacite; R, rhyolite

Note: All analyses in weight percent.

For case *a* only one mineral, with composition E, is in the extract. The line E-P-D is a lever with point P at the fulcrum. A daughter melt of composition D can be formed by extraction of E from parent melt P. Percentages of daughter and extract mineral can be determined by:

$$\%D = 100 \times E\,P/E\,D \quad and \quad \%E = 100 - \%D$$

The symbols E P and E D represent lengths of lines E-P and E-D. Percent E can also be considered percent fractionation. The line E-D is frequently referred to as a *control line* for mineral E. In case *b* two minerals, A and B, are involved in the extract. A bulk composition E must be removed from parent P to produce daughter D. Again, the E-P-D lever can be read to determine percentages of daughter and extract, but to determine percentages of A and B in the extract, the lever B-E-A must be read:

$$\%B = 100 \times E\,A\,/\,B\,A \qquad \%A = 100 - \%B$$

In case *c* the three minerals A, B, and C are in the extract. Their compositions connect to form the *extract triangle* A-B-C. No unique solution for E can be made from this diagram alone because the intersection of a line and a triangle in the same plane is a line. Hence, no levers can be read. We will return to a three-mineral extract later and discuss ways in which the problem can be solved.

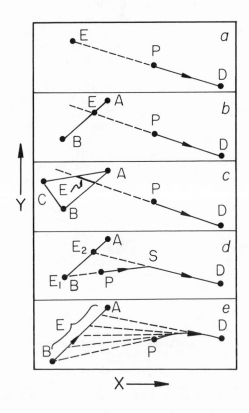

Figure 5.20 Stacked variation diagrams for hypothetical elements *X* and *Y*. Data can be in either weight or mol percent. *D*, daughter; *P*, parent; *E*, extract assemblage. Diagrams are: *a.* one-mineral extract assemblage; *b.* two minerals *(A* and *B)* in the extract assemblage; *c.* three minerals *(A-C)* in the extract assemblage; *(d)* extraction of first *B* and then *A + B;* and *e.* extraction of the solid solution *B-A.*

Cases *a-c* involve only one extract composition throughout the entire fractionation process, producing linear LLDs. It is apparent from Figure 5.19 that not all LLDs are single straight lines. Case *d* represents first extraction of B and then a mixture of B and A; this produces a break in slope in the LLD at point S. This situation is analogous to crystallization of one mineral in a primary-phase field, then coprecipitation of two minerals when the liquid path LP reaches the cotectic (e.g., Section 4.1.1.5). Under this situation a break in slope also occurs for LP on a phase diagram. The extract composition is first at composition E_1 and then at E_2. The lever B-P-S can be read to determine the percentages of B and melt when the melt is at composition S. The two levers B-E_2-A and E_2-S-D can be read to calculate the percentages of A, B, and melt when the melt is at composition D.

On Figure 5.19 three trends are curved: Al_2O_3, MgO, and Na_2O. Case *e* on Figure 5.20 demonstrates how such a curved trend could come about. The line A-B can represent a solid solution. Recall that crystallization of solid solutions produces curved paths for LP on ternary phase diagrams (see Section 4.1.3.2). Minerals A and B can also be immiscible minerals crystallizing when LP is moving along a curved cotectic (see Section 4.1.2.2). For whatever reason, the ratio A/(A+B) is continually increasing as fractionation proceeds, producing a curved LLD from P to D.

In many instances an equally plausible explanation for some of the relationships on Figure 5.19 is crystal accumulation. As an example, let us redefine points E, P, and D on Figure 5.20*a*. Composition P may represent an accumulative rock formed by the addition of cumulus crystals (primocrysts) of composition E to melt D. Reading the lever P D / E D, accumulation of about 45 percent primocrysts would account for rock P. Ragland and Arthur (1985) reported a situation similar to that shown on Figure 5.20*b*, in which orthopyroxene and plagioclase accumulated in a basaltic melt.

How do we distinguish between fractional crystallization and accumulation? Frequently we cannot, because they both were simultaneously operative. Textures, however, provide the best clue. For rock P to have been a true parental melt, it probably should now be aphyric; if it contains primocrysts, and especially true phenocrysts, a good chance exists that at least some accumulation has taken place. Because crystal size is the most important term in Stokes' Law (see Section 3.2.7.5), true phenocrysts have the best opportunity for crystal settling or floating. Only if perfect equilibrium crystallization were operative so that all primocrysts stayed in the crystal mush until crystallization ceased (see Section 3.2.4.4) would a phyric rock of composition P represent an original parental melt.

We can now continue with our evaluation of the hypothesis that fractional crystallization is responsible for the trends on Figure 5.19. We can go about this by using one of two methods. The first and simplest is based on two facts: (1) any bulk mineral extract from a melt of composition B must initially lie along a line extrapolated to lower SiO_2 values; and (2) no oxide can have a negative composition. Before proceeding with both these two methods, a word of warning is necessary. Had all trends on Figure 5.19 been linear, a single extraction assemblage could have explained the evolution of all five daughters from the parent basalt. However, the implication of the three curved trends is that the extraction assemblage must change. Because the lever rule requires a linear trend, the parent and daughter compositions must fall on straight lines for all elements. The most conservative approach is to evaluate adjacent samples, the lower SiO_2 sample designated as the parent and the other as the daughter. Accordingly, we will assume that sample B is the parent and BA is the daughter.

Three oxides have trends with positive correlations with SiO_2 for the more mafic rocks: Al_2O_3, Na_2O, and K_2O. However, K_2O extrapolated to lower SiO_2 goes to zero (at 46.5 percent SiO_2; a good exercise is to confirm this graphically) at a much higher SiO_2 value than the other two oxides. Because the extract cannot have negative K_2O, this places a constraint on the *minimum* percent fractionation possible. If we assume that K_2O is strongly incompatible in the early extraction assemblage (a reasonable assumption for a dry basalt melt), then the bulk composition and norm for an estimated extract are given in Table 5.4. Each oxide composition is obtained by extrapolating the appropriate line BA-B on Figure 5.19 to 46.5 percent SiO_2 and determining the oxide composition at that SiO_2 value, the SiO_2 composition where the extrapolated K_2O composition goes to zero. This extrapolation can be done more accurately using mathematics than graphics. Refer to the data in Table 5.3. First, percent SiO_2 (S_{ZK}) is calculated at zero percent K_2O:

$$(54.3 - 50.2) / (2.1 - 1.0) = (50.2 - S_{ZK}) / (1.0 - 0)$$

$$S_{ZK} = 46.5\%$$

Then the other oxides are calculated based on SiO_2 equal to 46.5 percent at zero percent K_2O. Taking MgO (M_{ZK}) as an example:

$$(54.3 - 50.2) / (7.4 - 3.7) = (50.2 - 46.5) / (M_{ZK} - 7.4)$$

$$M_{ZK} = 10.8\%$$

The norm in Table 5.4 is reasonable as an approximation to an extract assemblage in most respects. The fact that plagioclase, clinopyroxene *(di)*, and olivine are abundant in the norm is quite reasonable. This is a common assemblage that differentiates from basaltic magmas at relatively low pressure (Section 5.1.1.3). Recalculating *ab* and *an* to 100 percent yields an AN content of the extracted plagioclase of AN62, which is somewhat low, but not extremely so. AN-contents for plagioclases fractionating from basaltic melts are commonly in the AN70-AN85 range. Assuming a $FeO/(FeO+Fe_2O_3)$ oxidation ratio of 0.90, the Mg number for this assemblage is 0.65, which is in the expected range (about 0.6-0.7). Some *hy* can be accomodated in the clino-pyroxene structure (a great deal in the case of pigeonite), so 4.7 percent *hy* in the norm of a low-pressure extraction assemblage is typical.

TABLE 5.4 Extraction Assemblage of BA from B

SiO_2	46.5	*ab*	18.3
TiO_2	1.4	*an*	30.1
Al_2O_3	14.2	*di*	23.2
$Fe_2O_3^*$	11.5	*hy*	4.7
MgO	10.8	*ol*	19.3
CaO	11.6	*mt*	1.7
Na_2O	2.1	*il*	2.7
K_2O	0.0		

Analyses of B and BA are given in Table 5.3.

Depending solely on a norm as a means of identifying an extraction assemblage is useful as an approximation, but is not sufficiently exacting. We will now turn to the second method. Our first step is to examine the phenocryst assemblage. Phenocrysts are uncommon in these samples, but those present should provide valuable clues to which phases were extracted to produce the liquid lines of descent. If we cannot find an acceptable assemblage that satisfies the variation diagrams on Figure 5.19, we must look to other petrologic causes for the trends. For this example, a few phenocrysts of olivine, plagioclase, and augite are present in sample B and even fewer phenocrysts of plagioclase and augite are in BA.

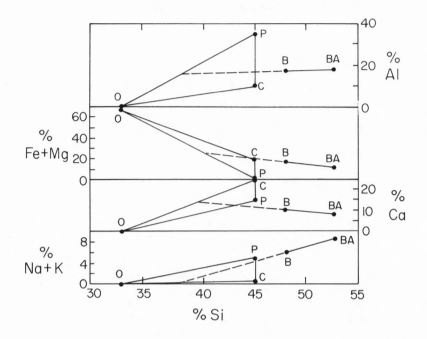

Figure 5.21 Stacked variation diagrams for fractional crystallization of a basalt parent *(B;* Table 5.3) to form a basaltic andesite daughter *(BA;* Table 5.3). All data in mol percent. Extract assemblage consists of: *O,* olivine; *P,* plagioclase; *C,* clinopyroxene.

Electron microprobe analyses of these phenocrysts allows us to examine the hypothesis of fractional crystallization in a much more exacting fashion than simply examining a norm. This can be done graphically by a technique referred to as the *principle of opposition* by Maaloe (1985) and is discussed at length in Cox and others (1979). It is shown on Figures 5.21 and 5.22. The method works well for three minerals, but not more. Before plotting these figures the data in Table 5.3 were recast as cation percent. *It is not absolutely necessary to do this; weight percent oxides can generally be used equally well.* In fact, weight percent oxides are used more often than cation percent, although the use of cation percent is quite common (e.g., Berg and Klewin, 1988). These data were recalculated to cation percent to combine Fe + Mg and Na + K. Because the $FeO/(FeO+Fe_2O_3)$ oxidation ratio was not determined in the bulk rocks and could not be determined in the minerals, combining cation percent Fe + Mg can minimize complications brought about by different values of this ratio from sample to sample. The alkalies were combined because of their tendencies to be affected by sub-solidus alteration (deuteric, hydrothermal, or weathering). Had we chosen RD as a parent and R as a daughter, the alkalies could not be combined because K_2O increases from RD to R, whereas Na_2O decreases.

xxx

If data are converted to cation percent, then a whole-rock compositional variable that has merit as an approximate measure of the behavior of Fe^{2+} in the silicate minerals only is:

$$F_S = Fe - 2Fe^{3+} - Ti$$

The formula for magnetite can be written as $FeO \cdot Fe_2O_3$, so two Fe^{3+} ions exist for each Fe^{2+} ion in magnetite. The formula for ilmenite is $FeO \cdot TiO_2$, so the ratio $Fe^{2+}/Ti = 1/1$. By subtracting $2Fe^{3+}$ and Ti from total Fe, the remainder (F_S) approximates the Fe^{2+} content in the silicate portion of the rock. F_S can then be used with Mg as a differentiation index, or simply used as a compositional variable. This assumes that most of the Ti and Fe^{3+} ions are in the oxides. If oxide abundances are low and minerals that contain some Ti and Fe^{3+} are abundant, such as biotite, augite, hornblende, and epidiote, then F_S is not an acceptable measure of Fe^{2+} in the silicates. In general, F_S is most useful in mafic rocks. Although it has limitations, it does point out the utility of cation percents in comparison with weight percent oxides. It does require, however, separate analyses for FeO and Fe_2O_3.

xxxxxxxxxxxxxxxxxxxxxxxxxxxxxxxxxxxxxxxxxxxxxxxxxxxxxxxxxxxxxxxx

We will now describe the principle of opposition. The compositional points on Figure 5.21 are defined as follows:

B basalt, Table 5.3
BA basaltic andesite, Table 5.3
P plagioclase phenocrysts
C clinopyroxene (augite) phenocrysts
O olivine phenocrysts

The line BA-B is extrapolated to lower SiO_2 values on each of the four plots until it crosses the triangle P-C-O. Each of the four triangles on Figure 5.21 is analogous to the triangle for situation *c* on Figure 5.20. The bulk composition of the extraction assemblage must lie along the extrapolated line BA-B and within the triangle. The intersection of any two of these lines provides a unique solution for the bulk composition; any additional lines provide a check.

To find the intersection of two lines that produces the unique solution, we must have some common basis for comparison. We do this by plotting the appropriate line representing the extrapolation for each element on an equilateral triangle (Fig. 5.22). Each line is plotted by determining where it intersects two sides of the appropriate scalene triangle, and then translating these points using the same relative distances to the equilateral triangle. The small, stippled triangle is referred to as the *polygon of error*. Perfect results are obtained when this triangle of error becomes a point; the triangle of error on Figure 5.22 is considered acceptably small. If the triangle of error is large, either (1) the wrong minerals were chosen, (2) too few minerals were chosen, (3) chemical compositions of the phenocrysts are inappropriate for the fractionating minerals, or (4) fractional crystallization either did not occur or occurred as one part of a multiple process. A numerical technique to perform the same analysis will be described in Section 5.2.3.2.

5.2.2.4 Log-log Plots and Constant-ratio Lines

Figure 5.19 shows a series of stacked Harker diagrams, which, as discussed earlier, are scattergrams primarily used to display trends (correlations) among the various oxide concentrations plotted on the six graphs. Had the database been much larger and a different symbol used for each lithology, we also could have determined which oxides discriminate best among the various lithologies. For instance, we might use closed circles to represent basalts, open circles for basaltic andesites, etc. In addition to using absolute concentrations as discriminators for different lithologic groups, petrologists frequently use ratios. Constant-ratio lines drawn on a bivariate scattergram allow us to (1) determine whether a ratio between the two variables or the absolute concentration of a variable is the best discriminator, and (2) determine whether a trend exists between the two variables (Fig. 5.23a).

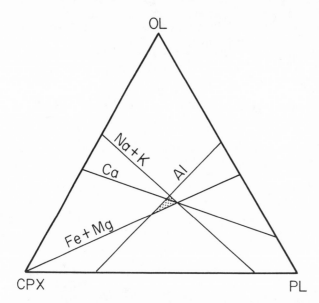

Figure 5.22 Equilateral extract triangle for Figure 5.21 based on principle of opposition. *Stippled area* is polygon of error.

    Three constant-ratio lines are plotted on Figure 5.23a: Rb/Ba = 0.2, 1.0, and 5.0. On an arithmetic plot, such as Figure 5.23a, constant-ratio lines must be straight lines that pass through the origin of the graph. Ratios between the two variables along a straight line vary if the line does not pass through the origin. Two groups of granitoids are plotted on the figure: (1) group L, relatively *leucocratic* (light colored), and (2) group M, relatively *melanocratic* (dark colored). The leucocratic rocks are dominantly granites, quartz monzonites, and some granodiorites, whereas the melanocratic rocks are primarily tonalites and some granodiorites. Which best discriminates between groups L and M, ppm Rb, ppm Ba, or the Rb/Ba ratio? The poorest of the three is Rb because the most overlap exists for Rb between the two groups. The best is the Rb/Ba ratio, because no overlap exists; all samples in group L have ratios less than about 1.2, whereas the ratio for all group M samples is greater than 1.2.
    A vague suggestion of a positive correlation exists between Ba and Rb on Figure 5.23a. One difficulty with this figure is a large number of points are clustered together in the lower left corner of the graph and the point density

decreases away from this corner. Trace elements commonly exhibit this type of positively skewed distribution and approach log normality (see Section 1.1.3.1). If both elements on a bivariate scattergram have positively skewed distributions, the practical result is uneven point density with clustering at relatively low values (Fig. 5.23a).

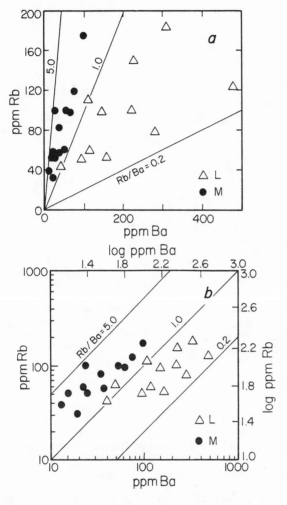

Figure 5.23
Scattergrams of Rb versus Ba showing constant-ratio lines for two groups of granitoids: *L (open triangles)* and *M (closed circles)*. *a,* arithmetic plot; *b,* log-log plot. The same constant-ratio lines are drawn on both graphs.

If we desire a more uniform point density, we can either take the logs of all the values and plot the resultant logs on arithmetic paper or we can plot the original data on *log-log paper* (Fig. 5.23b). Exactly the same data set is plotted on Figure 5.23b as on Figure 5.23a. This particular log-log plot is referred to as a *2 x 2 cycle* plot, each cycle covering one order of magnitude (i.e., 10 to 100 and 100 to 1000). On Figure 5.23b the log ppm scales are arithmetic, whereas the corresponding ppm scales are logarithmic. Thus a plot of logs on arithmetic paper would appear identical to a plot of raw data on log-log paper.

Constant-ratio lines are parallel to one another and have 45° slopes on log-log plots. These constant-ratio lines cannot intersect at the origin, as they do on an arithmetic plot, because they are parallel to one another and no origin exists on a log-log diagram. On this log-log plot the point density is more uniform and a positive correlation between Rb and Ba is more apparent than on the arithmetic plot. In general, log-log plots are useful when the data are positively skewed and/or their values range over several orders of magnitude.

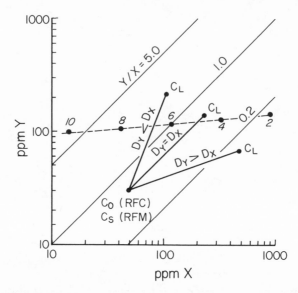

Figure 5.24 Log-log of hypothetical incompatible elements Y and X showing effects of Rayleigh fractional crystallization *(RFC; three solid lines C_O-C_L)* and Rayleigh fractional melting *(RFM; dashed line* for 2-10 percent partial melting; D_Y = 0.2, D_X = 0.02). C_O, composition of parent magma for *RFC*; C_S, composition of original source rock for *RFM* (C_O = C_S); C_L, - composition of liquid for *RFC*. Data can be in weight or molar units.

The trend on Figure 5.23b is not parallel to a constant-ratio line. Rather, the Rb/Ba ratio decreases in a rather systematic fashion, given considerable scatter, from slightly less than 5.0 to slightly greater than 0.2. We can examine this trend in the light of relative incompatibilities of Rb and Ba. Let us first determine the effect of Rayleigh fractional crystallization (RFC) on the trend. Equation 5.5 can be rewritten to include both elements as follows:

$$C_{LY} / C_{LX} = C_{OY} / C_{OX} \, F^{(DY - DX)} \tag{5.18}$$

where:

C_{LY}, concentration of element on Y-axis in daughter magma
C_{LX}, concentration of element on X-axis in daughter magma
C_{OY}, concentration of element on Y-axis in parent magma
C_{OX}, concentration of element on X-axis in parent magma
$DY = D_Y$, bulk distribution coefficient for element on Y-axis
$DX = D_X$, bulk distribution coefficient for element on X-axis
F, fraction melt remaining

During the argument that follows we will assume that both elements are behaving incompatibly. The same ultimate conclusions would be drawn had we assumed that both were compatible. If both elements are equally incompatible, then $D_Y = D_X$ and $C_{LY}/C_{LX} = C_{OY}/C_{OX}$. This means that any liquid line of descent (LLD) from the parent O to the daughter L must maintain a constant ratio equal to that of the original parent O (Fig. 5.24). As a result, an LLD for equally incompatible (or equally compatible) elements must be parallel to a constant-ratio line and both daughter and parent must have the same ratio. If $D_Y > D_X$ (element Y is more compatible than element X), then $C_{LY}/C_{LX} < C_{OY}/C_{OX}$ and LLD is not parallel to a constant-ratio but appears to have a lower slope (Fig. 5.24). It follows that if element Y is more incompatible and $D_Y < D_X$, then $C_{LY}/C_{LX} > C_{OY}/C_{OX}$ and LLD will appear to have a greater slope. Trends for fractional crystallization on log-log plots are straight lines (Cocherie, 1986).

As a consequence of these arguments, if the trend on Figure 5.23b were explained by RFC, Ba should be more incompatible than Rb. Assuming the sample with 38 ppm Rb and 13 ppm Ba represents C_O and the sample with 125 ppm Rb and 480 ppm Ba represents C_L after fractionation has almost completed ($F = 0.01$), then solving equation 5.18 in the form of equation 5.5:

$$0.260 = 2.92 \times 0.01^{(DY - DX)}$$

$$0.260 = 2.92 \exp [(D_Y - D_X) \ln 0.01]$$

$$(D_Y - D_X) = \ln 0.089 / \ln 0.01 = 0.53$$

Hence, the trend on Figure 5.23b can be explained by RFC of two incompatible elements in which the bulk distribution coefficient D for Rb is 0.53 greater than D for Ba. The trend will be linear if $D_Y - D_X$ remains constant. In this example Ba is clearly the more incompatible element; this is reasonable because with increasing fractionation its concentration increases in residual melts much more rapidly than does the concentration of Rb.

Can the trend on Figure 5.23b be explained by partial melting? Either incremental EBM or RFM can be modeled, but in this case it is simpler to consider RFM. Equation 5.6 can be rewritten to accommodate two elements on a bivariate scattergram:

$$C_{LY}/C_{LX} = (C_{SY}/C_{SX}) (D_X/D_Y) (1 - F)^{(1/DY - 1/DX)} \tag{5.19}$$

where C_{LY} and C_{LX} represent compositions of the two elements in the partial melt; C_{SY} and C_{SX} are their compositions in the original source rock; D_Y and D_X (or DY and DX) are their respective bulk distribution coefficients; and F is the fraction partial melt produced. Analogous to fractional crystallization, if $D_Y = D_X$, then $C_{LY}/C_{LX} = C_{SY}/C_{SX}$ and any partial melt must have the same ratio as the source rock; therefore, any trend of partial melts from the same source must be parallel to a constant-ratio line that passes through the point of the source composition.

The situation is more complex, however, if D_Y and D_X are different. Take the example of two incompatible elements. For fractional crystallization, if $D_Y < D_X$, then the ratio C_{OY}/C_{OX} is the lower limit for C_{LY}/C_{LX} and all compositions of residual liquids must fall on a line that is anchored on its lower end by point C_O (Fig. 5.24). Conversely, if $D_Y > D_X$, C_{OY}/C_{OX} is an upper limit for C_{LY}/C_{LX}.

The situation is quite different for fractional melting; in this case the ratio C_{LY}/C_{LX} can be higher or lower than C_{SY}/C_{SX} and the straight line representing the partial melts does not have to pass through point C_{SY}-C_{SX}. These relationships can be seen in Figure 5.24. The dashed line representing 2-10 percent partial melting was calculated by solving equation 5.6 for each element based on the following data:

$$C_{SY} = 30 \text{ ppm}$$
$$C_{SX} = 50 \text{ ppm}$$
$$D_Y = 0.2$$
$$D_X = .02$$

Whereas $C_{SY}/C_{SX} = 0.6$, C_{LY}/C_{LX} varies from less than 0.2 to over 5.0 (Fig. 5.24). In addition, the dashed line representing 2-10 percent partial melting never passes through the point 30,50. More complex models for partial melting are given by Shaw (1970) and Cocherie (1986).

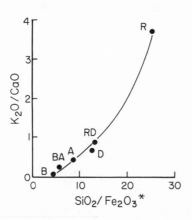

Figure 5.25 Four-oxide ratio-ratio diagram based on data in Table 5.3 and Figure 5.19. Abbreviations are explained on Table 5.3 and Figure 5.19. Calculation of curve is given in text.

These examples indicate that if no knowledge of either source or parental melt composition is available, a linear trend of trace-element analyses can be easily interpreted as either partial melting or fractional crystallization. Other evidence, such as the presence of an adequate extraction assemblage and age relationships, are needed to determine which process is the more likely. For example, if cross-cutting relationships indicate that the group M rocks (Fig. 5.23) are older, then RFC is more likely than RFM. In general, early partial melts will be relatively felsic and incompatible element enriched, whereas early residual liquids will be more mafic and incompatible element depleted.

5.2.2.5 Diagrams Using Ratios

The previous sections have considered various types of bivariate scattergrams, all of which plot the absolute concentration of one element against that of another. Scattergrams involving ratios are also widely used by petrologists. They are generally one of three types: (1) ratio-ratio diagrams involving four elements, (2) ratio-ratio diagrams for three elements, with one element being in the denominator of both ratios, and (3) ratio-absolute concentration diagrams. This third type can involve either three or two elements.

We will first examine ratio-ratio diagrams involving four elements. A good discussion of these diagrams can be found in Cox and others (1979). Trace rather than major elements are generally plotted on these diagrams because trace-element ratios commonly vary over a larger range than do major-element ratios. On theoretical grounds it is just as valid to plot major-element ratios as trace-element ratios on diagrams of this type, so we will again use the data in Table 5.3 and on Figure 5.19 as an example. Ratio-ratio diagrams of this type are commonly used to determine if some mixing process has taken place, such as contamination of country rocks or magma mixing. Mixing processes involving two end-member compositions produce straight-line trends on simple element-element plots, so we will only use the elements on Figure 5.19 that display linear trends with SiO_2: $Fe_2O_3^*$, CaO and K_2O. In Section 5.2.2.3 our hypothesis was that these trends are caused by fractional crystallization. We will now adopt a new hypothesis: a basaltic melt of composition B was mixed in various proportions with a rhyolitic melt of composition R to produce the four intermediate rock types.

Before plotting the ratio-ratio diagram we must first decide what elements to ratio with one another. Silica and K_2O positively correlate, as do $Fe_2O_3^*$ and CaO, so ratios involving these two pairs would exhibit very little variation. A better approach in this case is to work with ratios of elements with negative correlations because a greater variation in ratios will be achieved. Accordingly, the ratios K_2O/CaO and $SiO_2/Fe_2O_3^*$ taken from the data in Table 5.3 are plotted against one another on Figure 5.25. The curve on this figure is not capriciously drawn through the six points. It is a hyperbolic curve calculated from the two end-member compositions B and R to represent an ideal curve for mixing these two compositions in various proportions (Langmuir and others, 1977). It is not necessary that only end-member compositions be used to calculate the curve. In fact, it is a common practice to calculate the curve based on two intermediate samples and use the curve to predict the ratios for the two end-members. However, the greater the difference in ratios for the two samples used to calculate the curve, the better will be the fit.

The formula for this hyperbolic mixing curve is of the form:

$$(aY) + (bYX) + (cX) + d = 0 \qquad (5.20)$$

Rewritten to solve for Y given different values of X:

$$Y = [-d - (cX)] / [a + (bX)] \qquad (5.21)$$

where Y is the ratio on the Y-axis and X is the ratio on the X-axis. We need only determine the coefficients a, b, c, and d to construct the curve. Data from Table 5.3 needed to calculate these four coefficients are:

$X_B = SiO_2/Fe_2O_3^*$ for basalt = 4.83
$X_R = SiO_2/Fe_2O_3^*$ for rhyolite = 25.5
$Y_B = K_2O/CaO$ for basalt = 0.10
$Y_R = K_2O/CaO$ for rhyolite = 3.73
$F_B = Fe_2O_3^*$ for basalt = 10.4
$F_R = Fe_2O_3^*$ for rhyolite = 2.8
$C_B = CaO$ for basalt = 10.0
$C_R = CaO$ for rhyolite = 1.1

Note that the numerators of the ratios are not necessary to calculate the coefficients. The four coefficients are calculated by:

$$a = (C_B F_R X_R) - (C_R F_B X_B) = 658.74$$
$$b = (C_R F_B) - (C_B F_R) = -16.56$$
$$c = (C_B F_R Y_B) - (C_R F_B Y_R) = -39.87$$
$$d = (C_R F_B Y_R X_B) - (C_B F_R Y_B X_R) = 134.70$$

Solution of equation 5.21 then yields:

$$Y = [(39.87X) - 134.7] / [658.74 - (16.56X)]$$

Substituting different values for X and solving for Y allows construction of the mixing curve on Figure 5.25.

Langmuir and others (1977) have noted that two main criteria must be met if mixing between two end members is feasible: (1) the data points should fall along the ideal hyperbolic mixing curve; (2) on all ratio-ratio plots the same order of samples should be observed. In this example, the order B-BA-A-D-RD-R should be maintained on all graphs. The first criterion has been satisfied; data points on Figure 5.25 do fall along the ideal mixing curve. The second criterion can be tested with additional plots involving other elements.

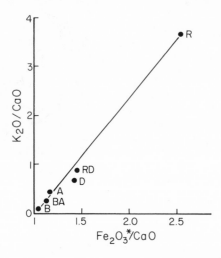

Figure 5.26 Three-oxide ratio-ratio diagram based on data from Table 5.3. Note that CaO is in the denominator of both ratios.

We can now introduce the second type of variation diagram using ratios: a plot involving three elements with the same element in the denominator of both ratios (Fig. 5.26). Such a plot should yield a linear, rather than hyperbolic, trend if mixing of two end members has occurred. The plot is indeed linear, confirming the fact that mixing is a reasonable hypothesis. A recent example of this type of diagram, involving Sr/Ca and Ba/Ca, is by Defant and Ragland (1988).

Recall, however, that trends involving other elements than the four on Figure 5.25 are not linear and cannot not be explained by mixing (Fig. 5.19). The fact that Figures 5.25 and 5.26 can be explained by mixing does not preclude other processes such as RFC or EBM. In fact, the apparent mixing trends on Figures 5.25 and 5.26 are probably fortuitous, because *all* the data are better explained by RFC (Section 5.2.2.3). We deliberately chose elements that were linearly correlated with SiO_2 on Figure 5.19 as an exercise, so the apparent mixing trends on Figures 5.25 and 5.26 are expected. Had all the trends been linear on Figure 5.19 and the above criteria met for several four- or three-element plots, then magma mixing or some other type of mixing process must be seriously considered.

In the last type of scattergram using ratios, an absolute concentration is plotted against a ratio. Many of these plots utilize three different elements, but others involve only two. This latter type is particularly useful to determine relative compatiblity (or incompatibility) of two elements; a typical example is a plot of K/Rb against K. They have been referred to as *process identification diagrams* because they are useful in distinguishing different magmatic processes (see, for example, Allegre and Minster, 1978; Camp and Roobol, 1989). Such diagrams occasionally have been criticized because they are analogous to scattergrams involving fractionation indices such as LI and NI; a trend is virtually assured because a variable is plotted against a ratio that contains that variable. The lever rule cannot be applied to these plots because they are generally curved. They seldom provide more information than can be obtained from a simple plot of two elements if constant-ratio lines are included on the graph (e.g., Figs. 5.23 and 5.24).

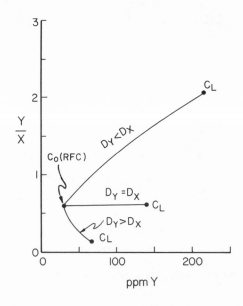

Figure 5.27 Ratio-element diagram for hypothetical elements Y and X. C_O, original melt composition for Rayleigh fractional crystallization *(RFC)*; C_L, liquid composition. Data can be in either weight or molar units.

Ratio-element diagrams, however, can be easier to interpret than corresponding element-element plots. For example, Figure 5.27 is a plot of Y/X against Y that corresponds to the Y-X trends on Figure 5.24 for Rayleigh fractional crystallization (RFC). The three trends on Figure 5.27 are the same as those shown on Figure 5.24. The more incompatible element can be easily determined from Figure 5.27. If the trend has a positive slope, Y is more incompatible $(D_Y < D_X)$; if the trend is horizontal, then the two elements are equally incompatible $(D_Y = D_X)$; if the trend is negative, X is more incompatible $(D_Y > D_X)$. Thus a plot of Rb/Ba versus Rb (not included) based on the data used to construct Figure 5.23 yields a negative slope because Ba is relatively incompatible.

5.2.2.6 Rare-earth Elements

This discussion is equally well placed under Section 5.2.1 on trace elements, but as rare-earth element (REE) compositions are traditionally plotted on a unique type of diagram, it is included here. REE compositions are typically "normalized" to those of an average chondritic meteorite and then plotted against atomic number. In this case to normalize a composition simply means to divide the composition by that of the average chondritic meteorite and produce a ratio C/C_M (Table 5.5). In this ratio C is the composition of the REE in the sample and C_M is the composition of that element in the average meteorite.

TABLE 5.5 Rare-Earth Element (REE) Data

Z	REE	r_c	C	C_M	C/C_M	K_D	40PL
57	La	1.26	24.2	0.367	65.9	0.14	102
58	Ce	1.22	53.7	0.957	56.1	0.14	87.0
59	Pr	1.22	6.50	0.137	47.4	--	--
60	Nd	1.20	28.5	0.711	40.1	0.08	64.1
61	Pm	--	--	--	--	--	--
62	Sm	1.17	6.70	0.231	29.0	0.08	46.4
63	Eu	1.15	1.95	0.087	22.4	0.32	31.7
64	Gd	1.14	6.55	0.306	21.4	0.10	33.9
65	Tb	1.12	1.08	0.058	18.6	--	--
66	Dy	1.11	6.39	0.381	16.8	0.09	26.7
67	Ho	1.10	1.33	0.0851	15.6	--	--
68	Er	1.08	3.70	0.249	14.9	0.08	23.8
69	Tm	1.07	0.51	0.0356	14.3	--	--
70	Yb	1.06	3.48	0.248	14.0	0.07	22.5
71	Lu	1.05	0.55	0.0381	14.4	0.08	23.0

r_c: ionic radius in angstroms for cubic co-ordinated, trivalent REE (Whittaker and Muntus, 1970)
C: REE concentration in USGS basaltic rock standard BCR-1 (Taylor and McLennan, 1985)
C_M: average REE concentration in type I carbonaceous chondrites (Evensen and others, 1978)
K_D: average K_D for plagioclase in mafic rocks (Henderson, 1982)
40PL: C/C_M in residual melt after 40 percent fractionation of plagioclase; original melt composition taken as BCR-1

This "normalization" is necessary because of the *Oddo-Harkins rule,* which states that in the cosmos elements of even atomic number are generally more abundant than elements of odd atomic number on either side. This relationship is seen in the REE data for the USGS basaltic rock standard BCR-1 and for an average type I carbonaceous chondrite (Table 5.5; see also Appendix C). REE elements with odd atomic (Z) numbers are less abundant than those with even Z-numbers on either side. The reasons for this relationship are beyond our concern here. The fact that this pattern in the cosmos is also observed on Earth suffices for our purposes. If we plot ppm REE against atomic number, a jagged, "saw-tooth" pattern emerges. This plot is not included, but it can be imagined by comparing Z-numbers with values for either C or C_M in Table 5.5. By calculating the chondrite-normalized ratio C/C_M, this jagged pattern is smoothed (Table 5.5, Fig. 5.28). A typical REE plot such as Figure 5.28 is simply one of convenience that enhances the visual effect by creating smooth rather than ragged patterns.

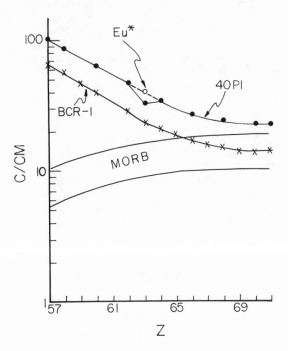

Figure 5.28 Rare-earth patterns based on chondrite-normalized concentrations (C/C_M) versus atomic number Z (Table 5.5). Range for MORB basalts is shown, in addition to basalt rock standard BCR-1 (parent) and daughter (40Pl) 40 percent plagioclase extraction from BCR-1. Note small Eu anomaly in pattern for daughter.

Ionic radii of REE decrease in a systematic manner with increasing atomic number (Table 5.5). As a result, geochemical properties of the relatively small, heavy rare-earth elements (HREE) are different from those of the relatively large, light rare-earth elements (LREE). An example of both an LREE (Ce) and an HREE (Yb) is included in the compilation of K_Ds (Table 5.1). For most minerals LREE are more incompatible than HREE. This difference is particularly striking for garnet, where K_D for Yb is about two orders of magnitude higher than K_D for Ce. Table 5.5 gives K_Ds for several REE in plagioclase feldspar from mafic rocks.

These differences in K_Ds between LREE and HREE for various minerals allow constraints to be placed on such processes as fractional crystallization and partial melting. The slope of an REE pattern is frequently indicative of a lithologic type, a petrogenetic process, or even a tectonic environment. A typical MORB pattern is slightly concave down with a low positive slope (Fig. 5.28). In contrast, many other basalts have a pattern that is slightly concave up with a negative slope (e.g., BCR-1; Fig. 5.28). Granitoid REE patterns are generally even steeper and overall more enriched (higher overall REE contents). In general, the negative slope steepens and the overall abundance of all REE increases progressively from mafic through intermediate to felsic rocks, although exceptions exist. Because La is the lightest REE and Lu is the heaviest, some petrologists use the ratio La/Lu as a measure of the steepness of the slope on a REE pattern; the Ce/Yb ratio is used as well. What can cause these differences in slope, curvature, and absolute abundances? They are generally attributed to differences in some combination of partial melting, fractional crystallization, or assimilation.

A qualitative example is illustrative. Compare two basaltic melts being formed at different depths in the mantle. The deeper melt (below about 20 kb or 60 km) is thought to be forming by partial melting of a garnet-bearing lherzolite, whereas the presumed source for the shallower melt is a spinel-bearing lherzolite (Fig. 5.6, Sections 5.1.1.4 and 5.1.1.5). K_Ds for HREE and LREE in spinel are not radically different, but they are for garnet (Table 5.1). Because the HREE are so compatible in garnet, they are retained in the source area relative to LREE at depths greater than about 60 km, but this phenomenon does exist at shallower depths. As a result, basaltic partial melts from greater depths, where garnet is stable, should have REE patterns with much steeper negative slopes than partial melts of lherzolite from shallower levels where spinel is stable. If HREE are retained in the source relative to LREE, then LREE are enriched in the partial melt relative to HREE. Interestingly, alkali-olivine basalts generally have steeper negative slopes on REE patterns than do tholeiitic basalts and are thought to form at greater depths than many tholeiites.

If during fractional crystallization an assemblage of minerals is being extracted in which all REE have similar K_Ds, then the slope for the REE pattern will not change, although the overall pattern is likely to rise to higher values of C/C_M. For instance, K_Ds for all REE in olivine and orthopyroxene are extremely low; REE in these two minerals approach perfect incompatibility. Extraction of either or both of these minerals would raise the overall pattern in the residual melts, but the slope of the pattern would not change. Likewise, accumulation of these two minerals in a cumulate rock would lower the overall pattern but not change the slope.

5.2.2.7 Europium Anomalies

Another topic concerning REE patterns deserves mention. In some REE patterns the element Eu (Z = 63) is either abnormally enriched or depleted compared with the elements on either side of it, Sm (Z = 62) and Gd (Z = 64). If Eu is abnormally enriched relative to Sm and Gd, this is referred to as a *positive Eu anomaly,* if depleted, a *negative Eu anomaly.* A negative Eu anomaly can be observed in the REE pattern for the residual melt after 40 percent plagioclase extraction from BCR-1 on Figure 5.28. Plagioclase was chosen for this extraction calculation because the role of feldspar is very important in Eu anomalies. Europium is the most compatible of all REE in plagioclase from mafic rocks (Table 5.5). This increased compatibility of Eu is even more pronounced in feldspars from more felsic rocks (Table 5.1).

Most REE exist in nature in the trivalent state, but if conditions are quite reducing, Eu will be in the divalent state. As such, it can substitute in the W-position of feldspars much more readily than can other REE. If conditions are quite oxidizing, Eu will be trivalent and no anomaly will occur. In many magmas conditions are sufficiently reducing, so a Eu anomaly can be seen on a REE pattern if feldspars have played a role in the genesis of that rock. Some of the strongest Eu anomalies have been in REE patterns of lunar basalts, where conditions are quite reducing.

Hence, the presence of a Eu anomaly provides evidence concerning the role of feldspar in a petrogenetic process. If the anomaly is positive, feldspar accumulation or assimilation of feldspar-rich material is normally proposed. A negative anomaly is considered to be indicative of felspar fractionation or partial melting of a source rock containing feldspar. In either case Eu is preferentially retained in the feldspar relative to surrounding REE, so a negative anomaly results in the partial or residual melt. In fact, other minerals exhibit Eu anomalies as well, although they are generally negative anomalies and much smaller than the positive anomalies observed in feldspars. These include clinopyroxene, garnet, and amphibole; these negative anomalies seem to be more pronounced in felsic rocks. Thus other minerals may play a role in Eu anomalies, so to explain every Eu anomaly solely on the basis of a feldspar is an oversimplification. Moreover, because these other minerals have negative rather than positive anomalies, their effect can at least partially offset that of feldspar.

A measure of the size of the Eu anomaly is Eu/Eu^*, where Eu^* is the interpolated concentration of Eu from its surrounding elements in the REE pattern. Eu^* can be obtained by ignoring the anomaly and interpolating the value of Eu where the curve between Sm and Gd intersects the Eu position in the pattern, assuming that the overall pattern is effectively linear between Sm and Gd. Because the Z-number for Eu is equidistant from Z-numbers for Sm and Gd, this interpolation can be accomplished by simply averaging the C/C_M values for Sm and Gd. For example, the data in Table 5.5 (last column) can be used to calculate Eu^*:

$$Eu^* = (46.4 + 33.9) / 2 = 40.2$$
$$Eu/Eu^* = 31.7 / 40.2 = 0.79$$

Thus if the anomaly is negative, $Eu/Eu^* < 1.0$; if positive, $Eu/Eu^* > 1.0$. In some cases Gd is not analyzed, so Sm and Tb are used for the interpolation to obtain Eu^*.

The Eu anomaly can be used to check quantitatively fractional crystallization based on major-element modeling (Sections 5.2.2.3 and 5.2.3.2). This can be done by a modification of equation 5.18, assuming no anomaly existed before fractionation. If this assumption is true, then C_{OY}/C_{OX} in equation 5.18 must be equal to 1.0 for C_{EU}/C_{EU^*}; therefore, this equation reduces to:

$$EU/Eu^* = F^{(D_{EU} - D_{EU^*})} \qquad (5.22)$$

where Ds are bulk distribution coefficients and F is the weight fraction remaining melt. If Sm and Gd are used to determine Eu^*, then D_{EU^*} is the average D for Sm and Gd (Table 5.1). The advantage of this equation is that no assumption has to be made about parent magma compositions, other than $Eu/Eu^* = 1.0$. For example, 50 percent fractionation of an extract assemblage with $D_{EU} = 0.7$ and $D_{EU^*} = 0.1$ would lead to a Eu/Eu^* ratio of 0.66.

5.2.2.8 Diagram for Exchange Reactions

In this and previous chapters we have primarily considered crystal-liquid equilibria, i.e., crystallization and partial melting of silicate melts. Other processes can affect chemical compositions of igneous rocks as well. A difficulty, however, is that many of these processes are not readily treated by graphical or numerical techniques, with the possible exception of contamination

312

(synonymous with assimilation) or magma mixing (Sections 5.2.2.3 and 5.2.3.1). One such process that is frequently difficult to evaluate by variation diagrams is hydrothermal or deuteric alteration. These are subsolidus processes that involve crystal-fluid equilibria. *Deuteric alteration* takes place at temperatures only slightly lower than the solidus, whereas *hydrothermal alteration* occurs at comparatively low temperatures, but considerably above the temperatures of weathering. Both types of postmagmatic processes frequently lead to *metasomatism,* a term widely used for metamorphic as well as igneous rocks. If metasomatism has affected a body of rock, then its final *chemical* composition is different from its original composition.

A typical metasomatic reaction affecting igneous rocks is an *exchange reaction,* whereby one cation dissolved in a fluid phase exchanges with another in a coexisting mineral. A common example is:

$$Na^+ + KAlSi_3O_8 = K^+ + NaAlSi_3O_8$$

The effect is for Na-rich feldspar to replace K-rich feldspar in one part of the system, while the reverse is occurring in another. Orville (1963) has shown that Na-rich feldspar will replace K-rich feldspar in hotter parts of the system, and vice versa in the cooler parts. Trace elements can be involved as well -- Rb and Ba commonly follow K in this process, whereas Sr and Ca (a minor element in alkali feldspars) can follow Na.

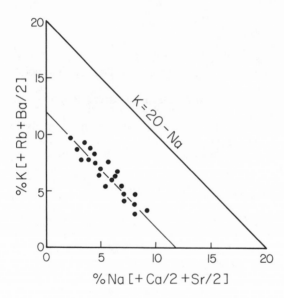

Figure 5.29 A plot of mol percentage K versus Na showing evidence for a possible exchange reaction. Possible substituting trace elements are given in brackets. Line with formula *K = 20 - Na* is for pure alkali feldspar.

How can we determine if some igneous rocks were affected by a metasomatic exchange reaction between K and Na in alkali feldspars? First, some evidence should exist in thin-section, such as Na-rich rims around K-feldspars or K-rich rims around Na-feldspars. Occasionally replacement textures appear as irregular, patch perthites. Second, evidence will be present on variation diagrams. If Na_2O exhibits a strong and negative correlation with K_2O, this plus petrographic observations would be evidence to suspect an alkali metasomatic exchange reaction. A test of this hypothesis is to plot Na_2O versus K_2O, or even better cation percent K against Na. Cation proportions can be used as well. Trace elements can be included, but if Ba, Sr, or Ca, because they are divalent, are added to either Na or K, their molar concentrations must be halved before including them with univalent ions.

A plot testing the hypothesis of a metasomatic exchange reaction in some granitoids is shown in Figure 5.29. The trace elements Rb and Ba are included in brackets with K, whereas Ca and Sr are similarly added to Na. The brackets denote an option and some evidence must exist for grouping the elements in this way. For example, K, Rb, and Ba should correlate strongly with one another, as should Na, Ca, and Sr. If any trace or minor element does not correlate with K or Na, it apparently was not involved in the exchange reaction and should be omitted. In fact, all trace elements can normally be omitted because they have little effect on the trend in comparison with the major elements K and Na.

Pure K-feldspar ($KAlSi_3O_8$) contains 20 cation percent K, and pure albite ($NaAlSi_3O_8$) contains 20 percent Na. Thus all alkali feldspar compositions should fall along the line with a slope of -1 and a formula K = 20 - Na. Compositions of granitoids undergoing metasomatic alkali exchange will not fall along this line because granitoids contain minerals other than alkali feldspars; their trend, however, should parallel this line and have a slope of -1 (Drummond and others, 1986).

Figure 5.29 offers no evidence as to which rocks have been affected by metasomatism and which have not. We are primarily dependent on petrography to answer this question. For example, if the relatively K-rich rocks exhibit replacement textures and the Na-rich rocks do not, then a K-rich fluid apparently interacted with a relatively Na-rich granitoid. Another possibility is that all rocks were affected by metasomatism acting along a temperature gradient such that the fluid was hotter and thus more Na-rich in some areas, while being cooler and more K-rich in other. In this case, replacement textures should be observed in all the rocks, although they should be different in the different rock types. Electron microprobe data for the composition of a mineral supposedly affected by the metasomatism compared with data for that same mineral, but apparently not metasomatized, are often critical to these arguments. Other evidence, such as that from stable isotopes (Drummond and others, 1986), can be invaluable.

5.2.3 Numerical Modeling

The numerical methods introduced herein are extensions of the graphical techniques in previous sections. In fact, graphical methods already discussed will achieve the same results as the numerical approach that follows shortly. The principal advantage to any numerical method is, given a properly programmed computer, speed and accuracy. In Section 5.2.3.2 we will deal with the numerical solution of a three-dimensional geometric problem to model fractional crystallization. Much more sophisticated programs involving the solution of an n-dimensional problem are available (e.g., Wright and Doherty, 1970; Bryan and others, 1969). These programs are too long to be included in this book, but they should be utilized for a more in-depth treatment of fractional crystallization. We are primarily concerned with general principles, and they can be most easily grasped by examining numerical methods in the light of graphics.

5.2.3.1 Linear Mixing Calculations

These calculations are simply numerical expressions of the lever rule. They assume a linear relationship exists between the rock compositions in question. For instance, if a linear mixing calculation is dealing with fractional crystallization, compositions of the extraction assemblage, the parent magma, and the daughter magma must be linearly related. This linear relationship, of course, is the lever. A basic tenet of linear mixing leads to the fact that rhyolite R (Fig. 5.19) cannot be considered as a daughter with basalt B as the parent because their compositions for some elements fall on opposite ends of a curve rather than a straight line. Any two adjacent samples on Figure 5.19 could be treated by linear mixing because the assumption of compositional linearity between them is reasonable.

TABLE 5.6 Definitions of Terms in Linear Mixing Equations

Process	Y	Z	X	N
fractional crystal-lization	residual liquid	parent magma	extract	fraction crystal-lized
accumu-lation	melt	cumulate	cumulus crystals	fraction crystals accumulated
partial melting	restite residuum	source rock	partial melt	fraction melted
two-member mixing	end-member A	hybrid rock	end-member B	fraction B in hybrid

Percent fractionation is fraction crystallized x 100.

Linear mixing is the graphical or mathematical means by which we treat such processes as restite unmixing (White and Chappell, 1977), assimilation, and magma mixing, all of which produce straight-line trends on variation diagrams. Numerous examples of linear mixing processes are discussed in the literature (e.g., Reidel and Fecht, 1987; Wiebe, 1987). In some instances fractional crystallization, crystal accumulation, or partial melting can also be treated by linear mixing calculations. The basic equation for all types of linear mixing is:

$$Z = NX + (1 - N)Y \tag{5.23}$$

Terms are explained in Table 5.6. A modified form of this equation was used by Weigand and Ragland (1970) to model fractional crystallization of some tholeiitic magmas (see also Drummond and others, 1988). For computational purposes it is important to write the expressions for Y, X and N:

$$Y = [Z - NX] / (1 - N)$$
$$X = [Z - Y(1 - N)] / N$$
$$N = (Z - Y) / (X - Y)$$

Equation 5.23 for Z is simply the formula for a weighted average, which is a numerical expression of the lever rule. Ultimately, equation 5.23 and its derivative formulae are nothing more than an expression of the lever rule:

X   Z   Y
|_____|_____|

N = Z Y / X Y

The definitions of X, Y, Z, and N depend on what process is being modeled. The choices are (1) fractional crystallization, (2) crystal accumulation, (3) partial melting, and (4) bulk contamination or magma mixing. Table 5.6 defines these terms for each of the three processes. Examine the lever above for each process in light of these definitions. One simply chooses a process to model and solves for one of the three variables X, Y, or Z, given values for the other two and N. Each element is treated individually. If X, Y, and Z are known, one can solve for N, but this situation is rare. A more common practice is to place constraints on N, as is explained below.

The extraction assemblage X for fractional crystallization includes not only the minerals involved but also any trapped liquid. Likewise, the partial melt X can include *cognate xenoliths* (also referred to as *autoliths*), which are rock fragments from the restite transported upward by the partial melt. During restite unmixing a magma is contaminated by cognate xenoliths from its own source area. In general, composition X refers to whatever (solid or liquid) is being removed from the parent melt, in the case of fractional crystallization, or from the source rock, for partial melting. Similarly, residual liquid Y includes not only the magma but also any primocrysts not removed by fractionation, or restite Y also contains any trapped melt. These complications can obviate linear mixing calculations if they are not taken into account. We will consider them further in the next section.

The utility of linear mixing calculations can best be understood through some examples. In the first example we will again model fractional crystallization of basalt B to form basaltic andesite BA (Table 5.3). In Section 5.2.2.3 we used as a constraint the fact that a solution with negative numbers is impossible, because a rock cannot have a negative concentration. The K_2O-SiO_2 plot on Figure 5.19 was used to constrain the composition of the extraction assemblage in modeling fractional crystallization of parent B to form daughter BA. This particular plot was used because of the positive K_2O-SiO_2 trend and the fact that an extrapolation of the line BA-B to lower SiO_2 values, in the direction of the extract composition, quickly intersects the 0 percent K_2O line. Numerically, this exercise can be carried out by:

$$N = (1.0 - 2.1) / (0 - 2.1) = 0.52$$

SiO_2: $X = [50.2 - 54.3 \times (1 - .52)] / .52 = 46.4\%$
TiO_2: $X = [1.1 - 0.8 \times (1 - .52)] / .52 = 1.4\%$
Al_2O_3: $X = [14.9 - 15.7 \times (1 - .52) / .52 = 14.2\%$

and so on for the remaining oxides. These compositions are the same, allowing for measuring errors in the graphical method, as those generated by graphical means in Table 5.4.

The second example involves partial melting of rhyodacite RD to leave a restite residuum of composition D (Table 5.3). We are again constrained by negative numbers, but an additional constraint exists. The maximum partial melting possible is normally about 40 percent, so we will assume that N cannot be greater than 0.4. A careful inspection of compositions RD and D on Figure 5.19 is useful at this point. In this case the concern is for negative values of oxides that correlate negatively with SiO_2 over this compositional range: Al_2O_3, $Fe_2O_3^*$, MgO, and CaO. These partial melts of RD must have SiO_2 concentrations higher than those of RD itself, so extrapolation of the line D-RD to higher SiO_2 values indicates that MgO is the first oxide to reach 0 percent concentration. Accordingly:

$$N = (0.9 - 1.7) / (0 - 1.7) = 0.47$$

Because we have assumed that N cannot be greater than 0.40, we will use this lower value to constrain the most differentiated melt possible by partial melting of RD:

$$SiO_2: \quad X = [66.2 - 64.9 \times (1 - .40)] / .40 = 68.2\%$$

The remaining oxide compositions are calculated similarly.

xxxxxxxxxxxxxxxxxxxxxxxxxxxxxxxxxxxxxxxxxxxxxxxxxxxxxxxxxxxxxxxxx

Igneous intrusions are three-dimensional bodies, yet unless numerous drill cores are available, we are generally confined to studying them in two dimensions, at the present level of erosion. A tacit assumption is often made that the exposed rocks are representative of the entire intrusion, an assumption that may or may not be valid. Petrogenetic conclusions drawn from modeling calculations, such as those described in these sections, frequently implicitly depend on this assumption. An example is illustrative. For many years petrologists have been puzzled as to why rocks of intermediate composition are not present in the Concord ring dike complex, North Carolina. Geochemical modeling calculations (Olsen and others, 1983) indicate that the magma which formed the syenite ring dike was derived from a magma similar in composition to the gabbro interior by fractional crystallization, yet no intermediate rocks are exposed. Williams and McSween (1989), based on geophysical data, postulate a large igneous body at depth, very likely of composition intermediate to the syenite and gabbro. Very likely this body was a magma chamber at one time, which was periodically tapped to produce the gabbro and syenite. This is an excellent example of the fact that rocks at the present level of erosion may not be representative of the entire magmatic system. Consequently, geochemical modeling calculations should be made with caution.

xxxxxxxxxxxxxxxxxxxxxxxxxxxxxxxxxxxxxxxxxxxxxxxxxxxxxxxxxxxxxxxxx

The final example will model a mixing process, such as magma mixing. In Section 5.2.2.5 we examined the hypothesis that all intermediate rocks in Table 5.3 were formed by mixing of magmas B and R. Let us examine this hypothesis further by testing whether the andesite A can be a hybrid rock formed by the mixing of basalt B and rhyolite R. Graphically, this can be visualized by drawing a straight line between points B and R on Figure 5.19 and observing the degree to which point A deviates from this line. If point A on each graph falls sufficiently close to line B-R, after taking into account the analytical error on compositions of A, then the hypothesis is reasonable. This same exercise can be accomplished numerically as follows. N is calculated on the basis of the differentiation index SiO_2:

$$N = (60.1 - 71.5) / (50.2 - 71.5) = 0.54$$

TiO_2: $Z = (.54 \times 1.1) + (1 - .54) \times 0.3 = 0.73$
Al_2O_3: $Z = (.54 \times 14.9) + (1 - .54) \times 14.1 = 14.5$

Again no need exists to continue with the remainder of the oxides. Although the value for TiO_2 agrees well with the concentration of TiO_2 in the andesite A, the value for Al_2O_3 does not. Assuming a conservative 4 percent as the analytical error on Al_2O_3, 14.5 falls outside the range 15.5-16.7. This is no surprise. As stated in Section 5.2.2.3, the curved trends for some oxides on Figure 5.19 preclude magma mixing of rhyolite and basalt to form any of the intermediate rock types. A useful exercise is to confirm graphically the calculations in the above three examples.

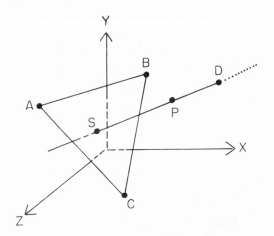

Figure 5.30 Schematics for fractional crystallization of three minerals *(A-C)* from a parent magma *(P)* to produce a daughter *(D)*. The solution is the point of intersection *(S)*. Elements or oxides *X-Z* can be in mol or weight percent.

5.2.3.2 Modeling of Fractional Crystallization

In Section 5.2.2.3 we applied the *principle of opposition* to fractional crystallization. This technique lends itself to a graphical solution much more readily than it does a numerical solution. A far more straightforward numerical approach is to visualize a three-dimensional, orthogonal graph in which each of of the axes is an element (X,Y, and Z; Fig. 5.30).

Compositions of the three minerals in the proposed extract assemblage form a triangle within the volume of the three-dimensional graph (A, B, and C; Fig. 5.30). The parent magma is represented on Figure 5.30 as point P, and the daughter composition is at point D. The liquid line of descent P-D can be extrapolated to point S, where it intersects the extract triangle A-B-C. The position of point S relative to positions of points A, B, and C yields percentages of minerals in the extract assemblage. If point S coincides with point A, B, or C, only one mineral was involved in the fractionation. If point S intersects one of the three sidelines of the triangle, then two minerals were involved. Moreover, because point S falls on a straight line extrapolated from D through P (a lever), the position of point S relative to D and P determines percent fractionation. Only three elements are required for a solution, so several three-element groups can be used as a check. A BASIC program to perform these calculations is included in Appendix B.

TABLE 5.7 Extraction Assemblage for Parent B and Daughter BA

Elements	% Plag.	% Cpx.	% Oliv.	% Fract.
Si-Al-FM	31	39	30	40
Si-Al-Ca	33	32	35	38
Si-Al-NK	31	33	36	38
Al-FM-CA	32	34	34	36
Al-FM-NK	31	36	32	38
FM-Ca-NK	40	28	32	40
$\bar{x}$	33	34	33	38
s	3.5	3.7	2.2	1.7

FM = Fe + Mg
NK = Na + K

Let us now consider some complications in this approach, the first being the effect of trapped liquid on fractional crystallization. Point S on Figure 5.30 represents the composition of the extraction assemblage, but does not take into account any trapped liquid in the "pore space" of these minerals. The composition of this trapped liquid must lie along the line P-D (liquid line of descent), so trapped liquid will not affect the solution for percentages of the three minerals in the extraction assemblage. It will, however, affect the percent fractionation, because the composition of the entire extraction assemblage, including trapped liquid and crystals, will lie somewhere between point S and point P. The ratio P D / S D, which is the algorithm in the BASIC program (Appendix B), will result in a percent fractionation that is too low. Hence percent fractionation as calculated by the program will always be a minimum value. If a reliable estimate of trapped liquid is known, percent fractionation can be easily corrected.

Another complication relates to the presence of primocrysts in daughter magma D (Fig. 5.30). Let us assume that daughter D contains some primocrysts that were members of the extraction assemblage that did not "escape." If relative amounts of these primocrysts are the same in daughter D as in the extraction assemblage, then the calculated percentages of minerals in the assemblage will not be affected by the presence of primocrysts in the daughter. Percent fractionation will be affected, however, because the composition of the daughter melt, as represented by the groundmass composition of D, must fall along a linear extrapolation of line P-D (represented by the dotted line in Fig. 5.30). As a result, the presence of primocrysts will also cause the percent fractionation to be too low. Hence, percent fractionation can always be considered a minimum value; this is because presence of trapped liquid in the extraction assemblage or primocrysts in the "daughter" will have the same effect.

The fact that percent fractionation will always be a minimum value may not be considered a serious problem, but *selective fractionation,* causing the relative amounts of primocrysts to be different from relative amounts of minerals actually being extracted, is of more concern. This process can occur because one mineral in the extraction assemblage is more easily extracted than another. For example, during gravity settling olivine, owing to its greater density, might be more effectively extracted than plagioclase. A result might be for daughter D to contain primarily plagioclase primocrysts, while the extraction assemblage contains a much higher percentage of olivine. As a result, percentages of minerals in any extraction assemblage calculated by the BASIC program would not be accurate. Little can be done about this problem except to avoid analyzing obviously phenocrystic rocks. If line P-D is truly a liquid line of descent (i.e., all rocks whose compositions fall along it are aphyric), then selective fractionation should pose no major problem.

As an example of the use of the BASIC program, the same data set used to illustrate the principle of opposition will also be used for this exercise. Again we will assume that basalt B (Table 5.3) is the parent and basaltic andesite BA is the daughter. Minerals involved in the extraction assemblage are plagioclase, clinopyroxene (augite), and olivine. Data input into the program are in mol percent, with Na and K summed, as are Fe and Mg. The reasons for this are explained in Section 5.2.2.3, but recall that the data can also be input in weight-percent oxide form. Results are presented in Table 5.7 and generally agree with those on Figure 5.22, as they must if the two procedures are carried out correctly. Six groups of three elements were used, then a mean and standard deviation were calculated for each mineral and for percent fractionation. The error is greatest for percent clinopyroxene and smallest for percent fractionation (Table 5.7). The program can also be used for an accumulation model (Appendix B).

5.2.3.3 An Independent Check

The process modeled is closed-system fractional crystallization, so an independent check of the results can be made by utilizing Rayleigh fractional crystallization (RFC). Although TiO_2 is not a trace element in these rocks, it is a minor element (Table 5.3). As no trace-element analyses are available, we did not include TiO_2 in the analysis summarized in Table 5.7 and will use it as a check. Equation 5.5 can be rewritten in the following manner:

$$D = [\ln (C_L/C_O) / \ln F] + 1 \qquad (5.24)$$

Terms are as defined for equation 5.5 (Section 5.2.1.2). A negative value for D means that the solution is impossible. Recall that a limiting factor for C_L/C_O, assuming perfect incompatibility, is $1/F$. Assuming 38 percent fractionation from Table 5.7, taking TiO_2 concentrations for B and BA from Table 5.3, and solving equation 5.24:

$$D = [\ln (0.8/1.1) / \ln 0.62] + 1 = 1.67$$

Assume that Ti is perfectly incompatible in olivine and plagioclase. Using a Ti partition coefficient for clinopyroxene of 0.8 (Henderson, 1982) and percentages of the three minerals in the extract from averages in Table 5.7:

$$D = (0.33 \times 0) + (0.34 \times 0.8) + (0.33 \times 0) = 0.27$$

Because no trapped liquid $(K_D = 1.0)$ was included in the calculation, D is actually too low. If, for example, the extract contained 30 percent trapped liquid, then:

$$D = (0.27 \times 0.70) + (1.0 \times 0.30) = 0.49$$

This calculation suggests that Ti should be incompatible, whereas $D = 1.67$ suggests the opposite. Furthermore, the decrease in TiO_2 with increasing fractionation (Fig. 5.19) indicates that Ti is compatible.

How do we resolve these two conflicting results? One way is to question whether we have chosen the correct K_D values. Certainly K_Ds for plagioclase and olivine are not exactly zero, but they are very small numbers. Although K_D for augite may be too low, in all probability it cannot be raised enough to account for this discrepancy. D can be raised by increasing the percent trapped liquid, but not sufficiently. Another way is to recall the argument that percent fractionation will always be a minimum value. If we raise the percent fractionation to 52 percent, based on the arguments for the example of fractional crystallization in Section 5.2.3.1, $F = 0.48$ and D is only lowered to 1.43.

The only reasonable conclusion is that another phase must be involved in the extraction, very likely a high-Ti phase such as ilmenite ($FeTiO_3$) or titanomagnetite (a solid solution between Fe_3O_4 and Fe_2TiO_4). Unfortunately, the program in Appendix B cannot accomodate a fourth mineral. A more sophisticated program (e.g., Bryan and others, 1969) is required for that. The lesson to be learned from this exercise is that in order for a hypothesis to have merit, *all* the data must fit, or a good reason must exist as to why they do not.

5.2.4 Igneous Rock Series Revisited

The importance of SiO_2 saturation is stressed in Section 1.3.3.1 as a means to distinguish between the two principal rock series, alkaline and subalkaline. Most alkaline rocks are *ne* normative, whereas most subalkaline rocks are *hy* normative. These fundamental differences in SiO_2 saturation are related to, among other factors, a low-pressure thermal divide (see Section 3.2.5.2). At relatively low pressures compositions of alkaline magmas tend to evolve away from this divide toward more SiO_2-deficient compositions, whereas subalkaline melts fractionate toward more SiO_2-enriched compositions. Within the subalkaline suite there are two main groups: tholeiitic and calcalkaline. Confusion exists, however, primarily because there is no general agreement on how to define a particular series on the basis of chemical or any other criteria. Three different sets of criteria exist, for example, to distinguish between calcalkaline and tholeiitic rocks. If the following discussion occasionally seems confusing, a good reason exists.

Oftentimes a classification scheme was developed for rocks in a particular region, then later applied (without the original author's intent) on a worldwide basis. This worldwide application of a method initially developed locally can produce results that are less than satisfactory. Although the principles and general relationships do not change, the precise location of a boundary curve on a variation diagram may change slightly from one area to another. Only when a worldwide data base is used can a generally applicable set of criteria be developed. Some examples of locally derived versus worldwide boundary curves are cited in the following discussion. A summary of chemical criteria as related to igneous rock series can be found in Ragland and Rogers (1984).

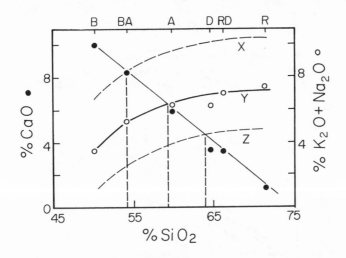

Figure 5.31 Schematics of a classification for igneous rock series using the method of Peacock (1931). Data from Table 5.3 are plotted. All data are in weight percent. *Open circles* are for Na$_2$O + K$_2$O (curve *Y)* and *closed circles* are for CaO. Curves *X* and *Z* are discussed in text.

A classification of basaltic rocks based on SiO$_2$ saturation is discussed in Section 1.3.3.2, while granitoids are classified according to Al$_2$O$_3$ saturation in Section 1.3.4.1. Other dominantly chemical schemes for subdividing individual rock types exist as well, such as I-type and S-type granitoids (Chappell and White, 1974), or magnetite-series and ilmenite-series granitoids (Ishihara, 1977). I-type granitoids are thought to be partial melts of igneous source rocks, whereas S-types are from sedimentary (or metasedimentary) source rocks. Interestingly, Al$_2$O$_3$ saturation is involved in the I-type versus S-type classification; all S-type granitoids are peraluminous, but most I-types are not. Magnetite-series granitoids, relative to those from the ilmenite series, are believed to form under relatively oxidizing conditions. This section, however, concentrates on rock series that contain rocks of diverse compositions.

5.2.4.1 Importance of Alkalies

One of the most important early contributions to the recognition and classi-fication of igneous rock series was that of Peacock (1931). His classification is still occasionally used today. Two of his terms, *alkalic* and *calc-alkalic,* are still widely used, although frequently in a different context and spelled differently (alkaline and calcalkaline). Two superimposed Harker plots on the same graph form the basis for his classification scheme: SiO_2 versus $Na_2O + K_2O$ and then against CaO. The critical value on this plot is the SiO_2 value at the point where best-fit curves through the two trends intersect. Four rock series are defined by this weight percent SiO_2 value:

alkalic	<51
alkali-calcic	51-56
calc-alkalic	56-61
calcic	>61

According to this scheme, the rock series of Table 5.3 and Figure 5.19 is calc-alkalic because the SiO_2 value at the point of intersection between the best-fit curve Y for $Na_2O + K_2O$ and the CaO curve is about 59.5 (Fig. 5.31).

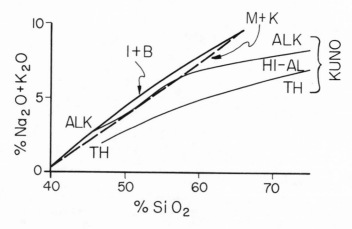

Fig 5.32 Alkali-silica plot showing fields for tholeiitic *(TH),* high-alumina *(HI-AL),* and alkaline *(ALK)* rock series according to various workers. Macdonald and Katsura (1964; *M+K)* and Irvine and Baragar (1971; *I+B)* divided the field into two areas, alkaline and tholeiitic, whereas Kuno (1968) proposed three. All data in weight percent.

The point of intersection of these two best-fit curves, from which the critical SiO_2 value is derived, is related more to changes in the alkalies than CaO. This is because the CaO curves of most rock series are generally similar, while the alkali contents, and thus curves, can be quite high in some series and low in others (Fig. 5.31). The higher hypothetical alkali curve X intersects the relatively stationary CaO curve at lower values of SiO_2 than does the lower alkali curve Z. Essentially, suites with the lowest overall alkali content are calcic and those with the highest are alkalic. The suite with alkali curve X would be alkali-calcic because the SiO_2 value at the point of intersection is about 54.3. Similar reasoning indicates that hypothetical suite Z is calcic.

Because alkalies are more important in classifying igneous rock suites than is CaO, one might use only the plot SiO_2 against $K_2O + Na_2O$. A similar plot is used to identify specific volcanic rock types, as well as rock series (Zanettin, 1984; Le Bas and others, 1986). Kuno (1968) was the first petrologist to use such a plot extensively for volcanic rocks from a convergent plate margin. He classified basaltic rocks associated with subduction in Japan, Manchuria, and Korea into three suites: *tholeiitic, high-alumina,* and *alkaline* (Fig. 5.32). Most petrologists now refer refer to Kuno's high-alumina basalts as calcalkaline, although he used the terms differently, as will be discussed in the next section. The term *high-alumina basalt* is still retained, however (e.g., Myers, 1988). The SiO_2 values on Figure 5.32 extend beyond 75 percent, so these three types of basalts presumably differentiated to form derivative magmas up to rhyolite in composition.   In fact, at the present-day level of exposure, andesites are generally the most common volcanic rocks, particularly among the calcalkaline rocks, at a convergent plate boundary such as an island arc. It is not clear whether these calcalkaline andesites are derivative melts from fractional crystallization of high-alumina basalts, or are primary melts.

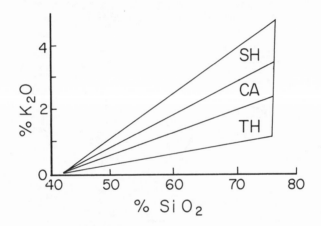

Figure 5.33 Classification scheme for igneous rock series modified after Gill (1970). *SH,* shoshonitic series; *CA,* calcalkaline series; *TH,* tholeiitic series. All data in weight percent.

Macdonald and Katsura (1964) used a similar alklali-silica plot to classify basaltic rocks in Hawaii, but they proposed only two suites: *tholeiitic* and *alkalic* (dashed line M+K, Fig. 5.32). The boundary between Kuno's high-alumina and alkaline basalts is quite close to the linear boundary between Macdonald and Karsura's tholeiitic and alkaline basalts. Hence, mafic rocks of a certain composition would be classified as tholeiitic by Macdonald and Katsura's method, but high-alumina by Kuno's. Which of these two methods is correct? If one is working on subduction-related basaltic rocks associated with a convergent plate margin, then Kuno's scheme is preferable. If the mafic rocks are related to "hot-spots" or divergent plate margins, Macdonald and Katsura's method is probably more applicable. Both classification schemes, however, are limited because they were not based on worldwide data.

A similar plot for volcanic rocks was compiled by Irvine and Baragar (1971) and is based on data from around the world. Their dividing curve between alkaline and tholeiitic rocks, relative to that of Macdonald and Katsura, is raised to higher alkali values and is concave down rather than linear (curve I+B, Fig. 5.32). Some rocks classified as alkaline by Macdonald and Katsura are tholeiitic by Irvine and Baragar. It is recommended that Irvine and Baragar's dividing curve be used.

Another method by which tholeiitic and alkaline basalts are distinguished is by the use of norms in conjunction with Yoder and Tilley's basalt tetrahedron. In essence, if a basalt is *hy* normative, it is tholeiitic, but if it is *ne* normative, it is an alkaline basalt. Some basalts are tholeiitic by the normative criterion, but an alkaline basalt if the alkali-silica plot is used (see Section 1.3.3.1). The opposite situation also exists. Again we can ask the question as to which scheme is preferable, and again no definite answer exists. Because the normative method can be more closely tied in with experimental petrology, some petrologists prefer it. Others prefer a simpler alkali-silica plot.

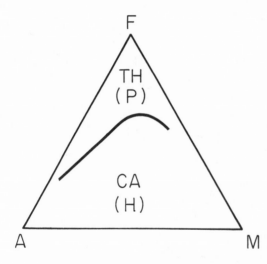

Figure 5.34 AFM diagram showing fields for two rock series: tholeiitic or pigeonitic *(TH, P)* and calcalkaline or hypersthenic *(CA, H)*. *A*, $Na_2O + K_2O$; *M*, MgO; *F*, $FeO + Fe_2O_3$. All data in weight percent.

Further research indicated that K_2O is more important than Na_2O in classifying rock series, so petrologists (e.g., Gill, 1970; Jakes and Gill, 1970) began using only SiO_2-K_2O plots to characterize subduction-related suites. Apparently K_2O is a better discriminator than Na_2O because K is more incompatible than Na, unless a mica or potash feldspar is present. A K_2O-SiO_2 plot has been used to classify individual rock types as well (Peccerillo and Taylor, 1976). Because K_2O is one of a large number of elements that is presumed to be generally incompatible in the lower crust and upper mantle, any incompatible element could be used. For instance, Gunn and others (1974) used K, the K/Na ratio, and many other incompatible trace elements to characterize the volcanic rocks on Martinique. Because K_2O is routinely analyzed as a part of any major-oxide analysis, it is convenient to use K_2O. Which is a better discriminator, K_2O or total alkalies? No clear-cut answer exists to this question and it is largely a matter of personal choice. Perhaps K_2O alone is preferable because K is normally more incompatible than Na and in general incompatible elements seem to discriminate better than other elements.

Based on K_2O-SiO_2 plots, a new suite was added to the suites already named. This suite is variously referred to as the *high-K calcalkaline, shoshonitic,* or *Andean* series (Fig. 5.33). It is intermediate between the normal calcalkaline and alkaline suites, but as the name above implies, can be considered a high-K subgroup within the calcalkaline series. Not all petrologists look upon these three terms as synonomous, but we will do so.

5.2.4.2 Variations on the Fe/Mg Ratio

Kuno (1968), who concluded that the alkali-silica relationship and Al_2O_3 content could be used to distinguish between three different subduction-related types of basalts, also found that the Fe/Mg ratio differs between calcalkaline and tholeiitic rock series. Herein lies a source of some of the confusion that exists today. Many modern petrologists equate the calcalkaline and high-alumina series, but Kuno did not do so. He specifically stated that high-alumina basalts could fractionate to form *either* tholeiitic or calcalkaline rock suites. He used an AFM diagram to distinguish between the tholeiitic and calcalkaline series, although he originally defined them as the *pigeonitic* and *hypersthenic* series, respectively, named after the types of pyroxene phenocrysts present. We now consider the pigeonitic series to be tholeiitic and the hypersthenic as calcalkaline.

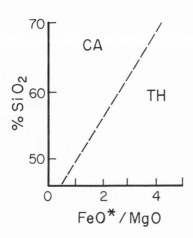

Figure 5.35 Classification of rock series into calcalkaline *(CA)* and tholeiitic *(TH)* after Miyashiro (1974). All data in weight percent.

This boundary separating compositional fields of these two series on an AFM diagram is referred to as the *critical iron enrichment curve* (Fig. 5.34). It approximately parallels a typical *Fe-enrichment* or tholeiitic differentiation trend (Fig. 5.18b). Schwarzer and Rogers (1974), however, have shown that alkaline basalts and their derivative magmas overlap this critical iron enrichment curve, thus an AFM plot is of little use in delineating the alkaline and subalkaline series.

Kuno worked primarily in the Far East, where, for example, a sample that he classified by one criterion as tholeiitic was generally tholeiitic by any other criterion. Because he set up the original criteria, this agreement is no surprise. Keep in mind that he was able to define his own boundaries on variation diagrams separating compositional fields for different suites; these boundaries were by necessity arbitrarily drawn. Had he initially worked in another part of the world, the boundaries very likely would have been at least slightly different. A good example of this was discussed in the previous section, where the boundary between the tholeiitic and alkaline fields set by Macdonald and Katsura (1964) for Hawaiian rocks was changed when a world-wide data base was used (Irvine and Baragar, 1971).

Workers such as Gill (1970) and Gunn and others (1974) relied primarily on the alkalies as a means of classifying rock series, whereas Kuno (1968) relied on the Fe/Mg ratio as determined by the critical iron enrichment curve on an AFM plot. Miyashiro (1974) suggested returning to some measure of Fe/Mg to delineate calcalkaline from tholeiitic rocks. His scheme for doing so is shown in Figure 5.35, a plot of FeO*/MgO against SiO_2. A classification of rocks by this method frequently does not agree with either the alkali-silica or the K_2O-SiO_2 plot. Furthermore, a classification of basalts, supposedly the parent rocks to many of these rock series, based on norms does not necessarily agree with one based on the alkali-silica plot or with Miyashiro's method. Must a rock fit all these various criteria to be classified as being in a particular series? If this were true, a considerable number of rocks could never be placed in any series. The next section will examine an approach that attempts to bring some order out of this chaos.

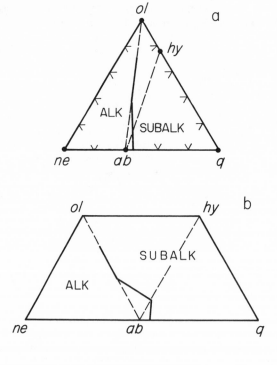

Figure 5.36 Classification of rock series into alkaline and subalkaline: *a.* after Irvine and Baragar (1971); *b.* after Figure 1.13, section 1.3.2.4. Refer also to Figure 1.12a. All data are molecular norms.

5.2.4.3 An Integrated Approach

The previous discussion leaves one with the feeling that a single, integrated scheme, based on a world-wide data base, is needed to classify igneous rock series. Irvine and Baragar (1971) have done this for volcanic rocks, which should be equally applicable for hypabyssal rocks. Although this method has also been used to characterize plutonic series, doing so is more dangerous. Subtle compositional differences do exist between plutonic rocks and their supposed volcanic equivalents, although these differences may not be statistically significant. No scheme comparable with that of Irvine and Baragar exists for all plutonic rocks.

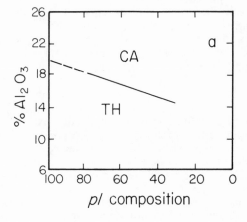

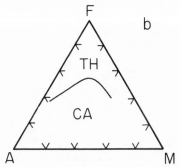

Figure 5.37 Two schemes for subdividing subalkaline rock series into calcalkaline and tholeiitic (after Irvine and Baragar, 1971). Use (a) for basaltic rocks and (b) for rocks more felsic than basalts. Note that the dividing curve between the calcalkaline and tholeiitic fields on (b) is different from that on Figure 5.34. All data are in weight percent, except pl, which is mol percent normative plagioclase composition, as defined in text.

In this section we will adopt a simplified, slightly modified version of Irvine and Baragars's classification. It will include only the larger subgroups. For a complete discussion, refer to their 1971 paper. Three major groups exist:

    I. Subalkaline rocks
        A. Tholeiitic basalt series
        B. Calcalkaline (spelled *calc-alkali* by
           Irvine and Baragar) rock series
    II. Alkaline rocks
        A. Alkali-olivine basalt series
        B. Extremely SiO_2-undersaturated rocks such as
           nephelinite, leucitite, and analcitite
    III. Peralkaline rocks, such as *pantellerite* and *comendite*

Subalkaline rocks are most abundant, and peralkaline are least abundant. Irvine and Baragar further divided Group IIA into a sodic and potassic series, but we will not examine this subdivision. The procedure will be presented in steps, which should be carried out in the order indicated:

1. The rock compositions must first be adjusted for secondary alteration. If a rock clearly has an anomalous chemical composition that appears to be related to alteration, it should be rejected. In general, altered rocks are higher in H_2O and CO_2, as well as being more oxidized with respect to Fe, than unaltered rocks. Of the major oxides, the alkalies are most commonly affected by alteration. Recalculation to 100 percent without H_2O+, H_2O-, and CO_2 is advisable. A problem remains, however, about how to deal with the iron oxidation ratio. This ratio can have a profound effect on a norm and thus on this classification scheme (see Section 1.3.1.4). Irvine and Baragar used the following equation for the upper limit on Fe_2O_3:

    $$\% \ Fe_2O_3 = \% \ TiO_2 + 1.5$$

    For any rock that contains more than this amount of Fe_2O_3, the "excess" is converted to FeO. If FeO and Fe_2O_3 are not determined separately, this equation can be used to partition Fe between these two oxides.

2. Molecular norms must be calculated. Irvine and Baragar's classification scheme depends on molecular rather than CIPW norms. The remaining variables are all weight percent oxides.

3. A peralkaline rock is alumina undersaturated and generally has *ac* in the norm (see Section 1.3.4.2). If the rock contains normative *ac,* it almost invariably can be classified as peralkaline. If the rock is not peralkaline, go to step 4.

4. Use Figure 5.36a to determine if the rock is subalkaline or alkaline. Because there are three scalene triangles on Figure 5.36a, refer to Section 1.3.2.3. This figure is essentially a projection from the *di* apex of the basalt tetrahedron onto the triangle *ne-q-ol*, although it is used herein to classify many volcanic rocks other than basalts. A variation on this figure has been used earlier (Fig. 5.10) to explain various theories concerning the origin of basaltic rocks. Alternatively, use Figure 5.36b (which avoids the problem of scalene triangles) by recalculating the appropriate three normative minerals to 100 percent. Figure 5.36a is quite similar to Figure 1.12a; Figures 5.36b and 1.13 are also very similar. Irvine and Baragar state that the alkali-silica plot (curve I+B, Fig. 5.32) can also be used to distinguish between subalkaline and alkaline rocks, although the *ne-q-ol* plot is probably more reliable. If the rock is subalkaline, go to step 5; if alkaline, step 6.

5. If the rocks are basaltic, use Figure 5.37a. Figure 5.37b can be used for basaltic rocks, but Figure 5.37a is preferred. Normative plagioclase composition is defined by Irvine and Baragar as:

    $$100 \ x \ an \ / \ [an + ab + (5/3)ne]$$

    This is a measure of the degree of fractionation. In general, calcalkaline rocks are more alumina-rich than tholeiitic rocks, in accordance with the belief of some petrologists that Kuno's high-alumina basalts can be considered as calcalkaline. If the rocks are more felsic than basalt, use Figure 5.37b. This is the familiar AFM diagram, with the boundary between the TH and CA fields slightly different from the boundary proposed by Kuno (Fig. 5.34). Because Irvine and Baragar's boundary was established from a worldwide data base, it is preferable.

6.   If the rock does not contain abundant nepheline, leucite, or analcite, and does not plot in the most SiO_2-undersaturated part of the *ne-ab-ol* field on Figure 5.36a or 5.36b (near the *ne-ol* sideline), consider the rock to be in the alkali-olivine basalt series. Basanites (see Fig. 1.14) are in the alkali-olivine basalt series. If further subdivision into a potassic or sodic series is desired, consult Irvine and Baragar (1971). In addition, see Section 1.3.3.3 for an alternative way of considering alkaline rocks based on the alkaline rock tetrahedron.

TABLE 5.8 Classification of Basalt in Table 1.1

Figure	Source	Classification
5.32	Kuno	high-alumina
5.32	Irvine and Baragar	tholeiitic
5.32	Macdonald and Katsura	tholeiitic
5.33	Gill	tholeiitic
5.34	Kuno	calcalkaline
5.35	Miyashiro	tholeiitic
5.36	Irvine and Baragar	subalkaline
5.37a	Irvine and Baragar	tholeiitic
5.37b	Irvine and Baragar	tholeiitic

An interesting exercise is to plot the basaltic composition in Table 1.1 on all discrimination diagrams (Figs. 5.32 to 5.37) to determine the degree of agreement. Results are shown in Table 5.8. In the strictest sense, Figure 5.37a, rather than 5.37b, is to be used for basaltic rocks. However, Kuno's AMF diagram (Fig. 5.34) classifies the basalt as calcalkaline, in disagreement with most other methods, so it is informative to plot the composition on Irvine and Baragar's AMF diagram (Fig. 5.37b). All plots agree (including Fig. 5.37b) that this basalt is in the tholeiitic series except the alkali-silica and AMF diagrams of Kuno. One would expect a high-alumina basalt to be calcalkaline, so these two discriminators agree with one another, although they are in conflict with those that classify the basalt as tholeiitic. Fortunately, the basalt is also classified as subalkaline (Fig. 5.36). If classifications were made democratically, the rock is tholeiitic by a 6-2 vote.

5.3 IGNEOUS ROCKS AND PLATE TECTONICS

The study of igneous petrology is rewarding for its own sake, but it can be even more rewarding when it interacts with other disciplines in geology. For example, it impacts on economic geology through igneous-related ore deposits and on environmental geology through our concern about volcanic hazards. Tectonics, however, is probably the area in which the greatest opportunity exists for interaction. A common modus operandi has been first to search for chemical, mineralogical, or physical traits in modern igneous rocks characteristic of a particular tectonic environment. The next step is to invoke the concept of *uniformitarianism* by applying these traits to some very old igneous rocks and then speculating about the ancient tectonic setting in which they occurred. Whether uniformitarianism can always be invoked is a debatable point. The older the rocks, the greater are the chances that it cannot. If the igneous rocks in question are Cenozoic in age, probably the assumption of uniformitarianism is reasonable. If they are Archean, the assumption may not be valid.

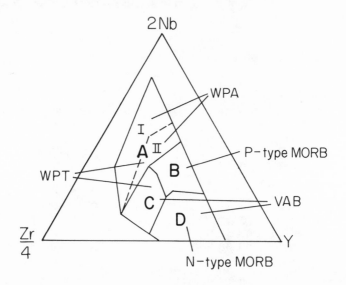

Figure 5.38 A diagram based on HFS, supposedly immobile trace elements to discriminate different tectonic settings (modified after Meschede, 1986). *AI, WPA* (within-plate alkali basalts); *AII, WPA* and *WPT* (within-plate tholeiites); *B,* P-type MORB (plume-related MORB); *C, WPT* and *VAB* (volcanic-arc basalts); *D, VAB* and N-type MORB (normal MORB).

5.3.1 Approaches

Three approaches are widely used: (1) compile as much descriptive information as possible (e.g., chemical, mineralogical, textural, structural, stratigraphic) about a particular rock type in different modern tectonic settings and then attempt to find a set of criteria among all this information that hopefully can be applied to its ancient equivalents; (2) assume that if some modern igneous rock situated in a particular environment has some quantitative signature, such as a unique trace-element or major-element distribution, its ancient equivalent with that same signature occurred in the same tectonic setting; and (3) make the same assumption as (2), but work with entire rock series rather than one rock type. For instance, Rogers (1982) has used approach (1), dealing primarily with basaltic rocks, but he also makes reference to particular rock series.

5.3.1.1 Basalts and High Field Strength Elements

The use of *immobile* elements is perhaps the best example of approach (2). Because many ancient volcanic rocks are altered, or even metamorphosed, their concentrations for many elements are probably different now than they were when the rock was formed. An obvious example is Fe_2O_3, which because of oxidation almost certainly is different in concentration now than it was originally. Fe_2O_3 should never be used as an indicator of ancient tectonic environments. Some elements, however, appear to be relatively immune to secondary alteration and even low-rank metamorphism, the *immobile elements*. During our earlier discussion of trace elements, the point was made that incompatible elements are generally more useful than compatible elements in modeling igneous processes. Many of these immobile elements are incompatible minor and trace elements, particularly high field strength (HFS) elements (Section 5.2.1.1). These include Ti, P, Zr, Hf, Ta, Th, Nb, Y, and the REE.

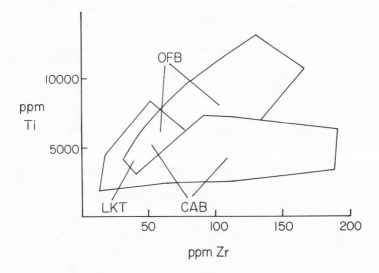

Figure 5.39 A scattergram based on two HFS, supposedly immobile trace elements used to differentiate between different tectonic settings (modified after Pearce and Cann, 1973). *CAB,* calcalkaline basalts; *OFB,* ocean-floor basalts (MORB); *LKT,* low-K tholeiites (arc tholeiites). Data are in ppm.

An early effort to use an incompatible element as a tectonic indicator was by Chayes (1965), who used TiO_2 to discriminate between intra-oceanic and continental-margin basalts. Intraoceanic basalts generally contain more TiO_2 than do those from the continental margin. Other studies, primarily based on basalts, include Floyd and Winchester (1978), Pearce and Norry (1979), Wood (1980), and Meschede (1986); the classic paper is by Pearce and Cann (1973). Another variation is by Mullen (1983), who used the minor elements MnO, TiO_2, and P_2O_5. Pearce and others (1984) used a similar technique for granitic rocks. Most of these papers employ either bivariate scattergrams or triangular diagrams to discriminate different tectonic environments. An example of each is shown in Figures 5.38 and 5.39. When applying these diagrams to ancient rocks, several limitations exist:

1.   The assumption of uniformitarianism may not hold. A Paleozoic basalt in a certain tectonic setting, for example, is assumed to have had the the same chemical composition as its modern counterpart. This may not be true.
2.   These supposedly immobile elements may not be immobile (e.g., Bartley, 1986). For relatively moderate levels of alteration and metamorphism apparently involving minimal metasomatism, immobility is probably a valid assumption. If the rock has been metasomatized, some petrographic evidence should be apparent (e.g., replacement textures, microveins of foreign material). Under these conditions even "immobile" elements may be mobile.
3.   Many of these chemical discriminators cannot subdivide, for instance, within-plate basalts. A within-plate setting, however, could be an oceanic island, a plume (hot-spot), or a continental volcanic center of uncertain origin. Hence some diagrams cannot discriminate on a sufficiently specific level.
4.   A tendency exists to equate rock series with tectonic settings. For example, if an immobile-element plot characterizes a suite as being calcalkaline, we tend to equate this with a convergent plate boundary. A good chance exists that this is true, but it is not an absolute certainty.

xxxxxxxxxxxxxxxxxxxxxxxxxxxxxxxxxxxxxxxxxxxxxxxxxxxxxxxxxxxxxxxxxxxx

When using these diagrams to identify tectonic settings from immobile element contents, if at all possible check any conclusions drawn from triangular plots with one or more bivariate scattergrams. Aside from the closure problem having a greater effect on triangular than on bivariate plots (Section 5.2.2.1), another problem exists. Most of the elements used on these diagrams are incompatible and their absolute concentrations are as important or even more important than their ratios among each other as discriminators. A simple example is illustrative. Composition of a rock containing 100 ppm each of X, Y, and Z will plot as a point exactly in the middle of an X-Y-Z triangular plot, but so will the composition of a rock containing 500 ppm, 1000 ppm, or any other equal concentration of all three elements. Bivariate scattergrams are not afflicted with this problem. Because absolute concentrations are plotted on a bivariate graph, if two rocks have different concentrations for X and Y, they will plot as different points. If the best discriminating parameters are X/(X+Y+Z), Y/(X+Y+Z), and Z/(X+Y+Z), then the triangular plot is ideal.

xxxxxxxxxxxxxxxxxxxxxxxxxxxxxxxxxxxxxxxxxxxxxxxxxxxxxxxxxxxxxxxxxxxx

5.3.1.2 Rock Series and Tectonic Environments

Approach (3) requires the consideration of entire igneous rock series. We will consider two examples of the manner in which igneous rock series are related to tectonic setting. It has long been known that certain rock series tend to occur in certain tectonic settings but not in others, *although no one-to-one correlation exists*. Kuno's (1968) classic recognition of the phenomenon later to become known as the *potash-depth relationship* is an example. In general, K_2O contents progressively increase from the tholeiitic through the high-alumina to the alkaline series in volcanoes from a island arc environment, although many exceptions exist. In this case K_2O is simply the most conveniently analyzed incompatible element; any incompatible element will exhibit a similar trend. This ties in well with the three rock series based on the K_2O-SiO_2 plot developed by Gill (1970).

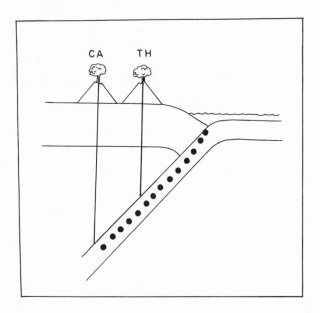

Figure 5.40 Schematic representation of subduction zone with tholeiitic volcano *(TH)* and, at greater depth, calcalkaline volcano *(CA)*. *Filled circles* represent earthquake focii in Benioff zone.

    Tholeiitic rocks tend to be on the outward, convex side of the island arc (nearest the trench).   High-alumina (calcalkaline, by some petrologists' definitions) rocks are more likely found on inward, concave side, farthest from the trench.   Continuing inward toward the continent, if high-K calcalkaline (shoshonitic; Gill, 1970) rocks are present, they will be closer to the continent, perhaps in the marginal sea between the island arc and the continent (as in a *back-arc basin*). Finally, in the marginal sea or even on the continent, alkaline rocks can dominate. Seldom do all four of these rock series develop in an island arc environment, but it is not unusual to find two or even three. Perhaps the most uncommon of the four series in a subduction setting is the alkaline series. Many rocks previously considered to be alkaline are now classified as shoshonitic.   A schematic depiction of a subduction zone with tholeiitic and calc-alkaline volcanos is shown in Figure 5.40.
    This systematic change in chemical compositions (particularly K_2O compositions) in volcanic rocks toward the continent has been correlated with depth to the top of the *Benioff zone,* as evidenced by seismic focii (e.g. Dickinson, 1975).   Presumably the Benioff zone marks the top of the down-

going lithospheric plate ("slab") in the subduction zone under the island arc. The term *potash-depth relationship* is derived from the approximate correlation between the K_2O content of lavas and depth to the Benioff zone. Reasons for this apparent correlation have engendered considerable speculation, particularly with regard to parental magma and source compositions, as well as the relative roles of partial melting, fractional crystallization, and contamination. Cawthorn (1977) provides some insight into these problems. Petrologists have used this relationship to estimate dip directions on ancient subduction zones, K_2O contents of rocks increasing in the direction of dip. Unfortunately, K_2O is one of the elements most susceptible to postmagmatic alteration.

The second example deals with igneous rock series and rift environments, whether they are oceanic (along midoceanic ridges) or continental. In continental rift systems *bimodal suites* are common, consisting of simultaneous eruptions of mafic and felsic lavas. Apparently two sources were partially melted to produce these suites: a mantle source for the basalts and a crustal source for the more felsic rocks (Section 5.1.2.2).

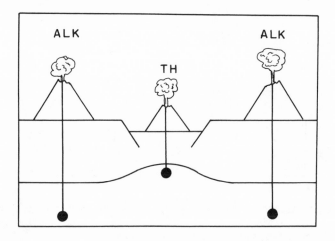

Figure 5.41 Schematic representation of a continental rift with a tholeiitic *(TH)* volcano in the rift and alkali-olivine basalt volcanos *(ALK)*, whose lavas are generated at deeper levels, on the flanks. Note crustal thinning beneath rift.

In many instances only basaltic rocks are found in continental rift environments; this is virtually always the case for oceanic rifts. The classic MORB is a tholeiitic basalt associated with rifting at the midoceanic ridges. However, oceanic-island basalts are more likely alkaline. Basalts erupting within continental rifts are generally tholeiitic, whereas those erupting on the flanks of the rift are relatively alkaline (e.g., Lipman and Mehnert, 1975). These relationships are shown schematically in Figure 5.41.

A reasonable explanation for these patterns is based on the belief that alkaline basalts (1) apparently form at greater depths than do tholeiitic basalts; and (2) represent smaller percentages of partial melting than do tholeiites. Partial melting will occur when the isotherm at a particular depth intersects the peridotite solidus. Because heat-flow values are generally higher within a rift system than some distance away, isotherms are nearer the surface under the rift system than on its flanks. The result seems to be for partial melting to occur at shallower depths under the rift than at a distance away, producing tholeiitic magmas in the rift zone and more alkaline magmas further away.

TABLE 5.9 Igneous Rock Series and Tectonic Settings

PLATE MARGIN			INTRAPLATE			
			Oceanic		Continental	
SETTING	conver-gent	diver-gent	mar-ginal sea	large ocean basin	rift	craton
SERIES	arc tholei-ite calc-alkaline shoshon-ite		low-K tholeiite (MORB)	tholeiite shoshonite (marginal sea only) alkaline bimodal (rift only)		alkaline ?

Modified after Condie (1976)

Only two examples are given of the manner in which igneous rock series and plate tectonics are related. In fact, a better correlation exists between igneous rock series and physicochemical conditions under which magmas are initially formed and subsequently modified, which may or may not have any one-to-one correspondence with any tectonic setting. For instance, Grove and Baker (1984) concluded that fractional crystallization of a basaltic magma at relatively low pressures could produce a tholeiitic trend, whereas fractionation of a similar magma at higher pressures could produce a calcalkaline trend. This is not to say that these two series could be formed no other way, but it is indicative of physicochemical control (see also Section 5.1.2.4). Presumably relatively low-pressure and high-pressure fractionation can occur in a variety of tectonic environments.

Nevertheless, Condie (1976) has provided a useful summary of the general correlation between various rock series and tectonic settings. Always keep in mind, however, that many exceptions exist. Table 7-2 (p. 147) from Condie has been modified and included as Table 5.9. A few points regarding

TABLE 5.10 Numerical Comparisons of Basalt Analyses

	a	b	c	d	e	f	g
SiO_2	49.2	3.23	49.3	0.007	51.5	0.152	48.7
TiO_2	1.9	1.03	2.4	0.513	1.2	0.007	1.29
Al_2O_3	15.8	2.13	14.6	0.274	16.3	0.006	16.6
FeO^*	10.7	2.33	11.4	0.148	10.4	0.009	10.1
MgO	6.6	2.11	7.4	0.080	5.9	0.090	6.63
CaO	10.0	1.46	10.6	0.001	9.8	0.083	10.7
Na_2O	2.7	0.76	2.2	0.180	2.5	0.044	2.83
K_2O	1.0	0.65	0.53	0.007	0.86	0.177	0.47
P_2O_5	0.33	0.25	0.26	0.013	0.21	0.000	0.20
χ^2				1.223		0.568	

	h	i	j	k	l	m
SiO_2	6.6	0.003	0.025	0.712	0.752	−0.155
TiO_2	54.2	0.485	1.160	−0.680	0.008	−0.592
Al_2O_3	13.5	−0.563	0.882	0.235	0.020	0.376
FeO^*	21.8	0.300	0.022	−0.129	0.080	0.153
MgO	32.0	0.379	0.143	−0.332	0.111	0.001
CaO	14.6	0.411	0.005	−0.137	0.379	0.479
Na_2O	28.1	−0.658	0.687	−0.263	0.188	0.171
K_2O	65.0	−0.723	0.008	−0.216	0.359	−0.815
P_2O_5	75.8	−0.280	0.058	−0.480	0.002	−0.520
SSD			2.990		1.899	

a: mean for 1996 basaltic rocks (Manson, 1967)
b: standard deviation for mean in column a
c: mean for 282 oceanic tholeiites (Manson, 1967)
$d = (g - c)^2/c$
e: mean for 946 continental tholeiites (Manson, 1967)
$f = (g - e)^2/e$
g: basalt analysis from Table 1.1
h: coefficient of variation in percent: $V = 100b / a$
i: Z-scores for column c: $Z = (c - a) / b$
j: deviations squared: $(m - i)^2$
k: Z-scores for column e: $Z = (e - a) / b$
l: deviations squared: $(m - k)^2$
m: Z-scores for column g: $Z = (g - a) / b$

Table 5.9 require some explanation. A question mark is in the column under *craton*. Although some cratonic volcanic centers contain alkaline rocks, occasionally tholeiitic or calcalkaline volcanic rocks are found on the craton that are not clearly associated with rifting or subduction, respectively.

The three types of tholeiites are listed in the table: (1) low-K tholeiites (MORB) found in large ocean basins and marginal seas, (2) arc tholeiites (also known as low-K tholeiites) generally confined to island-arc environments, and (3) continental tholeiites. For a given degree of fractionation, incompatible element contents (including K contents) are generally lowest in MORB and highest in continental basalts, although continental rift tholeiites can contain intermediate levels similar to those of arc tholeiites. Any shoshonites in a marginal sea will be there as a result of subduction associated with a nearby island arc.

5.3.2 Comparisons

One difficulty with using scattergrams of igneous-rock compositions to delineate tectonic environments is that only a few elements can be considered on any one diagram. It is possible to distinguish between rocks of different tectonic settings by simultaneously considering all the elements analyzed, rather than just a few. Sophisticated multivariate numerical methods, such as *discriminant function analysis* and *factor analysis* (Le Maitre, 1982), perform this function, but simpler techniques exist as well. Two examples are given in this section. They rely on transforming the raw concentration data into either the χ^2 statistical function (Sections 1.1.4.4 and 1.1.4.5) or Z-scores, which are explained later.

5.3.2.1 Comparisons using Chi-Squared Transformations

Suppose we have a major-element analysis for a basalt, such as the analysis in Table 1.1. Let us compare this basalt composition with those from the two mega-tectonic environments -- continents and oceans. Is this analysis more typical of an oceanic or a continental tholeiite? We can eliminate a comparison with alkali-olivine basalts because it has a substantial amount of *hy* in the norm (see Section 1.2.5.5).

We will first transform a major-oxide analysis to the χ^2 function to make this comparison (Table 5.10). The analysis for the basalt from Table 1.1 is in column g, and mean compositions for oceanic and continental tholeiites are in columns c and e, respectively. The equation for χ^2 is the summation of $(O-E)^2/E$, where O is the observed and E is the expected value. For this comparison the values in column g are observed, whereas those in columns c and e are "expected." Columns d and f provide the $(O-E)^2/E$ data, and the summations at the bottoms of these columns are the two χ^2 values. Because χ^2 for continental tholeiites is substantially lower than that for oceanic tholeiites, overall the analysis in column g is more similar to the mean for continental tholeiites.

The primary reason for the analysis in column g being more similar to continental tholeiites is TiO_2, whose deviation accounts for about 42 percent of χ^2 in the oceanic tholeiite. It has long been recognized that TiO_2 is substantially higher in oceanic than in continental tholeiites (e.g., Chayes, 1965). In general, χ^2 comparisons of this kind are dominated by the minor elements (or, if trace elements are included, both minor and trace elements). In part this is due to the fact that incompatible minor and trace elements are better discriminators, but it is also partly due to the calculation itself. Column a contains worldwide means for 1996 basaltic rocks; columns b and h respectively contain standard deviations and coefficients of variation for these means. The oxides with the three highest coefficients of variation *(Vs)* are the minor oxides TiO_2, K_2O, and P_2O_5. On average the term $(O-E)^2/E$ will be larger for elements with large overall *Vs*, so the minor elements generally tend to dominate χ^2. This is perfectly legitimate, and actually may be advantageous, but some petrologists prefer that the major and minor elements be weighed more equally.

5.3.2.2 Comparisons Using Z-Score Transformations

A technique to weigh major and minor elements more equally is to transform the analyses to *Z-scores,* also referred to as *standardized normal deviates* (only in statistics can a deviate be considered normal). The method is based on the assumption that all distributions are at least approximately normally distributed. If a distribution is clearly not normally distributed, it can be normalized by taking logs of the data (see Section 1.1.3.1), or some other appropriate mathematical conversion, before Z-scores are calculated. A Z-score is defined as:

$$Z = (X_i - \bar{x}) / s \tag{5.25}$$

where X_i is an individual value, $\bar{x}$ is assumed to be the overall mean (column a; Table 5.10), and s is the assumed overall standard deviation (column b). If Z-scores were calculated for all 1996 basaltic rocks whose analyses were used for the data in columns a and b, the arithmetic mean for all those Z-scores would be zero and the standard deviation would be 1.0. Units on Z are in standard deviations, so a mixed data set where values for some variables have been normalized and others have not can be transformed to Z-scores. Z-scores for columns c, e, and g are in columns i, k, and m, respectively. All the Z-scores in these columns are calculated from raw rather than normalized data.

For E in the χ^2 expression, when $X_E = \bar{x}$, $Z_E = 0$, and $(Z_O-Z_E)^2/Z_E = \infty$, thus the χ^2 transformation is not used for Z-scores. Only the numerator (sum of the squared deviations, SSD) is necessary:

$$SSD = \Sigma(Z_O - Z_E)^2 \tag{5.26}$$

The sum of the absolute deviations could be used as well. Squared deviations are given in columns j and l in Table 5.10 and SSD values are given at the bottom of these columns. Again, the relative SSD values indicate that the basalt in column g is more similar to continental than to oceanic tholeiites. As for χ^2, SSD is weighed heavily by TiO_2 for the comparison with the oceanic basalt (column j), but generally minor and major elements are weighed more equally for SSD than for χ^2. Examination of a large number of these comparisons confirms this observation.

As stated above, one technique is not necessarily more correct than the other. If minor elements are generally better discriminators, perhaps they should be emphasized and χ^2 used. On the other hand, a more equable weighting among the major and minor (plus trace elements, if included) elements may be desired. If so, Z-scores and SSD values should be used. The choice is left to the investigator.

5.3.3 Conclusion

This entire discussion of the correlation between tectonic settings and igneous rock series is only the briefest examination of a complex and fascinating problem. It is one of the most important aims of analytical petrology and is particularly interesting because it ties together, at least implicitly, many of the topics examined in this book. For instance, delineation of igneous rock series is indirectly dependent on thermal divides, which are best understood through experimental petrology. Trace-element and minor-element distributions are widely used in the recognition of tectonic settings. These distributions can be examined by means of crystal-melt partition coefficients, which are rooted, as is experimental petrology, in basic thermodynamic theory. Ultimately, analytical petrology depends on a good, sufficiently large database. These data can only be obtained through careful field and petrographic observations in concert with precise and accurate chemical analyses.

This book, however, should not end with the implication that the most important task facing us in the near future is gathering more, better chemical analyses. To the contrary, with a modern multielement technique in the hands of a good analyst, high quality data can be collected at such alarming rates that they can be assimilated only with great effort. In recent years we have tended to emphasize the collection of chemical analyses rather than, for example, physical properties of igneous rocks. This book is a reflection of that fact. It was written to collect between two covers many fundamental techniques in diverse areas necessary to interpret chemical data from igneous rocks. We are becoming increasingly aware, however, that we cannot understand the chemical nature of these rocks until more is known about their physical properties.

Unfortunately, to cover adequately even the most rudimentary aspects of all the topics discussed herein, something had to be omitted. The most glaring omission is a treatment of isotopes and their role in igneous processes. This in no way implies that they are of little significance. Isotopic ratios and incompatible elements are most useful in delineating source areas for magmas, as well as mixing of magmas from different sources. A suite of rocks may have quite diverse chemical compositions, but if their initial isotopic ratios are similar, a good chance exists that they are genetically related. In contrast, major-element chemistry of some rocks may be similar, but if their initial isotopic ratios are different, they probably are not related. Isotopic data are as useful in igneous petrology as are all other chemical data.

We also must examine closely some of our basic concepts about chemical compositions of igneous rocks. One final example is illustrative. The liquid line of descent is perhaps one of the most misused concepts in igneous petrology. For this reason the term *presumed* liquid line of descent is frequently used in this book. Only in aphyric glasses, or at the very least in fine-grained, aphyric rocks, should we have any confidence that the trend on a variation diagram represents a change in melt composition. The coarser grained the rocks, the greater is the chance that the trend is not a liquid line of descent, especially if it is linear. It could just as easily be a mixing line between a melt and accumulative crystals. Conversely, a magma chamber may have experienced crystal fractionation on a grand scale, but the larger, more euhedral crystals in any one phyric sample may be the result of equilibrium crystallization on a much smaller scale. If so, the bulk-rock composition of that phyric sample could truly represent a melt composition. Distinguishing between such a phyric sample and one that contains accumulative crystals may be difficult at best. Other examples could be cited, but this will suffice. Perhaps we should be as concerned about examining old tenets as we are about gathering new data.

APPENDIX A Calculation of a Molecular Norm

To calculate a molecular norm, follow steps 1-27 in the order given. An example has been worked following these steps in Section 1.2.5.5, along with conversion of a molecular norm to the equivalent of a CIPW norm. The following procedure assumes that weight percent oxides have already been converted to cation percent (see Section 1.2.3.3). Calculation of a standard molecular norm requires only 15-20 steps for most rocks. Fe^2 and Fe^3 are used as abbreviations for cation percents of Fe^{2+} and Fe^{3+}, respectively.

All but four normative minerals are pure compounds with fixed chemical compositions (see Table 1.9). These four are *di, hy, ol,* and *pl* (optional); they are all solid-solution series. Three of them, *di, hy,* and *ol,* are ferromagnesian silicates, so it is important that the $Mg/(Mg + Fe^2)$ ratio be the same for all three minerals in one norm (step 17).

The term *budget* refers to the amount of a particular cation available to "make" (i.e., to be converted to) one or more normative minerals. To *make provisionally* a mineral means initially to convert all of the appropriate cation to that mineral, but with the possibility that later part of that mineral must be converted to a new mineral to satisfy a deficiency. If a budget for a cation is a negative quantity (a deficiency), too much of that cation has already been used and another mineral must be made that requires less or none of the cation. If a budget has an excess after all minerals containing that cation have been made, then the amount remaining of that cation becomes the mineral representing its pure oxide (e.g., excess Si is *q*, excess Ti is *ru*, excess Al is *c*, and excess Fe^3 is *hm*). Calcium is an exception to this procedure; excess Ca is made into *wo* ($CaSiO_3$).

Many steps below include chemical equations. These equations are balanced according to equivalent rather than molecular units (see Section 1.2.2). They are used in two ways. First, they are used to make a normative mineral from its appropriate cations (e.g., steps 2, 3, 5, 6, and 7). Second, they are used to negate a deficiency in the budget of a cation by making a new normative mineral (e.g., steps 8B, 8D, 11C, and 12B). The new mineral and the deficient cation are on the right side of the equation. The appropriate concentrations of cations or minerals on the left side of the equation must be subtracted from their respective budgets, as they are used to make the new mineral and eliminate the deficiency. For example, assume the Fe^2 deficiency in step 11C were -2 percent:

$$2il + Ca + Si = 3sp + Fe^2$$

Then 6 percent *sp* would be made and 4, 2, and 2 percent *il*, Ca, and Si, respectively, are used in the process. These latter three values must be subtracted from their respective budgets.

In the equations for steps 8B, 8D, 11A, and 12B, Si is on the right side, but Si does not represent a deficiency. In these four cases the appropriate value for Si must be added to, rather than subtracted from, the Si budget. An example is:

$$\text{step 8B:} \quad 5ab + Fe^3 = 4ac + Al + Si$$

If the Al deficiency is -3 percent, then (1) 15 percent *ab* must be subtracted from provisional *ab;* (2) 3 percent Fe^3 must be subtracted from its budget; (3) 12 percent *ac* is made, (4) 3 percent Si must be added to its budget; and (4) the Al deficiency is eliminated.

Consequently, the following generalization always holds: any calculated quantity on the right side of any equation below is a product and thus a positive quantity, whereas any quantity on the left side is a reactant and is negative. In other words, products on the right side are new minerals formed, cations to be added to existing budgets, or positive values to eliminate deficiencies. In contrast, reactants on the left side are used up making the products and must be subtracted from their respective budgets or provisional quantities.

The 27 steps are:

1. Add Mn to Fe^2. If desired, add Ni to Fe^2, Rb to K, Sr and Ba to Ca, Cr and V to Fe^3, etc. These latter additions are seldom necessary.
2. Make *ap* from all P by:

$$P + 1.67Ca = 2.67\ ap$$

3. Make *cc* from all C by:

$$C + Ca = 2cc$$

4. If desired, make pyrite (FeS_2) from S, fluorite (CaF_2) from F, halite (NaCl) from Cl, etc. Be certain to subtract necessary quantities of other involved cations (Fe^2, Ca, and Na in the examples above) from their respective budgets. Since S, F, and Cl are frequently not analyzed, this step normally is not be required.
5. Provisionally make *il* from all Ti by:

$$Ti + Fe^2 = 2il$$

Generally this will also be the final amount of *il*.
6. Provisionally make *or* from all K by:

$$K + Al + 3Si = 5or$$

Except for extremely SiO_2 undersaturated rocks (*lc*-bearing; step 25), this will be the final *or* value.
7. Provisionally make *ab* from all Na by:

$$Na + Al + 3Si = 5ab$$

8. Check Al budget.
   A. If excess Al (i.e., the remaining Al is a positive quantity), as is typical, continue to step 9.
   B. If deficiency (remaining Al is a negative quantity), make provisional *ac* according to:

$$5ab + Fe^3 = 4ac + Al + Si$$

   where Al is the deficiency. In this case $an = 0$. All remaining Ca is used to make *wo, di,* and if necessary, *sp*.
   C. Check Fe^3 budget. If excess, go to step 11.
   D. For those rare situations in which an Fe^3 deficiency exists, make sodium metasilicate (*ns;* Na_2SiO_3) by:

$$8ac = 3ns + 2Fe^3 + 3Si$$

where $2Fe^3$ is deficiency. Because the rock will contain no *mt*, go to step 11 but omit steps 13-14.

9. Provisionally make *an* from all remaining Ca by:

$$Ca + 2Al + 2Si = 5an$$

10. Check Al budget again.
    A. If excess Al is present, report as *c* and continue to step 11.
    B. If deficiency, the rock is Al-undersaturated. Go to step 11.

11. Check Fe^2 budget.
    A. If excess Fe^2 exists and the rock is *c*-normative from step 10A, go to step 13. If excess Fe^2 and the rock is Al-undersaturated, calculate available Ca by:

    $$5an = Ca + 2Al + 2Si$$

    where $2Al$ is deficiency. Subtract this value of *an* from provisional *an* (step 9) to obtain final *an*. Go to step 12. Most rocks will have excess Fe^2 and steps 11B-D will not be necessary.
    B. If an Fe^2 deficiency exists and rock is *c*-normative, then as much *il* as possible is made from original Fe^2 by equation for step 5 and excess TiO_2 is reported as *ru*. For this step and steps 11C-11D, the remaining Fe^3 becomes *hm*; *mt* and *fs* equal zero. Go to step 16.
    C. If Fe^2 deficiency and rock is Al-undersaturated, calculate available Ca and final *an* using equation in step 11A. Then make *sp* according to:

    $$2il + Ca + Si = 3sp + Fe^2$$

    where Fe^2 is the deficiency. Go to step 12 but omit steps 13-15.
    D. If Fe^2 deficiency and rock is so Al-undersaturated that no *an* exists (i.e., the rock is *ac*-normative from step 8B), then ample Ca should be available and the above equation can be used to make *sp* and eliminate the Fe^2 deficiency. Omit steps 13-15.

12. Check *Ca* budget.
    A. If excess, make *wo* from all remaining Ca by:

    $$Ca + Si = 2wo$$

    B. If deficiency, which is rare, make *ru* according to:

    $$3sp = ru + Ca + Si$$

    where Ca is deficiency. *Di* and *wo* equal zero.

13. Provisionally make *mt* from all remaining Fe^3 by:

    $$Fe^3 + 0.5Fe^2 = 1.5mt$$

14. Check Fe^2 budget again.
    A. If excess, as is normal, continue to step 15.
    B. If deficiency, use all available Fe^2 to make *mt* by:

$$Fe^2 + 2Fe^3 = 3mt$$

Report excess Fe^3 as hm. In this case $fs = 0$. Go to step 16.

15. Make fs from all remaining Fe^2 by:

$$Fe^2 + Si = 2fs$$

16. Make en from all Mg according to:

$$Mg + Si = 2en$$

17. Calculate the $Mg/(Mg + Fe^2)$ ratio for the ferromagnesian silicates from $en/(en + fs)$. Maintain this ratio for di, hy, and ol when converting to "CIPW" norms.

18. If present, combine equal percentages of wo and $(en + fs)$ to make di.
    A. If no wo, report all $(en + fs)$ as provisional hy. Final $di = 0$.
    B. If excess $(en + fs)$ over wo, which is usually the case, report excess provisionally as hy; $wo = 0$.
    C. If excess wo over $(en + fs)$, report excess as wo.

19. Check Si budget.
    A. If excess, report as q and go to step 27.
    B. If deficiency and hy is available, provisionally make ol $(fo + fa)$ according to:

$$4hy = 3ol + Si$$

    where Si is the deficiency. $Q = 0$.
    C. If no hy is available, go to step 21. If no hy or ab are available, go to step 23.

20. Check hy budget.
    A. If excess, report as hy and go to step 27.
    B. If deficiency, use all available hy making ol and calculate new Si deficiency.

21. Attempt to eliminate this Si deficiency by provisionally making ne:

$$5ab = 3ne + 2Si$$

    where 2Si is the deficiency. Q and $hy = 0$.

22. Check ab budget.
    A. If excess, report as ab and go to step 27.
    B. If deficiency, use all original ab making ne and calculate new Si deficiency.

23. Attempt to eliminate this Si deficiency by provisionally making lc:

$$5or = 4lc + Si$$

    where Si is the deficiency. Q, hy and $ab = 0$.

24. Check or budget.
    A. If excess, report as or and go to step 27.
    B. If deficiency, which is rare, use all original or making lc and calculate new Si deficiency.

25. Attempt to eliminate this Si deficiency by making kp:

$$4lc = 3kp + Si$$

where Si is the deficiency. Q, hy, ab, and $or = 0$.

26. Check lc budget.

    A. If excess, report as lc and go to step 27.

    B. If deficiency, which is extremely rare, use all original lc making
        kp and calculate new Si deficiency. One possibility at this point
        is to admit defeat and report this Si deficiency as negative q.

27. If desired, report $(ab + an)$ as pl.

Finally, check to determine if the summation of normative minerals is within rounding error of the cation percent total. Both totals should be very close to 100.00 percent. If no mistakes are made and the chemical analysis is of a reasonably typical igneous rock, these summations normally will be within 99.9-100.1 percent. If a weight-percent norm similar to a CIPW norm is desired, refer to Section 1.2.5.6.

APPENDIX B A BASIC Program for Fractional Crystallization

The program involves three minerals in the extraction assemblage and three elements, which can be elements or oxides, in either mol or weight percent (see Section 5.2.3.2). If data are in mol percent, elements can be combined, such as Na + K or Fe + Mg. The program is the arithmetic solution of the graph shown on Figure 5.30, which can be interpreted in two ways. For fractional crystallization, the bulk composition S of the extraction assemblage A-B-C is removed from parent magma P to produce daughter magma D. Alternatively, for crystal accumulation, the three minerals A-B-C of bulk composition S are added to magma D, which produce accumulative rock P. The program is written in BASIC; for a general discussion of BASIC programming techniques, see Section 1.6.

The program is written in cryptic fashion, the most important reason for which is that with only slight modification it can be adapted to a hand-held calculator that is programmable in BASIC. Only lines 30 and 160-190 are absolutely required for data INPUT, so the remaining data-input lines can be omitted for adaption to a calculator. The current version was written for an IBM-compatible, desk-top computer, but the program can be adapted with little change to any computer. The data INPUT statements can be easily changed to read data from a file. If a printout of the results is required, change the PRINT statements in lines 490-550 to LPRINT statements.

In the following order, the program:

1. Prompts for names of minerals in the extraction (or accumulation) assemblage (lines 50-60);
2. Prompts for names of elements or oxides (lines 70-80)
3. Prints a table that explains order of data entry (lines 90-130)
4. Explains how to correct the previous incorrect data entry (line 140)
5. Warns the operator not to enter zeroes for concentrations (line 150)
6. Prompts for data entry (lines 160-190)
7. Provides second chance for correcting any data entry (lines 200-210 and the subroutine 610-650)
8. Performs the calculations (the algorithm, lines 230-490)
9. Prints the results (lines 510-570)
10. Offers opportunity to input new elements for same minerals (lines 580-590)
11. Ends (line 600)

The worked example that follows (data in mol percent) can be used to check the program. Note the value for Al in olivine, which is actually below detection. *It is very important not to enter zeroes for concentrations.* Doing so will cause a "division by zero" error. If some element's concentration is below detection, enter some trivially small number. Numbers 1-15 refer to order of data entry.

ELEMENT	PARENT	DAUGHTER	PLAG	CPX	OLIV
SI	1 - 47	4 - 52	7 - 45	10 - 45	13 - 33
AL	2 - 17	5 - 18	8 - 35	11 - 10	14 - .001
FE+MG	3 - 18	6 - 12	9 - .1	12 - 20	15 - 67

Output from the program is:

COMP. SI = 42.44482
COMP. AL = 16.08897
COMP. FE+MG = 23.46621

% PLAG = 29.76012
% CPX = 48.05102
% OLIV = 22.18886

% FRACT. = 52.32765

No provision is made in the program for rounding of numbers, but this could be easily added (see Section 1.6.2.3). Compositions of Si, Al, and Fe + Mg from the output refer to the bulk composition of the extraction (or accumulation) assemblage (the position of point S in X-Y-Z space; Fig. 5.30). Percentages are minerals in the extraction assemblage (position of point S within the extraction triangle A-B-C; Fig. 5.30), while % FRACTIONATION is percent fractional crystallization or percent crystals accumulated (P D / S D, expressed in percent; Fig. 5.30).

If a division by zero or overflow error message results from running the program, either zeroes were entered for data or an impossible solution was the result. Likewise, compositions for the three elements and percentages of the minerals should all be positive numbers and percent fractionation should be less than 100 (for either a fractionation or accumulation model). If *all* these criteria are not met, the program did not give a reasonable solution.

On Figure 5.30 the extrapolation of line D-P must intersect the extract triangle A-B-C to give a reasonable solution. On this figure they intersect at point S. If only one or two minerals are involved in the extraction, this intersection will be near an apex or somewhere along the sides of the extract triangle. Some uncertainty exists in the chemical analyses, both with regard to precision and how representative they are of the minerals and rocks in question. Hence, some uncertainty also exists about the exact location of both the extrapolated line and the triangle within X-Y-Z space, so it is possible that a one- or two-mineral fractionation process may not give a reasonable solution because the line and triangle do not quite intersect. If graphical solutions indicate that only one or two minerals are likely involved in the extraction, then this program should be used with caution. In fact, graphical solutions for only one or two minerals are so simple, this program is not even necessary (see Fig. 5.20).

```
10 CLS : PRINT
20 PRINT "THREE-MINERAL FRACTIONATION/ACCUMULATION:"
30 DIM V(16)
40 REM LINES 50-190 FOR DATA INPUT
50 PRINT : FOR I = 1 TO 3
60 PRINT "NAME OF MINERAL #";I; : INPUT M$(I) : NEXT I
70 PRINT : FOR I = 1 TO 3
80 PRINT "NAME OF ELEMENT #";I; : INPUT E$(I) : NEXT I
90 PRINT : PRINT "ENTER DATA IN FOLLOWING ORDER: ": PRINT
100 PRINT "ELEMENT","PARENT","DAUGHTER",M$(1),M$(2);TAB(71)M$(3)
110 PRINT E$(1),"1","4","7","10     13"
120 PRINT E$(2),"2","5","8","11     14"
130 PRINT E$(3),"3","6","9","12     15"
140 PRINT : PRINT "ENTER -9999 TO CORRECT PREVIOUS ENTRY"
```

```
150 PRINT : PRINT "NEVER ENTER ZEROES!"
160 PRINT : FOR I = 1 TO 15
170 PRINT "#";I; : INPUT V(I)
180 IF V(I) = -9999 THEN I = I-2
190 NEXT I
200 PRINT : INPUT "ALL DATA CORRECTLY ENTERED (Y/N)"; YN$
210 IF YN$ = "N" THEN GOSUB 610
220 REM LINES 210-470 FOR CALCULATIONS
230 M = V(11)-V(14) : N = V(10)*V(14)-V(11)*V(13)
240 O = V(14)*V(12)-V(11)*V(15)
250 P = V(8)-V(11) : Q = V(7)*V(11)-V(8)*V(10)
260 R = V(11)*V(9)-V(8)*V(12)
270 AP = (R*M-O*P)/(R*N-O*Q) : CP = (M-N*AP)/O
280 BP = (-1-V(7)*AP-V(9)*CP)/V(8)
290 A = -1/BP : B = -AP/BP : C = -CP/BP
300 E = (V(5)-V(2))/(V(4)-V(1)) : D = V(5)-E*V(4)
310 G = (V(5)-V(2))/(V(6)-V(3)) : F = V(5)-G*V(6)
320 Y = (A-B*D/E-C*F/G)/(1-B/E-C/G)
330 X = (Y-D)/E : Z = (Y-F)/G
340 L1 = SQR((V(7)-V(10))^2+(V(8)-V(11))^2+(V(9)-V(12))^2)
350 L2 = SQR((V(10)-V(13))^2+(V(11)-V(14))^2+(V(12)-V(15))^2)
360 L3 = SQR((V(13)-V(7))^2+(V(14)-V(8))^2+(V(15)-V(9))^2)
370 L4 = SQR((V(7)-X)^2+(V(8)-Y)^2+(V(9)-Z)^2)
380 L5 = SQR((V(10)-X)^2+(V(11)-Y)^2+(V(12)-Z)^2)
390 L6 = SQR((V(13)-X)^2+(V(14)-Y)^2+(V(15)-Z)^2)
400 C1 = (L5^2-L4^2-L1^2)/(-2*L4*L1)
410 A1 = .5*L4*L1*SIN(SQR(1-C1^2))
420 C2 = (L6^2-L2^2-L5^2)/(-2*L2*L5)
430 A2 = .5*L2*L5*SIN(SQR(1-C2^2))
440 C3 = (L4^2-L6^2-L3^2)/(-2*L6*L3)
450 A3 = .5*L6*L3*SIN(SQR(1-C3^2))
460 SU = A1+A2+A3 : PA = A2*100/SU
470 PB = A3*100/SU : PC = A1*100/SU
480 L7 = SQR((V(4)-X)^2+(V(5)-Y)^2+(V(6)-Z)^2)
490 L8 = SQR((V(4)-V(1))^2+(V(5)-V(2))^2+(V(6)-V(3))^2)
500 REM LINES 490-550 FOR PRINTOUT OF RESULTS
510 PRINT : PRINT "COMP. ";E$(1);" = ";X
520     PRINT "COMP. ";E$(2);" = ";Y
530     PRINT "COMP. ";E$(3);" = ";Z
540 PRINT : PRINT "% ";M$(1);" = ";PA
550     PRINT "% ";M$(2);" = ";PB
560     PRINT "% ";M$(3);" = ";PC
570 PRINT : PRINT "% FRACTIONATION = ";L8*100/L7
580 PRINT : INPUT "CONTINUE WITH SAME MINERALS (Y/N)";YN$
590 IF YN$ = "Y" THEN 70
600 END
610 PRINT : INPUT "ENTRY # FOR INCORRECT VALUE";Z
620 PRINT "CORRECT VALUE FOR #";Z; : INPUT X(Z)
630 PRINT : INPUT "MORE CORRECTIONS (Y/N)";YN$
640 IF YN$ = "Y" THEN 610
650 RETURN
```

APPENDIX C Abundances in the Solar System and Atomic Weights

Z	Symbol	gfw	log RA	Z	Symbol	gfw	log RA
1	H	1.08	10.50	44	Ru	101.07	0.28
2	He	4.00	9.34	45	Rh	102.91	−0.40
3	Li	6.94	1.69	46	Pd	106.40	0.11
4	Be	9.01	−0.09	47	Ag	107.87	−0.35
5	B	10.81	−0.03	48	Cd	112.41	0.17
6	C	12.01	7.07	49	In	114.82	−0.72
7	N	14.01	6.57	50	Sn	118.69	0.56
8	O	16.00	7.33	51	Sb	121.70	−0.50
9	F	19.00	3.39	52	Te	127.60	0.81
10	Ne	20.18	6.54	53	I	126.90	0.04
11	Na	22.99	4.78	54	Xe	131.30	0.73
12	Mg	24.30	6.01	55	Cs	132.91	−0.41
13	Al	26.98	4.92	56	Ba	137.33	0.68
14	Si	28.09	6.00	57	La	138.91	−0.35
15	P	30.97	3.98	58	Ce	140.12	0.07
16	S	32.06	5.70	59	Pr	140.10	−0.83
17	Cl	35.45	3.76	60	Nd	144.24	−0.11
18	Ar	39.95	5.07	62	Sm	150.40	−0.65
19	K	39.10	3.62	63	Eu	151.96	−1.07
20	Ca	40.08	4.86	64	Gd	157.25	−0.53
21	Sc	44.96	1.54	65	Tb	158.93	−1.26
22	Ti	47.90	3.44	66	Dy	162.50	−0.44
23	V	50.94	2.42	67	Ho	164.93	−1.10
24	Cr	52.00	4.10	68	Er	167.26	−0.65
25	Mn	54.94	3.97	69	Tm	168.93	−1.47
26	Fe	55.85	5.92	70	Yb	173.04	−0.67
27	Co	58.93	3.34	71	Lu	174.97	−1.44
28	Ni	58.70	4.68	72	Hf	178.49	−0.68
29	Cu	63.55	2.73	73	Ta	180.95	−1.68
30	Zn	65.38	3.06	74	W	183.85	−0.80
31	Ga	69.72	1.68	75	Re	186.21	−1.28
32	Ge	72.59	2.06	76	Os	190.20	−0.12
33	As	74.92	0.82	77	Ir	192.22	−0.14
34	Se	78.96	1.83	78	Pt	195.09	0.15
35	Br	79.90	1.13	79	Au	196.97	−0.69
36	Kr	83.80	1.67	80	Hg	200.59	−0.40
37	Rb	85.47	0.77	81	Tl	204.37	−0.72
38	Sr	87.62	1.43	82	Pb	207.20	0.60
39	Y	88.91	0.68	83	Bi	208.98	−0.84
40	Zr	91.22	1.45	90	Th	232.04	−1.24
41	Nb	92.91	0.15	92	U	238.03	−1.69
42	Mo	95.94	0.60				

gfw: gram formula weight or atomic weight ($^{12}C = 12.000$)

log RA: log relative abundance based on $Si = 10^6$ atoms (abundance data from Henderson, 1982)

APPENDIX D A BASIC Program for Bivariate Statistics

In Sections 1.1.3.3 and 1.1.3.4 linear correlation and regression were introduced. As observed in Chapter 5, however, many trends on variation diagrams are curved rather than linear. Least squares regression using a quadratic polynomial equation, which defines a parabola, generally provides a curve that fits a curved trend on a scattergram quite well. Some means should be available to fit both linear and curved trends and evaluate how well a curve or line fits the data. Accordingly, this program calculates the following:

I. Univariate statistics
II. Bivariate statistics
  A. Linear
    1. correlation
    2. regression
      a. reduced major axis (RMA)
      b. least squares regression
        (1) regressing Y on X (YOX)
        (2) regressing X on Y (XOY)
  B. Quadratic
    1. least squares regression
    2. "correlation"

All terms are defined in Sections 1.1.3.3 and 1.1.3.4. The SUBroutine on univariate statistics in the program calculates means and standard deviations for variables X and Y. A program to accomplish a similar task for one variable is given in Section 1.6.2.2. The X variable is assumed to be plotted on the horizontal axis of the graph, the Y variable on the vertical axis. Both the linear correlation coefficient (r) and the linear coefficient of determination (r^2) are calculated in step II.A.1. If the quadratic coefficient of determination is considerably higher than the linear r^2, then the quadratic fit is better than the linear fit. Linear regression coefficients are M and B for the equation $Y = MX + B$ (YOX or RMA) or $X = MY + B$ (XOY). All three types of linear regression are discussed in Section 1.1.3.4. Quadratic regression coefficients are A, B, and C and can be used in the equation $Y = A + BX + CX^2$.

In the previous outline linear correlation is listed before linear regression, but the opposite is true for quadratic regression and correlation. Linear regression coefficients can be calculated before or after correlation coefficients, but the quadratic "correlation coefficient" must be calculated last. In fact, a quadratic correlation coefficient similar to the linear correlation coefficient cannot be calculated, but a quadratic coefficient of determination can be. The quadratic regression coefficients are needed to calculate the quadratic coefficient of determination.

A sufficient number of REM statements are included in the program so it can be followed fairly easily. No variable contains more than two letters (or one letter plus a number) to accommodate some computers. No rounding routine is included, but can be easily added (see Section 1.6.2.3). As usual, the appropriate PRINT statements can be replaced with LPRINT statements if a printout of the results is desired. Alternatively, if an IBM-compatible computer is used, the PRT SC key prints the results on the screen. Two routines are included for data correction, one to correct a previous data by entering -9999, and the other to correct any entry after all X,Y pairs have been entered. After regression coefficients are calculated, another feature of the program allows the operator to enter a value for X and the program calculates Y (for RMA or YOX regressions). For the XOY regression, the operator enters Y and the program calculates X. These routines are useful for providing points to plot best-fit lines or curves on graphs. The program is written to be reasonably "user friendly." It could be drastically shortened if many of these special features were eliminated.

An example print-out, using the same data as the example in Sections 1.1.3.2 to 1.1.3.4, is given below. Within rounding errors, the following results agree wth the hand-calculated results. Input X,Y pairs and program output, followed by the program lines, are given below:

```
X:  1  3  2  5  4  1  5
Y:  2  5  6  810  4  7
```

MEAN FOR X = 3   STD. DEV. FOR X = 1.732051
MEAN FOR Y = 6   STD. DEV. FOR Y = 2.645751

LINEAR CORRELATION COEFFICIENT = .8001323
LINEAR COEFF. OF DETERMINATION = .6402118

LINEAR EQUATIONS ARE IN FORM Y = MX + B

M(RMA) = 1.527525   B(RMA) = 1.417424
M(YOX) = 1.222222   B(YOX) = 2.333334

LINEAR EQUATION IS IN FORM X = MY + B

M(XOY) = .5238096   B(XOY) = -.1428573

QUADRATIC COEFF. OF DETERMINATION = .717507

QUADRATIC EQUATION IS IN FORM $Y = A + BX + CX^2$

A = -.2753616   B = 3.657004   C = -.4057968

```
10 CLS : PRINT
20 PRINT "LINEAR AND QUADRATIC CORRELATION - REGRESSION:"
30 REM LINES 40-210 FOR DATA INPUT
40 PRINT : INPUT "MAXIMUM POSSIBLE NUMBER X,Y PAIRS";M
50 DIM X(M),Y(M)
60 PRINT : INPUT "MUMBER X,Y PAIRS THIS SET";N
70 PRINT : PRINT "TYPE -9999 TO CORRECT PREVIOUS ENTRY"
80 FOR I = 1 TO N
90 PRINT : PRINT "FOR PAIR #";I
100     INPUT "      X";X(I)
110 IF X(I) = -9999 THEN I = I-2 : GOTO 140
120     INPUT "      Y";Y(I)
130 IF Y(I) = -9999 THEN I = I-2
140 NEXT I
150 PRINT : INPUT "ALL DATA OK (Y/N)";YN$
160 IF YN$ = "Y" THEN 230
170 PRINT : INPUT "PAIR # TO CORRECT";D
180 INPUT "      X";X(D)
190 INPUT "      Y";Y(D)
```

```
200 PRINT : INPUT "MORE CORRECTIONS (Y/N)";YN$
210 IF YN$ = "Y" THEN 170
220 REM LINES 230-270 CALCULATE SUMMATIONS
230 SX=0 : SY=0 : QX=0 : QY=0 : CP =0 : XC=0 : XF=0 : CS=0
240 FOR I = 1 TO N
250 SX=SX+X(I) : SY=SY+Y(I) : QX=QX+X(I)^2 : QY=QY+Y(I)^2
260 CP=CP+X(I)*Y(I) : XC=XC+X(I)^3 : XF=XF+X(I)^4 : CS=CS+X(I)^2*Y(I)
270 NEXT I
280 PRINT : PRINT "CHOOSE ONE OF:"
290 PRINT
300 PRINT "          <1> LINEAR REGRESSION AND CORRELATION"
310 PRINT "          <2> QUADRATIC REGRESSION AND CORRELATION"
320 PRINT "          <3> QUIT THIS DATA SET"
330 PRINT : INPUT "YOUR CHOICE";Z
340 ON Z GOTO 350,650,850
350 GOSUB 890
360 REM LINES 370-480 CALCULATE & PRINT LINEAR CORREL. - REGR.
370 C=(CP-SX*SY/N)/(N-1) : R=C/(DX*DY) : DL=R^2
380 MR=DY/DX : IF R < 0 THEN MR = -MR
390 BR=AY-MR*AX
400 MY=C/DX^2 : BY=AY-MY*AX
410 MX=C/DY^2 : BX=AX-MX*AY
420 PRINT : PRINT "LINEAR CORRELATION COEFFICIENT =";R
430 PRINT "LINEAR COEFF. OF DETERMINATION =";DL
440 PRINT : PRINT "LINEAR EQUATIONS ARE IN FORM Y = MX + B"
450 PRINT : PRINT "M(RMA) =";MR; TAB(25) "B(RMA) =";BR
460 PRINT "M(YOX) =";MY; TAB(25) "B(YOX) =";BY
470 PRINT : PRINT "LINEAR EQUATION IS IN FORM X = MY + B"
480 PRINT : PRINT "M(XOY) =";MX ; TAB(25) "B(XOY) =";BX
490 REM LINES 500-640 CALCULATE Y FOR INPUT X, OR X FOR Y (LINEAR)
500 PRINT : PRINT "DO YOU WISH TO:"
510 PRINT : PRINT "     <1> ENTER X AND CALCULATE Y BASED ON RMA?"
520      PRINT "     <2> ENTER X AND CALCULATE Y BASED ON YOX?"
530      PRINT "     <3> ENTER Y AND CALCULATE X BASED ON XOY?"
540      PRINT "     <4> RETURN TO MAIN PROGRAM?"
550 PRINT : INPUT "YOUR CHOICE";Z
560 IF Z = 4 THEN 280
570 PRINT :PRINT "ENTER -9999 WHEN FINISHED CALCULATING X OR Y"
580 IF Z = 3 THEN 630
590 PRINT : INPUT "X";X : IF X = -9999 THEN 280
600 IF Z = 2 THEN 620
610 Y=MR*X+BR : PRINT "Y =";Y : GOTO 590
620 Y=MY*X+BY : PRINT "Y =";Y : GOTO 590
630 PRINT : INPUT "Y";Y : IF Y = -9999 THEN 280
640 X=MX*Y+BX : PRINT "X =";X : GOTO 630
650 GOSUB 890
660 REM LINES 670-780 CALCULATE & PRINT QUADRATIC CORREL. - REGR.
670 II=QX-SX^2/N : JJ=CP-SX*SY/N : KK=XC-SX*QX/N
680 LL=CS-QX*SY/N : MM = XF-QX^2/N
690 C1=(LL*II-JJ*KK)/(II*MM-KK^2)
700 B1=(JJ*MM-LL*KK)/(II*MM-KK^2)
710 A1=SY/N-B1*SX/N-C1*QX/N
720 F=0 : FOR I = 1 TO N
730 F=F+(A1+B1*X(I)+C1*X(I)^2-AY)^2
```

```
740 NEXT I
750 DQ = F/((N-1)*DY^2)
760 PRINT : PRINT "QUADRATIC COEFF. OF DETERMINATION =";DQ
770 PRINT : PRINT "QUADRATIC EQUATION IS IN FORM Y = A + BX + CX^2"
780 PRINT : PRINT "A =";A1;TAB(20)"B =";B1;TAB(40)"C =";C1
790 REM LINES 800-840 CALCULATE Y FOR INPUT X (QUADRATIC)
800 PRINT : INPUT "ENTER X AND CALCULATE Y (Y/N)";YN$
810 IF YN$ = "N" THEN 280
820 PRINT : PRINT "ENTER -9999 WHEN FINISHED CALCULATING Y"
830 PRINT : INPUT "X";X : IF X = -9999 THEN 280
840 Y = A1+B1*X+C1*X^2 : PRINT "Y=";Y : GOTO 830
850 PRINT : INPUT "ANOTHER DATA SET (Y/N)";YN$
860 IF YN$ = "Y" THEN 60
870 END
880 REM LINES 890-920 CALCULATE AND PRINT UNIVARIATE STATISTICS
890 AX=SX/N : DX=SQR((QX-SX^2/N)/(N-1))
900 AY=SY/N : DY=SQR((QY-SY^2/N)/(N-1))
910 PRINT : PRINT "MEAN FOR X =";AX; TAB(25) "STD. DEV. FOR X =";DX
920 PRINT "MEAN FOR Y =";AY; TAB(25) "STD. DEV. FOR Y =";DY
930 RETURN
```

APPENDIX E Compositions of Some Igneous Rocks and Minerals

PLUTONIC	GRA	GRD	TON	SYN	DIO	GAB	PER
SiO_2	71.3	66.1	61.5	58.6	57.5	50.1	42.3
TiO_2	0.31	0.54	0.73	0.84	0.95	1.12	0.63
Al_2O_3	14.3	15.7	16.5	16.6	16.7	15.5	4.23
Fe_2O_3	1.21	1.38	1.83	3.04	2.50	3.01	3.61
FeO	1.64	2.73	3.82	3.13	4.92	7.62	6.58
MnO	0.05	0.08	0.08	0.13	0.12	0.12	0.41
MgO	0.71	1.74	2.80	1.87	3.71	7.59	31.2
CaO	1.84	3.83	5.42	3.53	6.58	9.58	5.05
Na_2O	3.68	3.75	3.63	5.24	3.54	2.39	0.49
K_2O	4.07	2.73	2.07	4.95	1.76	0.93	0.34
P_2O_5	0.12	0.18	0.25	0.29	0.29	0.24	0.10

VOLCANIC	RHY	RYD	DAC	TRA	AND	BAS	BSN
SiO_2	72.8	65.6	65.0	61.2	57.9	49.2	44.3
TiO_2	0.28	0.60	0.58	0.70	0.87	1.84	2.51
Al_2O_3	13.3	15.0	15.9	17.0	17.0	15.7	14.7
Fe_2O_3	1.48	2.13	2.43	2.99	3.27	3.79	3.94
FeO	1.11	2.03	2.30	2.29	4.04	7.13	7.50
MnO	0.06	0.09	0.09	0.15	0.14	0.20	0.16
MgO	0.39	2.09	1.78	0.93	3.33	6.73	8.54
CaO	1.14	3.62	4.32	2.34	6.79	9.47	10.2
Na_2O	3.55	3.67	3.79	5.47	3.48	2.91	3.55
K_2O	4.30	3.00	2.17	4.98	1.62	1.10	1.96
P_2O_5	0.07	0.25	0.15	0.21	0.21	0.35	0.74

MINERALS	OLV	OPX	AUG	HBD	BIO	PLG	KFD
SiO_2	37.3	55.2	51.8	48.7	38.1	51.1	64.0
TiO_2	0.09	0.22	0.49	0.32	3.60	0.05	--
Al_2O_3	0.18	1.50	3.07	9.48	14.0	31.0	19.6
Fe_2O_3	1.60	0.84	1.38	2.33	3.98	0.28	0.32
FeO	21.6	11.9	7.21	9.12	20.2	0.12	--
MnO	0.27	0.28	0.17	0.23	0.09	0.01	--
MgO	38.1	28.1	16.0	14.4	7.96	0.27	0.16
CaO	0.38	1.93	19.2	11.9	0.90	14.0	0.33
Na_2O	0.03	--	0.27	1.16	0.50	3.30	3.02
K_2O	0.05	--	0.02	0.15	8.31	0.13	12.1

Rock compositions (excluding volatiles) from Le Maitre (1976); Mineral analyses from Deer and others (1962). All data in weight percent.

Abbreviations: GRA, granite; GRD, granodiorite; TON, tonalite; SYN, syenite; DIO, diorite; GAB, gabbro; PER, peridotite; RHY, rhyolite, RYD, rhyodacite; DAC, dacite; TRA, trachyte; AND, andesite; BAS, basalt; BSN, basanite; OLV, olivine (FO76); OPX, orthopyroxene (EN77); AUG, augite (WO40EN46); HBD, hornblende (Mg69); BIO, biotite (PH31); PLG, plagioclase (AN70); KFD, K-feldspar (OR71)

APPENDIX F Thermochemical Data for Some Igneous Minerals

Mineral	S^o_{298}	V^o_{298}	ΔH^o_f	ΔG^o_f	log K_f
ALBITE (HIGH)	226.40	100.43	−3924240	−3706507	649.367
$NaAlSi_3O_8$	±0.40	±0.09	±3640	±3660	±0.641
ANORTHITE	199.30	100.79	−4229100	−4003326	701.371
$CaAl_2Si_2O_8$	±0.30	±0.05	±3125	±3145	±0.551
CLINOENSTATITE	67.86	31.47	−1547750	−1460883	255.942
$MgSiO_3$	±0.42	±0.05	±1215	±1225	±0.215
DIOPSIDE	143.09	66.09	−3210760	−3036554	531.994
$CaMgSi_2O_6$	±0.84	±0.10	±9120	±9160	±1.605
FAYALITE	148.32	46.39	−1479360	−1379375	241.662
Fe_2SiO_4	±1.67	±0.09	±2410	±2470	±0.432
FORSTERITE	95.19	43.79	−2170370	−2051325	359.385
Mg_2SiO_4	±0.84	±0.03	±1325	±1345	±0.236
ILMENTITE	105.86	31.69	−1236622	−1159170	203.083
$FeTiO_3$	±1.25	±0.08	±1590	±1632	±0.286
JADEITE	133.47	60.4	−3029400	−2850834	499.456
$NaAlSi_2O_6$	±1.25	±0.1	±4180	±4230	±0.741
KALIOPHILITE	133.26	59.89	−2121920	−2005975	351.440
$KAlSiO_4$	±1.25	±0.05	±1435	±1450	±0.254
LEUCITE	200.20	88.39	−3038650	−2875890	503.846
$KAlSi_2O_6$	±1.70	±0.05	±2755	±2850	±0.499
MAGNETITE	146.14	44.524	−1115726	−1012566	177.398
Fe_3O_4	±0.42	±0.008	±2092	±2134	±0.374
MUSCOVITE	334.6	140.71	−5976740	−5600671	981.219
	±1.0	±0.18	±3225	±3290	±0.576
$KAl_2(AlSi_3O_{10})(OH)_2$					
NEPHELINE	124.35	54.16	−2092110	−1977498	346.449
$NaAlSiO_4$	±1.25	±0.06	±2420	±2450	±0.363
PYROPE	260.76	113.27	−6284620	−5932412	1039.341
$Mg_3Al_2Si_3O_{12}$	±10.00	±0.05	±6000	±7000	±1.226
QUARTZ	41.46	22.688	−910700	−856288	150.019
SiO_2	±0.20	±0.001	±1000	±1100	±0.193
SANIDINE	232.90	109.05	−3959560	−3739776	655.196
$KAlSi_3O_8$	±0.48	±0.10	±3370	±3400	±0.596
SPINEL	80.63	39.71	−2299320	−2174860	381.028
$MgAl_2O_4$	±0.42	±0.03	±750	±760	±0.133
TREMOLITE	548.90	272.92	−12355080	−11627910	2037.170
	±1.25	±0.73	±17320	±17360	±3.041
$Ca_2Mg_5(Si_8O_{11})(OH)_2$					

All data are at STP and are from Robie and others (1979).
Errors are expressed as two standard errors ($\pm 2e_s$).

Units:
S^o_{298} joules per mol·degree (j/mol°K)
V^o_{298} cubic centimeters per mole (cc/mol)
ΔH^o_f joules per mole (j/mol)
ΔG^o_f joules per mol (j/mol)

REFERENCES

Abbey, S., 1983, Studies in standard samples of silicate rocks and minerals 1969-1982: Geol. Survey Canada Paper 83-15.

Ahrens, L.H. and Taylor, S.R., 1961, Spectrochemical Analysis, 2nd edition: Reading, Mass., Addison Wesley.

Allegre, C.J. and Minster, J.F., 1978, Quantitative models of trace element behavior in magmatic processes: Earth and Planetary Sci. Letters, v. 38, p. 1-25.

Andersen, O., 1915, The system anorthite-forsterite-silica: Am. Jour. Sci., 4th series, v. 39, p. 407-454.

Arth, J.G., 1976, Behavior of trace elements during magmatic processes -- a summary of theoretical models and their applications: Jour. Research U.S. Geol. Survey, v. 4, p. 41-47.

Barker, D.S., 1983, Igneous Rocks: Englewood Cliffs, N.J., Prentice-Hall.

Barth, T.F.W., 1962, Theoretical Petrology, 2nd ed.: New York, John Wiley.

Bartley, J.M., 1986, Evaluation of REE mobility in low-grade metabasalts using mass-balance calculations: Norsk Geol. Tidsskrift, v. 66, 145-152.

Berg, J.H. and Klewin, K.W., 1988, High-MgO lavas from the Keweenawan midcontinent rift near Mamainse Point, Ontario: Geology, v. 16, p. 1003-1006.

Bowen, N.L., 1913, The melting phenomena of the plagioclase feldspars: Am. Jour. Sci., 4th series, v. 34, p. 577-599.

Bowen, N.L., 1915, The crystallization of haplobasaltic, haplodioritic, and related magma: Am. Jour. Sci., 4th series, v. 40, p. 161-185.

Bowen, N.L. and Andersen, O., 1914, The system $MgO-SiO_2$: Am. Jour. Sci., 4th series, v. 37, p. 487-500.

Bowen, N.L. and Schairer, J.F., 1935, The system $MgO-FeO-SiO_2$: Am. Jour. Sci., v. 29, p. 151-217.

Boyd, F.R. and England, J.L., 1960, The quartz-coesite transition: Jour. Geophys. Research, v. 65, p. 749-756.

Bryan, W.B., Finger, L.W., and Chayes, F., 1969, Estimating proportions in petrographic mixing equations by least squares approximation: Science, v. 163, p. 926-927.

von Bunsen, R.W.E., 1853, Uber die Processe der Bulkanische Gesteins-bildungen Islands: Annales de Chimie et de Physique, v. 38, p. 215-300.

Burnham, C.W., 1975, Water and magmas: a mixing model: Geochim. et Cosmochim. Acta, v. 39, p. 1077-1084.

Burnham, C.W., 1979a, The importance of volatile constituents: in H.H.S. Yoder, Jr., ed., The Evolution of Igneous Rocks: Fiftieth Anniversary Perspectives, Princeton, N.J., Princeton University Press, p. 439-482.

Burnham, C.W., 1979b, Magmas and hydrothermal fluids: in H.L. Barnes, ed., Geochemistry of Hydrothermal Ore Deposits, 2nd ed., New York, John Wiley, p. 71-136.

Burnham, C.W. and Davis, N.F., 1974, The role of H_2O in silicate melts: II. thermodynamic and phase relations in the system $NaAlSi_3O_8-H_2O$ to 10 kilobars and 1000°C: Am. Jour. Sci., v. 274, p. 902-940.

Burnham, C.W., Holloway, J.R., and Davis, N.F., 1969, Thermodynamic Properties of H_2O to 1000°C at 10,000 Bars: Geol. Soc. America Spec. Paper 132.

Camp, V.E. and Roobol, M.J., 1989, The Arabian continental alkali basalt province: Part I Evolution of Harrat Rahat, Kingdom of Saudi Arabia: Geol. Soc. America Bull., v. 101, p. 71-95.

Carmichael, I.S.E, Turner, F.J., and Verhoogen, J., 1974, Igneous Petrology: New York, McGraw-Hill.

Cawthorn, R.G., 1977, Petrological aspects of the correlation between potash content of orogenic magmas and earthquake depth: Mineralog. Mag., v. 41, p. 173-182.

Chappell, B.W., and White, A.J.R., 1974, Two contrasting granitic types: Pacific. Geol., v. 8, p. 173-174.

Chayes, F. 1964, Variance-covariance relations in Harker diagrams of volcanic rocks: Jour. Petrology, v. 5, p. 219-237.

Chayes, F., 1965, Titania and alumina content of oceanic and circumoceanic basalt: Mineralog. Mag. v. 34, p. 126-131.

Cocherie, A., 1986, Systematic use of trace element distribution patterns in log-log diagrams for plutonic suites: Geochim. et Cosmochim. Acta, v. 50, p. 2517-2522.

Condie, K.C., 1976, Plate Tectonics and Crustal Evolution: New York, Pergamon.

Cox, K.G., Bell, J.D., and Pankhurst, R.J., 1979, Interpretation of Igneous Rocks: London, Allen and Unwin.

Darwin, C.R., 1844, Geological Investigations on the Volcanic Islands during the Voyage of the H.M.S. Beagle: London, Smith, Elder.

Deer, W.A., Howie, R.A., and Zussman, J., 1962, Rock Forming Minerals: New York, John Wiley.

Deer, W.A., Howie, R.A., and Zussman, J., 1966, An Introduction to the Rock Forming Minerals: London, Longmans.

DePaolo D.J., 1981, Trace element and isotopic effects of combined wallrock assimilation and fractional crystallization: Earth and Planetary Sci. Letters, v. 53, p. 189-202.

Dickinson, W.R., 1975, Potash-depth (K-h) relations in continental margin and intra-oceanic magmatic arcs: Geology, v. 3, p. 53-56.

Dowdy, S. and Wearden, S., 1983, Statistics for Research: New York, John Wiley.

Drummond, M.S., Ragland, P.C., and Wesolowski, D., 1986, An example of trondhjemite genesis by means of alkali metasomatism: Rockford Granite, Alabama Appalachians: Contr. Mineralogy and Petrology, v. 93, p. 98-113.

Drummond, M.S., Weslowski, D., and Allison, D.T., 1988, Generation, diversification, and emplacement of the Rockford granite, Alabama Appalachians: Mineralogic, Petrologic, Isotopic (C and O), and P-T constraints: Jour. Petrology, v. 29, p. 869-897.

Eggler, D.H., 1974, Effect of CO_2 on the melting of peridotite: Carnegie Inst. Washington Yearbook, v. 73., p. 215-224.

Eggler, D.H., Mysen, B.O., Hoering, T.C., and Holloway, J.R., 1979, The solubility of carbon monoxide in silicate melts at high pressures and its effect on silicate phase relations: Earth and Planetary Sci. Letters, v.43, p. 321-330.

Ehlers, E.G., 1987, The Interpretation of Geological Phase Diagrams: New York, Dover Publications.

Elthon, D. and Scarfe, C.M., 1984, High pressure phase equilibria of a high-magnesia basalt and the genesis of primary oceanic basalts: Am. Mineralogist, v. 69, p. 1-15.

Ernst, W.G., 1976, Petrologic Phase Equilibria: San Francisco, W.H. Freeman.

Evensen, N.M., Hamilton, P.J., and O'Nions, R.K., 1978, Rare-earth abundances in chondritic meteorites: Geochim. et Cosmochim. Acta, v. 42, p. 1199-1212.

Fairbairn and others, 1951, A cooperative investigation of precision and accuracy in chemical, spectrochemical, and modal analyses of silicate rocks: U.S. Geol. Survey Bull. 980, p. 1-71.

Flanagan, F.J., 1976, 1972 compilation of data on USGS standards: U.S. Geol. Survey Prof. Paper 840, p. 131-183.

Floyd, P.A. and Winchester, J.A., 1978, Identification and discrimination of altered and metamorphosed volcanic rocks using immobile elements: Chem. Geology, v. 21, p. 291-306.

Fyfe, W.S., Turner, F.J., and Verhoogen, J., 1959, Metamorphic Reactions and Metamorphic Facies: Geol. Soc. America Mem. 73.

Gibbon, D.L. and Wyllie, P.J., 1969, Experimental studies of igneous rock series: the Farrington complex, North Carolina, and the Star Mountain rhyolite, Texas: Jour. Geology, v. 77, p. 221-239.

Gill, J.B., 1970, Geochemistry of Viti Levu, Fiji, and its evolution as an island arc: Contr. Mineralogy and Petrology, v. 27, p. 179-203.

Goldschmidt, V.M., 1912, Die Kontakmetamorphose des KristianiaGebietes: Oslo, Norske Videnskaps-Akad. Skr. 1.

Green, D.H. and Ringwood, A.E., 1967, The genesis of basaltic magmas: Contr. Mineralogy and Petrology, v. 15, p. 103-190.

Green, T. H. and Ringwood, A.E., 1968, Genesis of the calc-alkaline igneous rock suite: Contr. Mineralogy and Petrology, v. 18, p. 105-162.

Greenwood, H.J., 1967, Wollastonite: stability in H_2O-CO_2 mixtures and occurrence in a contact metamorphic aureole near Salmo, British Columbia, Canada: Am. Mineralogist, v. 52, p. 1669-1680.

Grieg, J.W., 1927, Immiscibility in silicate melts: Am. Jour. Sci., 5th series, v. 13, p. 1-44.

Grove, T.L. and Baker, M.B., 1984, Phase equilibrium controls on the tholeiitic versus calc-alkaline differentiation trends: Jour. Geophys. Research, v. 89, no. B5, p. 3253-3274.

Grove, T.L. and Bryan, W.B., 1983, Fractionation of pyroxene-phyric MORB at low pressure: an experimental study: Contr. Mineralogy and Petrology, v. 84, p. 293-309.

Gunn, B.M., Roobol, M.J., and Smith, A.L., 1974, Petrochemistry of Pelean-type volcanoes of Martinique: Geol. Soc. America Bull., v. 85, p. 1023-1030.

Harker, A., 1909, The Natural History of Igneous Rocks: New York, Macmillan.

Harker, R.I. and Tuttle, O.F., 1956, Experimental data on the P_{CO_2}-T curve for the reaction calcite + quartz = wollastonite + CO_2, Part I: Am. Jour. Sci., v. 265, p. 239-256.

Helgeson, H.C., Delany, J.M., Nesbitt, H.W., and Bird, D.K., 1978, Summary and critique of the thermodynamic properties of rock-forming minerals: Am. Jour. Sci., v. 278A, p. 1-229.

Henderson P., 1982, Inorganic Geochemistry: Oxford, Pergamon Press.

Hildreth, W., 1981, Gradients in silicic magma chambers: implications for lithopsheric magmatism: Jour. Geophys. Research, v.10, p. 10153-10192.

Irvine, T.N. and Baragar, W.R.A., 1971, A guide to the chemical classification of the common volcanic rocks: Canadian Jour. Earth Sci., v. 8, p. 523-548.

Ishihara, S., 1977, The magnetite-series and ilmenite-series rocks: Mining Geol., v. 27, p. 293-305.

Ito, K. and Kennedy, G.C., 1967, Melting and phase relations in a natural peridotite to 40 kilobars: Am. Jour. Sci., v. 265, p. 519-538.

Jakes, P. and Gill, J., 1970, Rare earth elements and the island arc tholeiitic series: Earth Planetary Sci. Letters, v. 9, p. 17-28.

Kennedy, G.C. and Holser, W.T., 1966, Pressure-volume-temperature and phase relations of water and carbon dioxide: Geol. Soc. America Mem. 97, p. 371-384.

Kern, R. and Weisbrod, A., 1967, Thermodynamics for Geologists: San Francisco, Freeman, Cooper.

Kuno, H., 1968, Differentiation of basaltic magmas: in H.H. Hess and A. Poldervaart, eds., Basalts, v. 2, New York, Wiley-Interscience, p. 623-688.

Kushiro, I., 1973, The system diopside-anorthite-albite: determination of compositions of coexisting phases: Carnegie Inst. Washington Year Book, v. 72, p. 502-507.

Kushiro, I. and Yoder, H.S., Jr., 1966, Anorthite-forsterite and anorthite-enstatite reactions and their bearing on the basalteclogite transformation: Jour. Petrology, v. 7, p. 337-362.

Langmuir, C.H., Vocke, R.D., Hanson, G.N., and Hart, S.R., 1977, A general mixing equation applied to the petrogenesis of basalts from Iceland and Reykjanes Ridge: Earth Planetary Sci. Letters, v. 37, p. 380-392.

Larsen, E.S., 1938, Some new variation diagrams for groups of igneous rocks: Jour. Geology, v. 46, p. 505-520.

Le Bas, M.J., LeMaitre, M.J., Streckeisen, R.W., and Zanettin, B., 1986, A chemical classification of volcanic rocks based on the total alkali-silica diagram: Jour. Petrology, v. 27, p. 745-750.

Le Maitre, R.W., 1982, Numerical Petrology: Amsterdam, Elsevier.

Lipman, P.W. and Mehnert, H.H., 1975, Late Cenozoic basaltic volcanism and development of the Rio Grande Depression in the Southern Rocky Mountains: Geol. Soc. America Mem. 144, p. 119-154.

Luth, W.C., Johns, R.H., and Tuttle, O.F., 1964, The granite system at pressures of 4 to 40 kilobars: Jour. Geophys. Research, v. 69, p. 759-773.

Maaloe, S., 1985, Principles of Igneous Petrology: Berlin, Springer-Verlag.

Maaloe, S., and Johnston, A.D., 1986, Geochemical aspects of some accumulation models for primary magmas: Contr. Mineralogy and Petrology, v. 93, p. 449-458.

McBirney, A.R., 1979, Effects of assimilation: in H.S. Yoder, Jr., ed., The Evolution of Igneous Rocks, Princeton, N.J., Princeton University Press, p. 307-338.

McBirney, A.R., 1984, Igneous Petrology: San Francisco, Freeman, Cooper.

McBirney, A.R. and Noyes, R.M., 1979, Crystallization and layering of the Skaergaard intrusion: Jour. Petrology, v. 20, p. 487-554.

Macdonald, G.A. and Katsura, T., 1964, Chemical composition of Hawaiian lavas: Jour. Petrology, v. 5, p. 82-133.

Marsh, B. D., 1988, Crystal capture, sorting, and retention in convecting magma, Geol. Soc. America Bull., v. 100, p. 1720-1737.

Mason, B. and Moore, C.B., 1982, Principles of Geochemistry, 4th ed.: New York, John Wiley.

Maxwell, J.A., 1968, Rock and Mineral Analysis: New York, Interscience.

Meschede, M., 1986, A method of discriminating between different types of mid-ocean ridge basalts and continental tholeiites with the Nb-Zr-Y diagram: Chem. Geology, v. 56, p. 207-218.

Miyashiro, A., 1974, Volcanic rock series in island arcs and active continental margins: Am. Jour. Sci., v. 274, p. 321-355.

Morse, S.A., 1980, Basalts and Phase Diagrams: New York, Springer-Verlag.

Morse, S.A., 1968, Feldspars: Carnagie Inst. Washington Yearbook, v. 67, p. 120-126.

Mullen, E.D., 1983, MnO/TiO$_2$/P$_2$O$_5$: a minor element discriminant for basaltic rocks of oceanic environments and its implications for petrogenesis: Earth and Planetary Sci. Letters, v. 63, p. 53-62.

Mysen, B.O. and Boettcher, A.L., 1975, Melting of a hydrous mantle: II. geochemistry of crystals and liquids formed by anatexis of mantle peridotite at high pressures and high temperatures as a function of controlled activities of water, hydrogen, and carbon dioxide: Jour. Petrology, v. 16, p. 549-593.

Mysen, B.O. and Virgo, D., 1985a, Raman spectra and structure of fluorine- and water-bearing silicate glasses and melts: in R.L. Snyder, R.A. Condrate, and P.F. Johnson, eds., Advances in Materials Characterization II, New York, Plenum Publishing, p. 43-55.

Mysen, B.O. and Virgo, D., 1985b, Iron-bearing silicate melts: relations between pressure and redox equilibria: Phys. Chem. Minerals, v. 12, p. 191-200.

Myers, J.D., 1988, Possible petrogenetic relations between low- and high-MgO Aleutian basalts: Geol. Soc. America Bull., v. 100, p. 1040-1053.

Niggli, P., 1954, Rocks and Mineral Deposits: San Francisco, W.H. Freeman.

Nordstrom, D.K. and Munoz, J.L., 1986, Geochemical Thermodynamics: Palo Alto, Blackwell Scientific, 477 p.

O'Hara, M.J., 1968, The bearing of phase equilibria studies in synthetic and natural systems on the origin and evolution of basic and ultrabasic rocks: Earth-Sci. Rev., v. 4, p. 69-133.

O'Hara, M.J., 1977, Geochemical evolution during fractional crystallization of a periodically refilled majma chamber: Nature, v. 266, p. 503-507.

Olsen, B.A., McSween, H.Y., Jr., and Sando, T.W., 1983, Petrogenesis of the Concord gabbro-syenite complex, North Carolina: Am. Mineralogist, v. 68, p. 315-333.

Orville, P.M., 1963, Alkali ion exchange between vapor and feldspar phases: Am. Jour. Sci., v. 261, p. 210-237.

Osborn, E.F., 1942, The system CaSiO$_3$-diopside-anorthite: Am. Jour. Sci., v. 240, p. 751-788.

Peacock, M.A., 1931, Classification of igneous rock series: Jour. Geology, v. 39, p. 54-67.

Pearce, J.A. and Cann, J.R., 1973, Tectonic setting of basic volcanic rocks determined using trace element analyses: Earth and Planetary Sci. Letters, v. 19, p. 290-300.

Pearce, J.A., Harris, N.B.W., and Tindle, A.G., 1984, Trace element discrimination diagrams for the tectonic interpretation of granitic rocks: Jour. Petrology, v. 25, p. 956-983.

Pearce, J.A. and Norry, M.J., 1979, Petrogenetic implications of Ti, Zr, Y, and Nb variations in volcanic rocks: Contr. Mineralogy and Petrology, v. 69, p. 33-47.

Philpotts, A.R., 1976, Silicate liquid immiscibility: its probable extent and petrogenetic significance: Am. Jour. Sci., v. 276, p. 1147-1177.

Philpotts, A.R., 1982, Compositions of immiscible liquids in volcanic rocks: Contr. Mineralogy and Petrology, v. 80, p. 201-218.

Philpotts, A. R., and Reichenbach, I., 1985, Differentiation of Mesozoic basalts of the Hartford Basin, Connecticut: Geol. Soc. America Bull., v. 96, p. 1131-1139.

Piwinskii, A.J. and Wyllie, P.J., 1970, Experimental studies of igneous rock series: felsic body suite from the Needle Point pluton, Wallowa batholith, Oregon: Jour. Geology, v. 78, p. 52-76.

Potts, P.J., 1987, A Handbook of Silicate Rock Analysis: London, Blackie.

Powell, R., 1978, Equilibrium Thermodynamics in Petrology: London, Harper and Row.

Presnall, D.C., Dixon, J.R., O'Donnell, T.H., and Dixon, S.A., 1979, Generation of mid-ocean ridge tholeiites: Jour. Petrology, v. 20, p. 3-35.

Presnall, D.C., Dixon, S.A., O'Donnell, T.H., Brenner, N.L., Schrock, R.L., and Dycus, D.W., 1978, Liquidus phase relations on the join diopside-forsterite-anorthite from 1 atm to 20 kbar: their bearing on the generation and crystallization of basaltic magma: Contr. Mineralogy and Petrology, v. 66, p. 203-220.

Presnall, D.C. and Hoover, J.D.,1984, Composition and depth of origin of primary mid-ocean ridge basalts: Cont. Mineralogy and Petrology, v. 87, p. 170-178.

Ragland, P.C., 1970, Composition and structural state of the potassic phase in perthites as related to petrogenesis of a granitic pluton: Lithos, v. 3, p. 167-189.

Ragland, P.C. and Butler, J.R., 1972, Crystallization of the West Farrington pluton, North Carolina, U.S.A.: Jour. Petrology, v. 13, p. 381-404.

Ragland, P.C. and Rogers, J.J.W., 1984, Basalts: New York, Van Nostrand Reinhold.

Ragland, P.C. and Arthur, J.A., 1985, Petrology of the Boyds diabase sheet, north Culpeper basin, Maryland: in G.R. Robinson, Jr., and A.J. Froelich, eds., Proceedings of the Second USGS Workshop on the Early Mesozoic Basins of the Eastern United States, U.S. Geol. Survey Circular 946, p. 91-99.

Rankama, K. and Sahama, T.G., 1950, Geochemistry: Chicago, University of Chicago Press.

Richardson, S.M. and McSween, H.Y., Jr., 1989, Geochemistry: Pathways and Processes: Englewood Cliffs, N.J., Prentice-Hall.

Robie, R.A, Hemingway, B.S., and Fisher, J.R., 1979, Thermodynamic properties of minerals and related substances at 298.15K and 1 bar (10^5 pascals) pressure and at higher temperatures: U,S. Geol. Survey Bull. 1452.

Robie, R.A. and Waldbaum, D.R., 1968, Thermodynamic properties of minerals and related substances at 298.15°K (25.0°C) and one atmosphere (1.013 bars) pressure and at higher temperatures: U.S. Geol. Survey Bull. 1259.

Roedder, E., 1979, Silicate liquid immiscibility in magmas: in H.S. Yoder, Jr., ed., The Evolution of Igneous Rocks, Princeton, N.J., Princeton University Press, p. 15-57.

Rogers, J.J.W., 1982, Criteria for recognizing environments of formation of volcanic suites; application of these criteria to volcanic suites in the Carolina Slate Belt: Geol. Soc. America Spec. Paper 191, p. 99-107.

Rosenhauer, M. and Eggler, D.H., 1975, Solution of H_2O and CO_2 in diopside melt: Carnegie Inst. Washington Yearbook, v. 74, p. 474-479.

Sandell, E.B., 1959, Colorimetric Determination of Traces of Metals, 3rd edition: New York, Interscience.

Schairer, J.F., 1955, The ternary systems leucite-corundum-spinel and leucite-forsterite-spinel: Jour. Am. Ceramics Soc., v. 38, p. 153-158.

Schairer, J.F. and Bowen, N.L., 1955, The system K_2O-Al_2O_3-SiO_2: Am. Jour. Sci. v. 253, p. 681-746.

Schairer, J.F. and Bowen, N.L., 1956, The system Na_2O-Al_2O_3-SiO_2: Am. Jour. Sci., v. 254, p. 129-195.

Schairer, J.F. and Yoder, H.S., Jr., 1960, The nature of residual from crystal-
    lization, with data on the system nepheline-diopside-silica: Am. Jour.
    Sci., v. 258A, p. 273-283.

Schairer, J.F. and Yoder, H.S., Jr., 1964, Crystal and liquid trends in simplified
    alkali basalts: Carnegie Inst. Washington Yearbook, v. 63, p. 64-74.

Schairer, J.F. and Yoder, H.S., Jr., 1967, The system albite-anorthite-forsterite at
    1 atmosphere: Carnegie Inst. Washington Yearbook, v. 66, p. 204-209.

Schwarzer, R.R. and Rogers, J.J.W., 1974, A worldwide comparison of alkali
    olivine basalts and their differentiation trends: Earth and Planetary Sci.
    Letters, v. 23, p. 286-296.

Shand, S.J., 1927, The Eruptive Rocks, New York, John Wiley.

Shapiro, L., 1975, Rapid analysis of silicate, carbonate, and phosphate rocks --
    revised edition: U.S. Geol. Survey Bull. 1401.

Shapiro, L. and Brannock, W.W., 1962, Rapid analysis of silicate, carbonate,
    and phosphate rocks: U.S. Geol. Survey Bull. 1144-A.

Shaw, D.M., 1970, Trace element fractionation during anatexis: Geochim. et
    Cosmochim. Acta, v. 34, p. 237-243.

Simpson, G.G., Roe, A., and Lewontin, R., 1960, Quantitative Zoology: New
    York, Harcourt-Brace.

Sood, M.K., 1981, Modern Igneous Petrology: New York, John Wiley.

Stolper, E., 1980, A phase diagram for mid-ocean ridge basalts: preliminary
    results and implications for petrogenesis: Contr. Mineralogy and
    Petrology, v. 74, p. 13-27.

Streckeisen, A., 1976, To each plutonic rock its proper name: Earth-Sci. Rev., v.
    12, p. 1-33.

Taylor, J.R., 1982, An Introduction to Error Analysis: New York, University
    Science Books.

Taylor, S.R. and McLennan, S.M., 1985, The Continental Crust: Its Composition
    and Evolution: Oxford, Blackwell Scientific.

Thompson, R.N., 1972, The 1-atmosphere melting patterns of some basaltic
    volcanic series: Am. Jour. Sci., 272, p. 901-932.

Thompson, R.N. and Tilley, C.E., 1969, Melting and crystallization relations of
    Kilauean basalts of Hawaii: the lavas of the 1959-1960 Kilauea eruption:
    Earth and Planetary Sci. Letters, v. 5, p. 469-477.

Thornton, C.P. and Tuttle, O.F., 1960, Chemistry of igneous rocks. I. differ-
    entiation index: Am. Jour. Sci., v. 258, p. 664-684.

Tilley, C.E., Yoder, H.S., Jr., and Schairer, J.F., 1967, Melting relations of
    volcanic rock series: Carnagie Inst. Washington Yearbook, v. 65, p. 260-
    269.

Tuttle, O.F. and Bowen, N.L., 1958, Origin of granite in the light of
    experimental studies in the system $NaAlSi_3O_8$-$KAlSi_3O_8$-SiO_2-H_2O:
    Geol. Soc. America Mem. 74.

Walker, D., Schibata, T., and DeLong, S.E., 1979, Abyssal tholeiites from the
    Oceanographer Fracture Zone, II. phase equilibria and mixing: Contr.
    Mineralogy and Petrology, v. 70, p. 111-125.

Watson, E.B., 1976, Two-liquid partition coefficients: Experimental data and
    geochemical implications: Cont. Mineralogy and Petrology, v. 56, p.
    119-134.

Weigand, P. W. and Ragland, P. C., 1970, Geochemistry of Mesozoic dolerite
    dikes from eastern North America: Contr. to Mineralogy and Petrology,
    v. 29, p. 195-214.

White, A.J.R. and Chappell, B.W., 1977, Ultrametamorphism and granitoid
    genesis: Tectonophysics, v. 43, p. 7-22.

Whitney, J.A., 1988, The origin of granite: the role and source of water in the evolution of granitic magmas: Geol. Soc. America Bull., v. 100, p. 1886-1897.

Whittaker, E.J.W. and Muntus, R., 1970, Ionic radii for use in geochemistry: Geochim. et Cosmochim. Acta, v. 34, p. 945-956.

Wiebe, R.A., 1979, Fractionation and liquid immiscibility in an anorhtositic pluton of the Nain complex, Labrador, Jour. Petrology, v.20, p. 239-269.

Williams, R.T. and McSween, H.Y., Jr., 1989, Geometry of the Concord, North Carolina, intrusive complex: a synthesic of potential field modeling and petrologic data: Geology, v. 17, p. 42-45.

Winkler, H.G.F., 1979, Petrogenesis of Metamorphic Rocks, 5th edition: New York, Springer-Verlag.

Wood, B.J. and Fraser, D.G., 1976, Elementary Thermodynamics for Geologists: Oxford, Oxford University Press.

Wood, D.A., 1980, The application of a Th-Hf-Ta diagram to problems of tectonomagmatic classification and to establishing the nature of crustal contamination of basaltic lavas of the British Tertiary Volcanic Province: Earth and Planetary Sci. Letters, v. 50, p. 11-30.

Wright, T.L. and Doherty, P.C., 1970, A linear programming and least squares computer method for solving petrologic mixing problems: Geol. Soc. America Bull., v. 81, p. 1995-2008.

Wyllie, P.J., 1979, Magmas and volatile components: Am. Mineralogist, v. 64, p. 469-500.

Wyllie, P.J. and Tuttle, O.F., 1964, Experimental investigations of silicate systems containing two volatile components. Part III. The effect of SO_3, P_2O_5, HCl, and Li_2O, in addition to H_2O, on the melting temperatures of albite and granite: Am. Jour. Sci., v. 262, p. 930-939.

Yoder, H.S., Jr. and Tilley, C.E., 1962, Origin of basalt magmas: an experimental study of natural and synthetic rock systems: Jour. Petrology, v. 3, p. 342-532.

Zen, E., 1966, Construction of pressure-temperarure diagrams for multicomponent systems afterthe method of Schreinemakers -- a geometric approach: U.S. Geol. Survey Bull. 1225.

SUBJECT INDEX

A/CNK *41, 66*
accuracy *28-32*
activity *124-125*
Alkemade Theorem *209-210*
aluminosilicates *73*
alteration *132, 235-236,*
298, 312, 328
amphibolite *268*
alalcitite *327*
analytical methods *8-13*
electrochemical *9-10*
spectrochemical *9-10*
"wet chemical" *8*
andesite *253, 293-294,*
298
anion number *41*
antiperthites *187*
aphyric rocks *153-154,*
158, 319
arithmetic mean *16*
atomic absorption *10,*
24-28
calibration curve
26-28
instrumental drift *26*
matrix effects *26*
atomic percent *40*
autoliths *315*

back-arc basin *333*
basalts
AB-AN-DI *219-220*
AB-AN-DI-FO *244-245*
chemical analyses
4, 294, 336
classification *62-63*
effect of CO_2 *270-272*
effect of H_2O *270-272*
norm *48-49*
plate tectonics
331-332, 337-338
silica saturation
62-63
tetrahedron *62-63*
basanites *62-63, 329*
BASIC programs *79-93*
examples *83-93*
loops *83-93*
mathematics *82-83*
SUBroutines *88-90*
variables *80-81*
Benioff zone *333-334*

bimodal suites
268, 334-335
biotite, formula *43-44*
Bowen trend *291*
Bragg's Law *12*
buffers *140, 256*

calorimetry *99*
carbon dioxide
116, 123-124, 269-270
cation percent *40*
central tendency *16*
chemical equilibrium *95*
chemical potential
126-129
chi-squared *27, 31*
337-338
Clapeyron equation
109-112, 134-135
139-140, 147, 197,
258, 265
Clausius - Clapeyron
equation *116*
closure problem *287-289*
coefficient
activity *124*
correlation *19-21*
determination *21*
fugacity *122-123*
regression *21-24*
variation *17*
cognate xenoliths *315*
color index *245*
colorimetry *10*
combinatorial formula *194*
comendite *327*
components, definition *95*
constant-ratio lines *255*
contact principle
144-145, 212-214
control lines *255*
conversion factors *34*
conversions *32-52*
weight/mole *38-41*
weight/weight *32-36*
coordination numbers
74-75
correlation *19-21*
cotectic
biresorptional *213*
coprecipitational *212*
monoresorptional *213*

cotectic
    quaternary *239*
    ternary *202*
covariance *20*
critical point *135-136*
cryptic layering *182*
crystal mush *146, 171,*
    173, 228
crystallization
    dry *253-261*
    equilibrium
        159, 204-205
    eutectic *152*
    fractional *159,*
        171-173, 181-182,
        205, 317-319
    quaternary system
        240-241
    Rayleigh fractional
        159, 277-279
    reaction *163-166*
        189-190, 228
crystals
    cumulus *232, 244*
    intratelluric *154,*
        244
    shapes *154*
crystal path *162,*
    166-168, 206, 214-215

dacite *294*
decarbonation reactions
    115-116
degenerate systems
    197-198
degrees of freedom
    16, 129-130
deyhdration reactions
    114-115
diabase *244*
diagram
    AFM *288, 324-328*
    binary eutectic
    equilibrium *144, 191,*
        234-236
    exchange reaction
        311-313
    Harker *289, 293*
    P-T *108-116, 190-198*
    P-T-G *192-193*
    P-T-X *145-148*
    ratio-element *306-307*
    ratio-ratio *304-306*

diagram
    stability, *see* dia-
        gram, equilibrium
    triangular *58-61,*
        288, 332
differentiation indices
    see indices, fraction-
    ation
dipole molecule *120-136*
dispersion,
    measures *16-19*
distribution coefficient
    see coefficient, parti-
    tion
distribution
    log-normal *15*
    normal *14-15*
dolerite *244*

eclogite, norm *51-52*
electromagnetic spectrum
    9
electron microprobe *10,12*
elements
    compatible *273*
    HFS *276, 331-332*
    hygromagmatophile *275*
    immobile *331-332*
    incompatible *273*
    LFS *276*
    major *5*
    minor *5*
    rare-earth *276*
        308-311
    trace *5, 272-286*
emission
    spectroscopy *10-11*
enantiomorphic forms *197*
endothermic reaction *97*
enthalpy *99-101*
entropy *101-103*
entropy, K-feldspars *102*
equilibrium boundary *109*
equilibrium constant
    105-107
equilibrium batch melting
    160, 282-286
equivalent *37*
error analysis, linear
    EBM *283*
    enthalpy *100*
    entropy *103*
    examples *70-72*

error analysis, linear
    formulation *67-70*
    free energy *104*
    norms *71*
    P-T diagrams *113*
    simple approach *68*
europium anomalies
    310-311
eutectic
    binary *146, 148-157*
    quaternary *236-241*
    recognition *165-166*
    ternary *199-207*
exchange reaction
    107, 311-313
exothermic reaction *97*
experiments
    buffered *255-256*
    high P-T *257-261*
    methodology *140-141*
    non-buffered *254-255*
    one-atmosphere
       254-256
exsolution *175-176,*
    186-190, 224
extensive parameters
    95-96
extraction assemblage *297*
extract triangle *295*

Fe, oxidation states
    6, 36
Fenner trend *291*
filter pressing *173*
flame photometry *10-11*
flow differentiation *173*
forbidden zone *149*
fractionation
    high- and low-P
       257-259, 270-272
    polybaric *245*
free energy of mixing *128*
free energy,
    Gibbs *103-107*
freezing point depression
    150, 264
fugacity *122-123*

gabbro, *316*
gamma-ray spectrometry
    10, 13
gasses, ideal and real
    119-121
geobarometry *107, 117-119*

geohygrometer *263*
geothermometry
    107-108, 117-119
Gibbs surfaces *192-193*
gram-formula weight *32*
granitoids
    alumina saturation
       66-67
    I- and S-type *331*
    magnetite and
       ilmenite series *331*
    system OR-Ab-Q-H_2O
       224-226
gravity settling and
    floating *171-172*
groundmass *154*
guest, exsolution *186*

haplobasalt *241*
haplogranite
    norm *47-48*
    phase rule *130-131*
    system *245-251*
harzburgite *260*
heat of formation *99*
heat of reaction
    elevated temperature
       100-101
    standard temperature
       99-100
Henry's Law *124-125*
Hess' Law of
    Heat Summation *100-101*
host, exsolution *186*

ICP analysis *10, 11*
immiscibility gap *184*
incompatible normative
    minerals *54-55*
index
    differentiation *291*
    felsic *291*
    fractionation *289*
    Larsen *290*
    mafic *71-72, 254-255,*
       290
    Nockolds *290*
    solidification *291*
intensive parameters *95*
intercept, P-T diagrams
    112-113
ionic potential *276*
isothermal sections *206*
join, def. *151*

K-feldspars, entropy *102*
Kilauea volcano *253-255*

layered intrusion *232*
layering
    cryptic *182*
    unimodal *232*
leucitite *327*
leucocratic rocks *300*
lever rule *141-144,*
    180-181, 229-234,
    292-299
lherzolite *260*
liquid immiscibility
    174-176
liquid line of descent
    250, 255, 294-296
liquid path *162,*
    166-168
liquidus, def. *146*
log-log paper *301*
loss on ignition *6*

M-ratio *40-41*
magma chamber, closed
    153
magma, parent and
    daughter *275*
mantle
    depleted and
        enriched *260*
    metamorphic facies
        261-263
mass spectrometry *10, 13*
matrix, igneous rock *154*
melanocratic rocks *300*
melt hop *170, 205*
melting
    basalt *253-259*
    congruent *164-165*
    dynamic *160, 284*
    equilibrium *160,*
        163, 204-205
    equilibrium batch
        160, 282-286
    fractional *160,*
        170-171, 205
    incongruent *164-165*
    partial *154-155*
    peridotite *259-261*
    point *146*
    quaternary system
        240-241

melting
    Rayleigh fractional
        160, 279-282
    tonalite *266-268*
mesostasis *232*
metamorphism *124, 235*
Mg number *40-41*
mineral formulae *41-44*
mineral groups *75-79*
minerals
    formulae *77-78*
    incompatible
        normative *54-55*
    normative *45, 51*
    rock-forming *72-79*
minimum melt region *249*
minimum point *184*
mixing line *255, 339*
mode *44*
mol fraction, def. *95*
molar ratios *40-41*
MORB *244, 253, 261, 263,*
    309, 330-331,
    334-335, 337
muscovite, in norm *50-51*

NBO/T *77*
near-liquidus phases *154*
nephelinite *327*
network formers and
    modifiers *73-74*
neutron activation *10, 13*
normal distribution *13-14*
norms and modes *44*
norms *44-52*
    alkaline rocks *64-66*
    alumina saturation *67*
    basalt *48-49*
    CIPW *45*
    eclogite *51-52*
    haplogranite *47-48*
    effect of alkalies
        56-57
    error analysis *71*
    molecular to "CIPW"
        49-50
    molecular *45*
    oxidation
        effects *56-57*
    tailor-made *50-52*
    use of *249-251*
numerical
    modeling *313-320*

Oddo-Harkins rule *308*
olivine, formula *42, 43*
oxidation,
    effect on norms
        57-63
    iron *6, 36*
    europium *310*
oxides
    mol percent *40*
    names of *7*

pantellerite *327*
parameter of state *95*
parameters, univariate
    16-19
partial molar quantity
    126
partition coefficients
    107-108, 117-119,
    179-180, 273-277
parts per million *5*
path, crystal and liquid
    162, 166-168, 206,
    214-215
percent deviation *17*
peritectic
    binary *159-173*
        189-190
    recognition *165-166*
    ternary *207-211*
perthites *187*
phase boundary *109*
phase contact *232*
phase rule *129-131*
    binary system *149*
    CaO-SiO$_2$-CO$_2$ *235*
    Gibbs *125-126*
    haplogranite
        130-131
    proof *129-130*
    quaternary system
        239-240
    ternary system
        202-203
    unary system *134*
phases, definition, *95*
phenocrysts *153,*
    158-160, 219,
    244-245, 292-293,
    296-299, 325
phyric rocks *158, 158,*
    161, 244
picrite *260*
piercing point *248, 248*

piston cylinder *140*
plate tectonics, rock
    series *332-337*
point
    critical, def. *133*
    eutectic, def, *146*
    freezing *146*
    invariant, def. *134*
    melting *146*
    peritectic *159-173,*
        189-190
    piercing *243, 248*
    reaction, *see*
        point, peritectic
polygon of error *299*
polymerization *76-77*
potash-depth relationship
    333-334
precision *28-32*
precision, total *30*
pressure effects
    binary systems
        169-170
    solvus *185*
    AB-AN-DI-FO *243-244*
    OR-AB-AN-Q-H$_2$O
        248-248
primary magmas *154, 261*
primary-phase field *202*
primocrysts *154, 244*
principle of opposition
    298
projections, quaternary
    system *236-239*
pseudosystems *156, 209,*
    216-217, 221, 236,
    241, 246, 248
pyroxenite *220*

QAPF *55-56*
quenching furnace *140*

radioactivity *10-13*
Raoult's law *124-125*
rapid analysis *9*
Rayleigh
    fractional crystal-
        lization *159,*
        277-279
    fractional melting
        160, 279-282
reaction point *see* point,
    peritectic

regression
    least squares 22-24
    linear 22
    quadratic 22
    reduced major axis
        23-24
reproducibility 29
residuals 23, 27
restite 163-164, 218,
    220, 268, 281, 314-316
restite unmixing 314
rhyodacite 294
rhyolite 294
rifts 334-335
rock hops 166-168
rock series
    alkalies 322-325
    alkaline 62,
        157-165, 320-329,
        333-334
    Andean 325
    calcalkaline 53, 67,
        293, 320-329, 333-
        335
    chemical analyses 294
    classification
        326-329
    Fe/Mg ratio 325-326
    high-alumina 323, 329
        333
    high-K calcalkaline
        325, 333
    hypersthenic 325
    metaluminous 66
    peralkaline 62, 66,
        327, 328
    peraluminous 66, 321
    pigeonitic 325
    plate tectonics
        329-338
    shoshonitic 325, 333,
        335
    tholeiitic 53,
        320-329, 334-335
    subalkaline 53, 62,
        157-165, 320,
        325-329

sample preparation, AA 25
saturation 52-67
    alumina in
        granitoids 66
    norms 54-55
    rock series 61-62

saturation
    silica 53-54
    silica in alkaline
        rocks 64-66
    silica in basalts
        62-63
scalene triangles
    59-61, 227
Schreinemakers' rules
    194-198
second boiling 265
silica, polymorphs
    136-137
silicates
    basic groups 75-76
    mineral groups 78-79
    structures 75-76
slope, P-T diagrams
    112-113
solid solution
    partial 183-184
    binary 177-182
solidus, def. 146
solubility, relative 204
solution
    ideal 124-124, 203
    real 124-125
    regular 185
solvus, def. 174
spontaneity, criterion
    104-105, 128-129
stability field 109
standard deviation 16
standard error 18
standard state 99
standardization
    of data 30
standards 25
stoichiometric compounds
    32
Stoke's Law 172
subduction zones 333-334
subsolidus reactions
    183-190
supercritical fluid
    135-136
system
    AB-AN-DI 215-220
    AB-AN-DI-FO 241-245
    AB-OR-H_2O 185-187
    AN-DI 155-157
    Ca0-SiO_2-CO_2 234-236
    CMAS 262-263
    FO-Q 174-175

system
  FO-SP-LC *201-207*
  H_2O *133-135*
  LC-Q *160-165*
  $MgO-SiO_2$ *144*
  NE-Q *157-159*
  OR-AB-AN-Q-H_2O
    245-251
  OR-AB-Q-H_2O *221-226*
  petrogeny's residua
    225
  SiO_2 *136-140*

ternary
  invariant points
    201-202, 210-211
  isothermal sections
    207
tetrahedron
  alkaline rock *64-65*
  alumina *72-73*
  basalt *62-63*
  silica *72-73*
thermal divide
  binary *157-158*
  ternary *207-209*
thermal valley *146, 202*
thermodynamics
  first law *97-101*
  second law *101-103*
  third law *102*
tholeiites *62-63*
  definition *63*
  olivine *63*
  quartz *62*
tie line *151, 161, 163,*
    177, 181, 188, 206-
    207, 213, 217,
    227-228, 234-236

tonalite, melting *266-268*
transformations *138*
transition loops *177-182,*
    188-189
triple point *134*
trondhjemite *225, 250*
Tschermak's molecule *155, 263*

uniformitarianism *330*
unmixing, *see* ex-
  solution

van der Waals
  constants *120*
  equation *121*
Van't Hoff reaction
  isotherm *106*
variables, statistical *14*
variance, *16*
volatiles, effect *263-272*

water
  adsorbed *6*
  Clapeyron equation
    139-140
  effect *263-265*
  phase rule *134*
  structural *6*
WXYZ system *72-74*

X-ray diffraction *10, 12*
X-ray fluorescence *10, 12*

zoned crystals *172*
zoned intrusions *169, 173*
Z-transformations *338*